34.2 2008

PALÉORIENT

Revue pluridisciplinaire
de préhistoire et protohistoire
de l'Asie du Sud-Ouest
et de l'Asie centrale

*Pluridisciplinary Review
of Prehistory and Protohistory
of Southwestern and Central Asia*

CNRS ÉDITIONS
15, rue Malebranche - 75005 Paris

Paléorient

Revue pluridisciplinaire
de préhistoire et protohistoire
de l'Asie du Sud-Ouest
et de l'Asie centrale

Pluridisciplinary Review
of Prehistory and Protohistory
of Southwestern and Central Asia

Les manuscrits, les ouvrages pour recensions, la correspondance des auteurs devront être adressés à :

CNRS – Paléorient
Maison René Ginouvès, UMS 844
21, allée de l'Université
F-92023 NANTERRE Cedex

Tél. : 33+(0)1.46.69.24.08
Fax : 33+(0)1.46.69.24.33
Courriel : paleorient@mae.u-paris10.fr
Site Internet : www.mae.u-paris10.fr/paleo_index.htm

Du volume 3 (1975-1977) au volume 6 (1980), *Paléorient* a fait l'objet d'un volume annuel.
À partir du volume 7/1, *Paléorient* est à nouveau publié en deux numéros annuels qui peuvent être fournis régulièrement sur commande permanente.

Paléorient *volumes 3 (1975-1977) to 6 (1980) have been published as single annual volumes.*
From volume 7/1 (1981) Paléorient *is once again published with two issues per year.*
Standing orders can be delivered regularly.

Pour toute information relative à la diffusion de nos ouvrages, merci de bien vouloir contacter notre service lecteurs :

For any information concerning the distribution of our publications please contact:

CNRS ÉDITIONS
15, rue Malebranche, F-75005 PARIS
Tél. : 01.53.10.27.00 - Fax : 01.53.10.27.27
Courriel : cnrseditions@cnrseditions.fr
Site Internet : www.cnrseditions.fr

Illustration de la couverture : Photo Y. Lehmann

ISBN : 978-2-271-06856-9 – ISSN : 0513-9345

SOMMAIRE DU VOLUME 34.2 – 2008

CONTENTS OF VOLUME 34.2 – 2008

ARTICLES

RECENSIONS

Cultural Relationships beyond the Iranian Plateau: The Helmand Civilization, Baluchistan and the Indus Valley in the 3rd Millennium BCE

E. Cortesi, M. Tosi, A. Lazzari and M. Vidale

Abstract: *This paper presents a systematic review of the archaeological evidence for cultural interaction between the Helmand and the Indus during the 3rd millennium BCE. A series of artefacts found at Shahr-i Sokhta and nearby sites (Iranian Seistan) that were presumably imported from Baluchistan and the Indus domain are discussed, together with finds from the French excavations at Mundigak (Kandahar, Afghanistan) that might have the same origin. Other artefacts and the involved technologies bear witness to the local adaptation of south-eastern manufactures and practices in the protohistoric Sistan culture. While the objects datable to the first centuries of the 3rd millennium BCE fall in the so called "domestic universe" and reflect common household activities, in the centuries that follow we see a shift to the sharing of luxury objects and activities concerning the display of a superior social status; but this might be fruit of a general transformation of the archaeological record of Shahr-i Sokhta and its formation processes.*

Résumé: *Cet article propose un ré-examen systématique des documents archéologiques qui témoignent d'interactions culturelles entre le Helmand et le bassin de l'Indus au cours du IIIe millénaire avant J.-C. Une série d'objets, selon toute vraisemblance, importés du Baluchistan et de la vallée de l'Indus, trouvés aussi bien à Shahr-i Sokhta que sur les sites avoisinants du Séistan iranien, sont discutés ici, en parallèle avec d'autres, recueillis par la mission française à Mundigak (Kandahar, Afghanistan). D'autres artefacts ainsi que leurs technologies montrent clairement dans la culture protohistorique du Séistan l'adaptation de coutumes et de pratiques en provenance du Sud-Est. Alors que les objets qui peuvent être datés des premiers siècles du IIIe millénaire avant J.-C. appartiennent à la « sphère domestique » et reflètent des activités quotidiennes, dans les siècles qui suivent on observe un glissement vers des objets de luxe et vers des activités permettant d'afficher un statut social supérieur; on pourrait y voir le résultat d'une transformation générale des assemblages de Shahr-I Sokhta et des processus qui y ont présidé.*

Keywords: *Sistan, Baluchistan, Indus Valley, 3rd millennium BCE, Interrelationships, Pottery, Seals, Terracotta figurines, Shell artefacts.*
Mots-clés: *Séistan, Baluchistan, Vallée de l'Indus, 3e millénaire BCE, Interaction culturelle, Poterie, Scellements, Figurines d'argile, Coquillages.*

 Manuscrit reçu le 1er juillet 2008, accepté le 25 novembre 2008

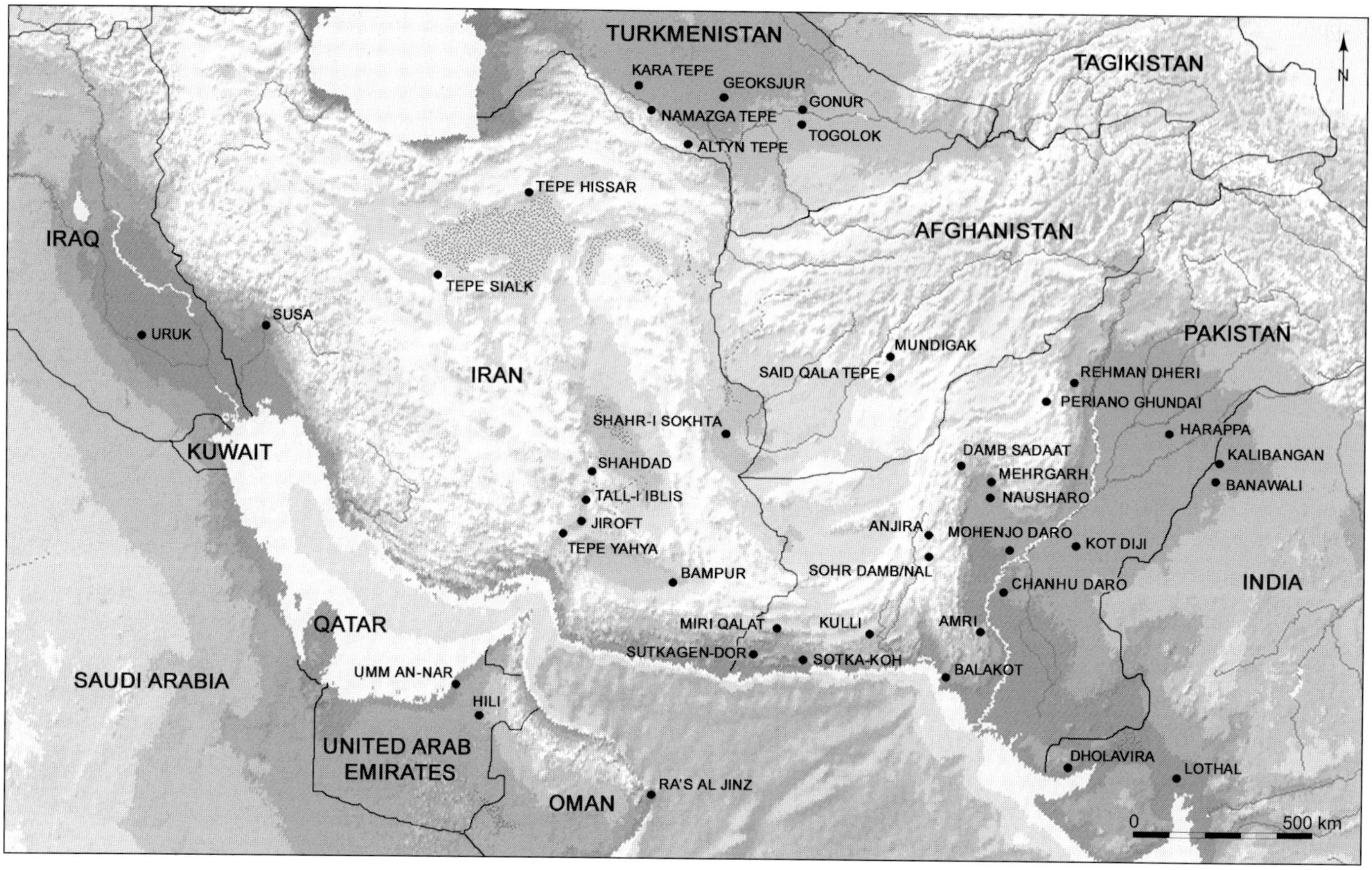

Fig. 1 – *Middle Asia map with the location of Shahr-i Sokhta and main protohistoric centres. (Source: Microsoft Encarta.)*

INTERACTION SPHERES IN THE INDO-IRANIAN BORDERLANDS

In 2003, Laurent Flusch, curator of a provocative exhibition in Lausanne, reported the hypothetical discovery, in future deposits dated to the 21st century AD, of metal brooches as the sole archaeological remnant of the contemporary production, trade and consumption of denims.[1] Below the pictures exhibited one sees a fictitious map of the world with the assumed archaeological locations of similar finds: the map simulates a dense cluster in North America followed by their spread to other regions of the planet. In current archaeological terms it would be the representation of a large "interaction sphere". Putting aside the contention that the entire planet has been turned into a global interaction sphere mainly pivoted in the northern American continent, we may briefly review the semantics of this term and its application to the prehistory and protohistory of Middle and South Asia.

J.R. Caldwell[2] originally applied the concept of interaction spheres to the study of late North American prehistory. According to L.R. Binford,[3] an interaction sphere is archaeologically defined by the evidence of a widespread exchange of material objects in a well delimited social context, and indicates an intensive process of active inter-cultural communication. J. Shaffer[4] systematically applied the concept to the general framework of South Asian prehistory. He stated that the basic preconditions for the identification of such historical contexts are the presence of similar artefacts occurring in wide exchange or trade networks, joined to the detailed reconstruction of the relevant social contexts. Later, K.C. Chang used this

1. FLUSCH, 2003: 98.

2. CALDWELL, 1964.
3. BINFORD, 1965: 208.
4. SHAFFER, 1992 and 1978: 133.

approach for his spectacular study of the mosaic of regional cultures of Late Neolithic China.[5]

In South Asian archaeology, the concept was used at a macro-regional scale for the first reconstruction of wide-range historical trends (fig. 1). While correlating various materials found at Shahr-i Sokhta and Tepe Yahya, C.C. Lamberg-Karlovsky and M. Tosi[6] recognized on the Iranian Plateau, at the end of the 4th millennium BCE, two main interaction spheres: a Jemdet Nasr one (late 4th-early 3rd millennium BCE), including southern Mesopotamia, the Diyala region, the Gulf and south-eastern Iranian Plateau, and extending to Tepe Yahya, somehow related to the so-called Uruk expansion and its later developments;[7] and a northern one, extending from the Kopet Dagh piedmont southward to the Helmand centres and Quetta valley.[8] After the middle of the 3rd millennium BCE, the authors observed the growth of new forms of interaction centred along the southern sea routes and what we presently see as the highly structured societies of the Magan coast centres. Ultimately they observed "a tendency towards greater nucleation and disintegration of the earlier pattern of cultural interconnections".[9] At the end of the 3rd millennium BCE, the main interaction spheres of southern Eurasia (from Mesopotamia to Central Asia, the Gulf and the Indus valley) shrunk in size and underwent a severe localization process apparently resulting in the growth of isolated polities.[10]

R.P. Wright applied a similar approach, but at a regional scale, to the study of the diffusion in the first half of the 3rd millennium BCE of the fine-ware production labelled Faiz Muhammad Gray.[11] Focussing on the lands artificially severed by the present southern boundary between Pakistan and Iran (in northern and north-eastern Baluchistan) and with reference to the "world systems" models proposed by Wallerstein,[12] Wright preliminarily analysed the production and exchange of these ceramics in terms of fields of interaction between core regions and borderlands.[13] Five overlapping production-distribution areas, with distinctive chemical patterns, were identified as pivots of interaction spheres of variable extent, involving local peer communities. *Contra* Wallerstein, Wright suggested that such spheres were actively promoting technical innovation and information exchange, and therefore the overlapping peripheries, rather than the centre, represented privileged points of archaeological observation.[14]

But the most influential application of a similar model to the 3rd millennium BCE archaeology of South Asian countries was P. Kohl's definition of an "intercultural style" commonly encountered in precious chlorite vessels carved with a complex set of icons and symbols. Kohl proposed that this class of luxury goods, found in distant regions such as Mesopotamia, south-eastern Iran (particularly in the Kerman Region) and the southern coasts of the Gulf shared the same manufacturing technology, aesthetics and iconography, as the marker of a widespread, interregional circuits of exchange. The high compositional variability of the raw materials, as observed in the initial mineralogical and chemical analyses, supported the hypothesis of several micro-regions of procurement and transformations and of a decentred production-distribution system.[15]

In 1978, the Islamic Revolution interrupted field research by foreign missions across the Iranian Plateau. As an immediate result, prehistoric-protohistoric research declined in Iran, but was intensified in the surrounding countries, initially in Pakistan and in the Gulf countries. When in 1989 the decline and collapse of the Soviet Union re-designed the geopolitics of Eurasia, a new phase of field archaeology in Central Asia was disclosed to for western scholars.

In the meantime, Iranian archaeologists had a new cycle of large-scale excavations in their country. Pre-Islamic archaeology suddenly became again very popular. Among the new results stand out the re-opening of the excavations at Shahr-i Sokhta in 1997[16] and the discovery of the Halil Rud Civilization, following the massive plundering of its rich graveyards and the beginning of regular excavations at the early urban site of Konar Sandal near Jiroft and nearby locations.[17] The chlorite carved artefacts illegally excavated at the cemetery of Mathoutabad and other sites, although missing at the time reliable contexts of provenience, suggested that the "intercultural style" is the actual marker of the local intensive production of such elite objects for local use, even before representing a widespread semantic system for international protohistoric trade. In a sense, the Halil Rud style was intercultural in Middle Asia in the 3rd millennium BCE in a fashion similar to the Egyptian style became intercultural across the Mediterranean during the first half of the 1st millennium BCE: it was perceived as highly

5. CHANG, 1986: 242.
6. LAMBERG-KARLOVSKY and TOSI, 1973.
7. ALGAZE, 1993.
8. LAMBERG-KARLOVSKY and TOSI, 1973: 52.
9. *Ibid.*: 52.
10. *Ibid.*: figs. 1 and 2.
11. WRIGHT, 1987 and 2002.
12. WALLERSTEIN, 1974.
13. WRIGHT, 2002.
14. *Ibid.*: 406.
15. KOHL, 1975, 1978 and 1979.
16. SAJJADI, 2003 and 2004; SAJJADI *et al.*, 2004.
17. MADJIDZADEH, 2003 and 2007; LAMBERG-KARLOVSKY, 2004; VIDALE, 2007.

prestigious by several contemporaneous cultures. It is too early to say to what extent it gave birth to local interpretations or variations, as the Egyptian style did in the Orientalizing craft productions of the 8th and 7th century BCE.

While Tepe Yahya, in period IVA, that is in the second half of the 3rd millennium BCE, was one of the manufacturing centres for these chlorite artefacts, and was certainly connected with the most important urban centres of the Halil Rud valley, many other centres of production await discovery in the nearby valleys and in the Konar Sandal area as well. Ongoing research on the subject shows that this industry in the Halil Rud basin goes back at least to the early 4th millennium BCE[18] and most probably represents an archaic indigenous economic adaptation. Furthermore, elaborately carved vessels found in the west bearing cuneiform inscriptions (mostly datable to the Akkadian period) suggest that such objects, in many cases, were taken as war booty dedicated to deities or exchanged as gifts among royal houses.[19] The impression is that the vessels with the royal dedications are those with the most elaborate mythological depictions, as if the Mesopotamian lords wished to state their authority on the culture that had produced such very complex and alien ideological representations. In other words, the general pattern of the production and distribution of chlorite goods in the area and in the period concerned needs to be conceived differently. Such fast and radical transformations of the concept of intercultural style (from a loosely defined economic and semantic sphere to the historical contingencies of complex civilizations) provide a paramount example of the need for investigating in depth the interaction spheres so far hypothesized in the prehistory of South Asian regions and the Gulf, trespassing upon their original definitions.

As very effectively stated by Lamberg-Karlovsky,[20] in the original formulation of interaction spheres as archaeological tools: "The distribution of artefacts stand as proxies for cultural interaction". As a consequence, in many cases little room was left for enquiring into the institutions, agencies and processes that materially affected and patterned social evolution. The next step is necessarily "the quest for an understanding of what brought different polities into contact".[21] Going back to the above-quoted case of contemporary denims, we immediately see why the interpretation of artefact distribution in terms of "cultural interaction" is hopelessly reductive. Denims were originally invented and traded as working class garments, to become unexpectedly a symbol of libertarian ideologies and new ways of life. At the same time, they were adopted by social and political groups actively opposing the political role of western countries and the expansion of their capitalistic markets. In the Lausanne catalogue, denims belong to a famous and powerful multi-national brand, but we all know that they have been independently produced by a multitude of independent factories all over the world. The brand itself is illegally copied for the mass production of counterfeit items, particularly in Third World countries. Confronted by such a nest of conflicting issues, their plain "archaeological" distribution, as reported in the Lausanne catalogue, conveys little or no historical meaning.

In Middle Asia, we are confronted with the nature of the boundary, or boundaries, separating the scattered early urban civilizations on the easternmost fringe of the Iranian plateau and in the Indus valley during the 3rd millennium BCE. An ideal straight line connecting Shahr-i Sokhta in the Iranian Seistan and Mundigak in Kandahar (Afghanistan), two of the larger early urban poles of the Helmand civilization, through the Registan desert, might represent, very generically, the north-western cultural boundary of the Indus core area. But the actual geopolitical and chronological context of such boundaries, to be looked for southwards, is still archaeologically unknown. Similarly field research on the Makran coast[22] had identified Sotka-Koh and Sutkagen-Dor, two Indus fortified outposts or forts aimed at controlling in security the eastern gates to the Gulf trade. Later, survey and excavations by French teams in Kech Makran[23] identified at the small site of Miri Qalat, Pakistan, specimens of Bevelled Rim Bowls, marking the present easternmost limit of diffusion of one of the most diagnostic ceramic types of the Uruk-related phenomenon (late 4th-early 3rd millennium BCE) as well as, for the second half of the 3rd millennium BCE, the traces of an Indus settlement. Besides these points on our maps, the actual geopolitical and chronological context of the Indus boundary is still archaeologically unknown.

The recent discovery in the Halil Rud Basin (Jiroft, Kerman, Iran) of a late 4th millennium administrative unit, endowed with large amounts of Bevelled Rim Bowls, stone weights and jar stoppers, suggests that we need to reconsider

18. M. Vidale, unpublished data from Mathoutabad.

19. MARCHESI, 2007.

20. LAMBERG-KARLOVSKY, 2008.

21. The interpretation of ancient social interaction in terms of "shared social fields" or the "social networks" recently advocated in prehistory (see GAMBLE, 1999) and similar abstract models does not seem particularly productive. See LAMBERG-KARLOVSKY, 2008 for a general discussion.

22. DALES, 1962a and 1962b; DALES and LIPO, 1992.

23. BESENVAL, 1994, 1997a, 1997b and 2000; BESENVAL and DIDIER, 2004 and 2005; DIDIER, 2007.

the so-called Uruk expansion and its later developments in a new light and scale. Similarly, the presence at Konar Sandal of objects and goods in Indus style or imported from the Indus valley[24] re-emphasize the importance of the Indus stamp seal impression found by Lamberg-Karlovsky at Tepe Yahya.[25] In such a perspective the critical constrain is that south-eastern Iran, particularly in the core regions along the fluvial corridors, remains largely unexplored. Before the "Jiroft revolution", archaeologists had to ground their assumption of local developments and interaction processes upon De Cardi's test trenches at Bampur in the Jazmurian depression and other small sites,[26] while current field exploration not far from Bampur reports the existence of imposing early urban site clusters extending for dozens of ha.[27]

HELMAND *VERSUS* INDUS

In a brief reply to R. Dittmann on the role played by the Iranian plateau as a geographical "continuum" during the 3rd millennium BCE between Mesopotamia and the Indus Valley,[28] U. Franke strongly stressed the present limitations of the currently available archaeological record.[29] Her list of the artefacts considered as positive material evidence of contacts between the Plateau and the Indo-Pakistani Subcontinent[30] is indeed very short. It includes

> "Only few Nal sherds, most of them found in Shahr-i Sokhta[31] and one from Tepe Yahya;[32] a canister jar, from Shahr-i Sokhta;[33] two so-called *Kot Dijian* jars, one from the Central Quarters of Shahr-i Sokhta and the latter from Mundigak[34] and last but not least an unfired TC (*terracotta*) female figurine, like the Zhob type found in Shahr-i Sokhta."[35]

While we agree that the material evidence of exchange and contacts is indeed scanty, the aim of this paper is to enlarge and substantiate this preliminary list, defining in greater detail the material evidence for ancient relationships among the Helmand civilization, Baluchistan and the Indus valley, during the Early and Middle Bronze Age, in a time range between the early 3rd millennium BCE and the third quarter of the same millennium. Our main focus is the archaeological record from Shahr-i Sokhta and Iranian Sistan, but we will also consider some relevant finds from Mundigak.[36]

On two occasions[37] Franke has stressed the need of solving a major disagreement on the protohistoric chronology of the Helmand civilization. In fact, J.-F. Jarrige[38] and other scholars, on the basis of their sound knowledge of the ceramic sequences of the region and of specific ceramic comparisons, have considered Shahr-i Sokhta Period III as contemporary with the Kot-Dijian phase of the Indus core area (*i.e.*, with Period II at Harappa, ascribed by a consistent series of calibrated ^{14}C datings to the range 2800-2600 BCE).[39] For example, in 1987 R. Biscione after specific information by S. Pracchia proposed to link a group of plain bowls with an upper S-shaped wall contour found at the ceramic production site of Lal Shah near Mehrgarh with apparently similar bowls from late Period III in Sistan and with Kot-Dijian production in general.[40]

In contrast, other Italian archaeologists, on the basis of a preliminary series of calibrated ^{14}C datings and other radiometric determinations from the settlement area of Shahr-i Sokhta and nearby sites, would rather place Period III, phase 4 between 2500 BCE and the 24th century BCE, and the ensuing phase 3 (very poorly represented in the settlement and unreported, so far, in the graveyard) between approximately 2350 and 2200 BCE[41] (see fig. 2). As Period III deposits, in the topmost levels of Shahr-i Sokhta, are to a great extent heavily damaged by wind and water erosion, the architecture often being abraded down to the foundations,[42] the relevant absolute readings are presently limited to five datings from the Eastern Residential Area and the layers underlying the Burnt Building, two

24. Y. Madjidzadeh, H. Pittman and M. Vidale, ongoing research.
25. LAMBERG-KARLOVSKY, 1972: pl. 2B; PITTMAN, 2001: fig. 10: 63.
26. DE CARDI, 1967, 1968 and 1983.
27. M. Mortazavi, personal communication.
28. DITTMANN, 2002.
29. FRANKE, 2002: 346.
30. *Ibid.*: 349.
31. AMIET and TOSI, 1978: 14-15.
32. LAMBERG-KARLOVSKY, 1970: pl. 34.
33. BISCIONE *et al.*, 1974: fig. 8.
34. CASAL, 1961: fig. 253.
35. TOSI, 1983: fig. 61.
36. Reported in CASAL, 1961.
37. FRANKE, 2002: 351 and 2008: 666.
38. JARRIGE, 1993: 152 and 162.
39. KENOYER, 1991a, 1991b and 1991c; KENOYER and MEADOW, 2000.
40. This claim, we feel, is based upon wrong ceramic associations: for example, Biscione compares some overfired potsherds with a saddle-shaped upper rim and stepped painted pattern from Tepe Rud-i Biyaban 2, Period IV, with Kot-Dijian fragments from the Indus valley having similar rims, but the painted designs are significantly different, and the Indus specimens are painted on a thick red slip, while the sherds from Seistan are the late bichrome (red-and-black) version of the coarse polychrome pottery most distinctive of the Helmand civilization. See BISCIONE, 1990: 399, figs 1-2 and comments in MUGAVERO and VIDALE, 2003 and 2006.
41. SALVATORI and TOSI, 2005: fig. 1.
42. As described by SALVATORI and VIDALE, 1997.

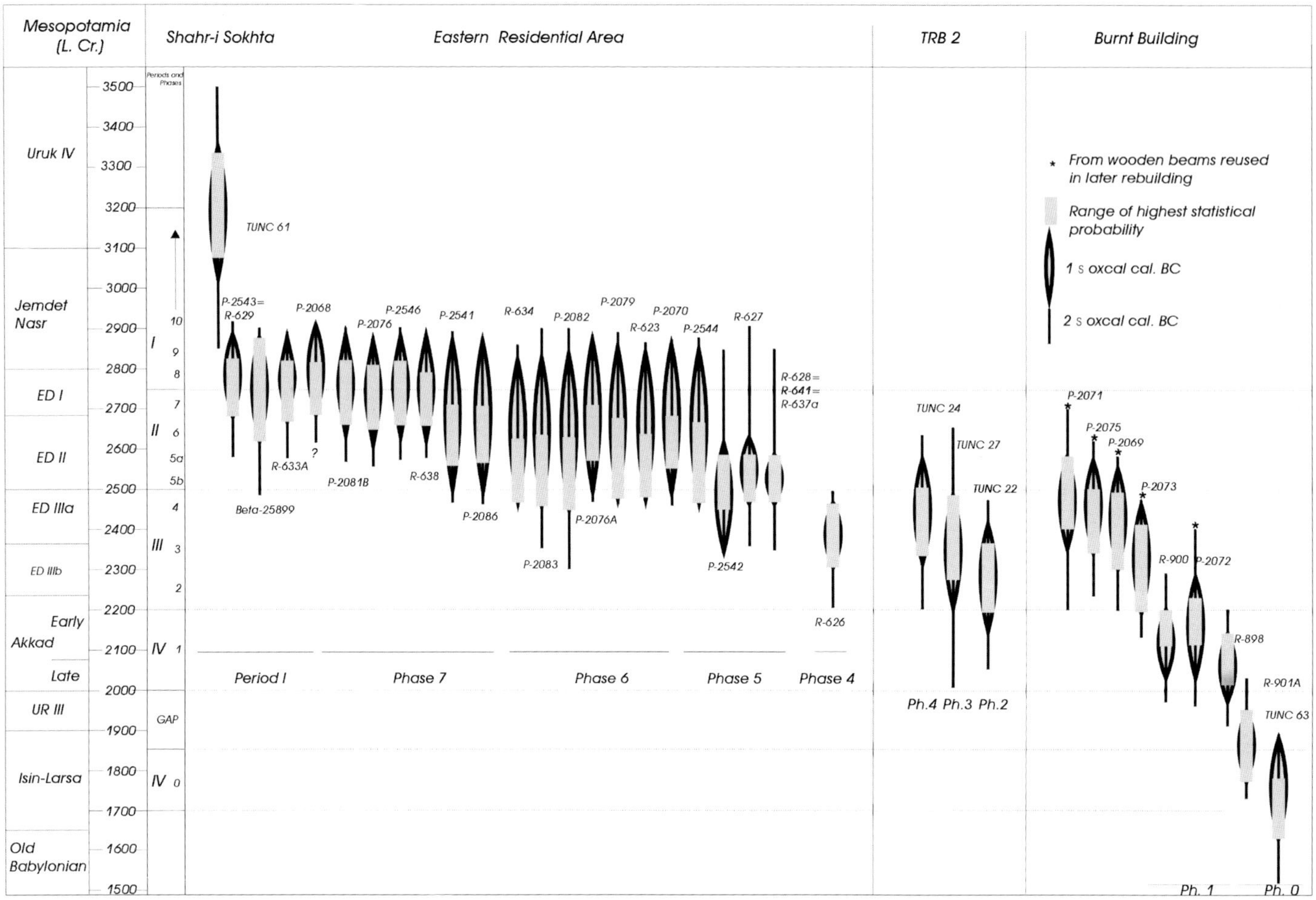

Fig. 2 – *Shahr-i Sokhta, C[14] chronology (SALVATORI and TOSI, 2005: fig. 1).*

radiometric determinations and a palaeomagnetic one from Tepe Rud-i Biyaban 2.[43] Furthermore, considering the recent results from Sohr Damb/Nal excavations,[44] Franke stated that "[...] evidence from Baluchistan, particularly Miri Qalat IIIc-IV and Sohr Damb III-IV, throws new light on this question, in favour of the later dating".[45]

If we accept the first viewpoint, both Shahr-i Sokhta and Mundigak would have disappeared, at least as large urban centres, immediately after 2600 BCE, without meaningful interaction with the urban formation processes of the Indus Integration Era (*ca* 2600-1900 BCE).[46] In the second perspective, the urban peaks of both civilizations would be roughly contemporary, and we should make an effort to define precisely the historical dynamics of the "interaction spheres" between the major civilization of the eastern Iranian Plateau and the Indus world.

CERAMIC LINKS WITH THE SUBCONTINENT: THE NAL CERAMICS

The record of possible ceramic imports or vessels imitating types from Baluchistan and the Indus basin found at Shahr-i Sokhta includes ten fragments and vessels (both from the graveyard and the settlement) close to the Nal production as well as a more limited and heterogeneous group of four ceramic finds similar to other types of products from the same

43. SALVATORI and TOSI, 2005: 288.
44. FRANKE, 2003-2004, 2005 and 2007; FRANKE and IBRAHIM, 2005.
45. FRANKE, 2008: 666.
46. JARRIGE, 1996: 864; JARRIGE and QUIVRON, 2008: 80.

Table 1 – *List and recovery contest of the Nal-like ceramic finds so far on record at Shahr-i Sokhta.*

Number	Description	Provenance	Period	Phase	References
MAI 8724 fig. 3: 1	Fragment of cylinder jar painted with metopes and stepped lines	XDV 11, South	I	10	AMIET and TOSI, 1978: fig. 4a: 23
MAI 8725 fig. 3: 2	Fragment of biconical jar with painted frieze including a pipal leaf	XCIIa, 8	I	10	*Ibid.* fig. 4c: 23
MAI 7041 fig. 4: 3 fig. 5: 3	Cylinder-like jar with well-distinguished neck and a painted frieze of triangles	G.413/5	I	10-9	TUCCI, 1977: fig. 28; PIPERNO and SALVATORI, 2007: fig. 493
1520.5 fig. 4: 6	Cylinder-like jar with reticular motifs on the shoulder and on the body	G.1520.5	I	10-9	SAJJADI, 2003: fig. 19
1700.4 fig. 4: 5	Cylinder-like jar with angular motifs on the shoulder and frieze of lozenges on the outer wall	G.1700.4	I	10-9	*Ibid.*
MAI 8692 fig. 6: 1	Fragment of cylinder-conical bowl with a frieze of zig-zag lines and semicircles	XX, 22	I	9	
Draw Dep.CS.3161 fig. 4: 1	Fragment of globular jar with a frieze of zig-zag lines and semicircles on the outer wall	XX	I	9-8	BISCIONE, 1984: 10.15b; LAMBERG-KARLOVSKY and TOSI, 1973: fig. 54; TOSI, 1983: fig. 38
MAI 2302 fig. 4: 2 fig. 6: 2	Carinated sub-cylindrical beaker with an outer frieze of triangles	Surface	I	?	SARIANIDI, 1983: fig. 12
MAI 6935 fig. 4: 4 fig. 5: 4	Cylinder-like jar with well-distinguished neck and an outer frieze with a vegetal design	G.62/1	I	8-7(?)	PIPERNO and SALVATORI, 2007: fig. 236
MAI 6917 fig. 4: 7 fig. 5: 1	Cylinder jar with saddle-like shoulder decorated with a simple radial motif	G.55inf./1	II	7	*Ibid.* fig. 224
MAI 6127 fig. 4: 8	Cylinder jar with saddle-like shoulder, and an outer frieze of stepped lines	G.10/2	II	5a	BISCIONE *et al.*, 1974: fig.8; PIPERNO, 1977: fig. 121a; D'AMORE, 1999: fig. 107: 139; PIPERNO and SALVATORI, 2007: fig. 47

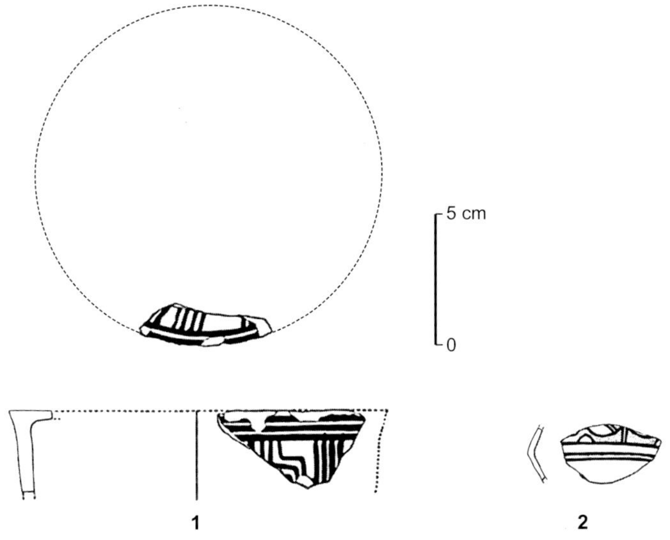

Fig. 3 – *Shahr-i Sokhta. The Nal group specimens.* ***1****, MAI 8724, 1,7 cm high (AMIET and TOSI, 1978: fig. 4a: 23), Period I, phase 10;* ***2****, MAI 8725, 2 cm high (ibid., 1978: fig. 4c: 23), Period I, phase 10.*

areas. Table 1 and fig. 3 present the complete list with the precise context of recovery of the ten Nal-like finds so far recovered.

MAI 8724,[47] dated to Period I, phase 10 (fig. 3: 1) was recovered on the south side of room XDV, artificial cut 11. This is one of the oldest deposits so far excavated at Shahr-i Sokhta. On the outer wall, metopal fields bounded by thick, vertical and horizontal lines include series of stepped lines. On the shoulder the remnants of further parallel radiating lines or possibly more metopes[48] are visible. Stepped lines are frequently encountered in Nal ceramic decoration;[49] for instance in Sohr Damb/Nal[50] or, again, along the Porali river in Jhalawan.[51]

47. AMIET and TOSI, 1978: fig. 4a: 23.

48. The originally Greek word metopē (from meta- + opē opening; akin to Greek ōps eye, face) indicates the space between two triglyphs of a Doric frieze often adorned with carved work (according to Merrian-Webster Online Dictionary). Here we used the term to describe the geometric space between two equal decorative elements often formed by one or more parallel vertical painted lines.

49. See for instance FAIRSERVIS, 1975: fig. 40, pl. 13 and figs. 41: 35.

50. HARGREAVES, 1929: pl. XVII: 20-21, pl. XVIII: 16.

51. See RAIKES, 1968: fig. 5: 1, 3, 6; fig. 7: 3; fig. 8: 2, 18; fig. 9: 9, 28; fig. 10; fig. 11: 26; fig. 13: 10.

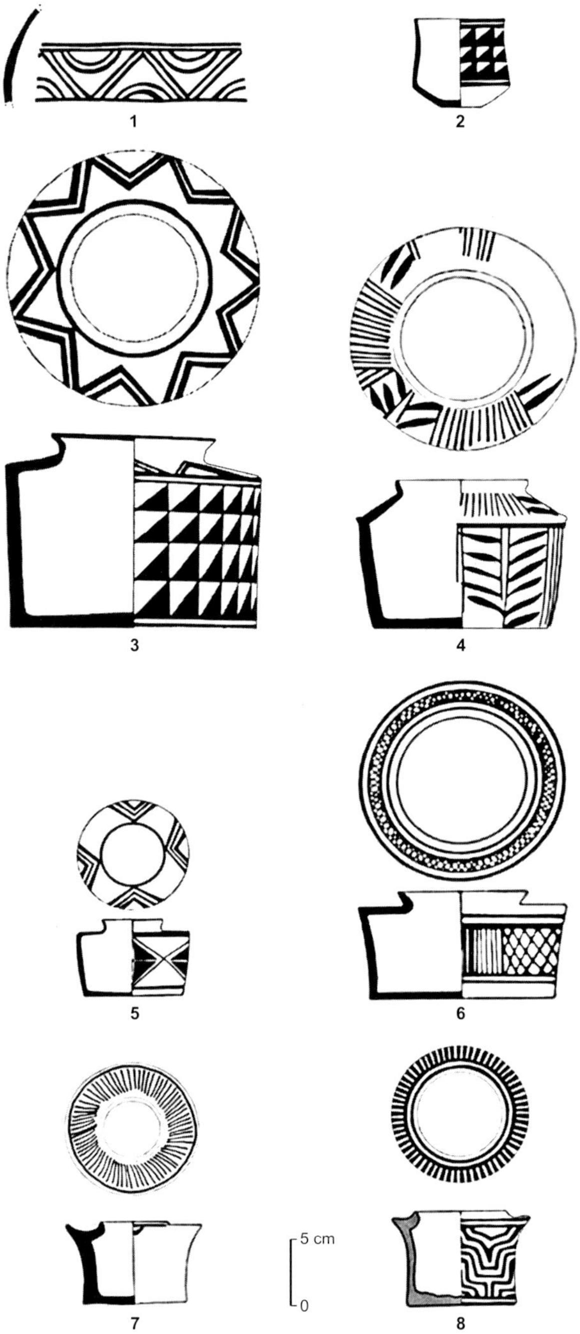

Fig. 4 – *Shahr-i Sokhta, Nal-related ceramics.* ***1***, *Draw Dep. CS. 3161, 9 cm high (BISCIONE, 1984: fig. 10.15b), Period I, phases 9-8;* ***2***, *MAI 2302, 7 cm high (SARIANIDI, 1983: fig. 12) Period I (?);* ***3***, *MAI 7041, 15 cm high (PIPERNO and SALVATORI, 2007: fig. 493), Period I, phases 9-8;* ***4***, *MAI 6935, 11,7 cm high (ibid.: fig. 236), late Period I, phases 8-7(?);* ***5***, *1700.4, 8,2 cm high (SAJJADI, 2003: fig. 19), Period I, phases 10-9;* ***6***, *1520.5, 7,5 cm high (ibid.), Period I, phases 10-9;* ***7***, *MAI 6917, 6,4 cm high (PIPERNO and SALVATORI, 2007: fig. 224), Period II, phase 7;* ***8***, *MAI 6127, 7,2 cm high (ibid.: fig. 47), Period II, phase 5a.*

Fig. 5 – *Shahr-i Sokhta, Nal-like canisters jars from the graveyard.* ***1***, *MAI 6917, Period II, phase 7 (see also fig. 4.7);* ***2***, *MAI 6935, Period II, phase 7(?) (see also fig. 4.4);* ***3***, *MAI 7041, Period I, phases 10-9 (see also fig. 4.3). No scale.*

MAI 8725 (fig. 3: 2) is a fragment of the wall of a biconical jar with a polychrome frieze including a *pipal* leaf embellished by a geometric filling. Found in room XCIIa, 8, it dates back to Period I, phase 10.[52] Hargreaves[53] described similar motifs on Nal wares found at Sohr Damb/Nal. Franke has recently published similar vessels found in her recent excavations at the same site.[54]

MAI 7041 (figs. 4: 6 and 5: 3) (Period I, phase 10/9) comes from G. (grave) 413.[55] The motif on the outer wall, with black and white triangles (see also MAI 2302 below), creates a bichrome effect and may be matched with a specimen from Jhalawan.[56] On the shoulder there are two thick parallel lines forming a zig-zag motif running around the neck. The vessel is deemed by Salvatori and Tosi as "surely imported".[57]

Vessel 1520.5 (fig. 4: 6) is a cylinder-like jar with a slightly concave profile, a raised, outward projecting rim and a horizontal plain shoulder. The outer wall is divided by series of vertical strokes into a sequence of metopes filled with a cross-hatching motif. On the shoulder there is a similar frieze delimited by three or more concentric lines. This motif is not common in the Nal tradition but it appears on some vessels

52. AMIET and TOSI, 1978: fig. 4c: 23.
53. HARGREAVES, 1929: pl. XVII: 24-25; pl. XXI: 1-4.
54. FRANKE-VOGT, 2005: figs. 2 and 60.
55. TUCCI, 1977: fig. 28; SALVATORI and TOSI, 2005: 283, fig. 2,2; PIPERNO and SALVATORI, 2007: fig. 493.
56. RAIKES, 1968: fig. 6: 14.
57. SALVATORI and TOSI, 2005: 283.

from Sohr Damb/Nal[58] and the similar forms appear in the publication of the recent excavations at the site.[59]

G. 1700.4 (fig. 4: 5) is another, similar cylinder-shaped jar with a raised, short neck and horizontal shoulder. Four angular motifs appear on the shoulder and a frieze of lozenges subdivided by cross-designs into four solid-filled triangles runs on the outer wall. Although the grave furnishings from this grave were not fully published, such motifs belong generically to the geometrical tradition of Period I, and sometimes they occur also on Nal wares.[60]

MAI 8692 (fig. 6: 1) is a fragment of a cylinder-conical bowl with a frieze of zig-zag lines and semicircles identical to the jar fragment Dep. CS. 3161 (fig. 4: 1), part of a globular jar bearing a wide painted band on the maximum expansion. Both may be ascribed to a late Period I context. The form of MAI 8692 is quite unusual at Shahr-i Sokhta and the vessel might have been imported. While this painted motif is very common in Nal wares,[61] this vessel may be compared with the Nal-like specimen found at Tepe Yahya, dated to the late 4th millennium BCE.[62]

MAI 2302 (fig. 4: 2) is a carinated sub-cylindrical beaker found on surface. The decoration is the same described for MAI 7041 jar (see above). The vessel has no stratigraphic context and was published by V. Sarianidi[63] as an import from Southern Turkmenia, but it is more likely a Nal-like product.[64] The discussion is anyhow open, and might be solved only by future analytical studies.

MAI 6935 (figs. 4: 4 and 5: 2) is a cylinder-shaped jar with a sloping shoulder, a short everted rim and outer frieze of metopes filled with a vegetal design. The vessel bears the same vegetable motif all around the mouth. Similar designs are well known in the Nal tradition, although U. Franke[65] considers this canister shape quite unusual for typical Sohr Damb forms.[66]

MAI 6917 (figs. 4: 7 and 5: 1) is a smaller cylindrical jar with a saddle-like shoulder decorated with a simple radial motif; the mouth is formed by a short everted rim. The decoration of the outer body included festoon-like designs but has almost completely vanished.

58. HARGREAVES, 1929: pl. XIX: 9.

59. FRANKE-VOGT, 2005: fig. 3.

60. See HARGREAVES, 1929: pl. XVII: 48.

61. *Ibid.*: pl. XVII: 46-48, 51; FAIRSERVIS, 1975: figs. 41, 33-35, 43-44; fig. 40; MARSHALL, 1931, III: pl. XCIII, 16N, 20N; RAIKES, 1968: fig. 13: 14.

62. LAMBERG-KARLOVSKY, 1970: pl. 34.

63. SARIANIDI, 1983: fig. 12.

64. HARGREAVES, 1929: pl. XVI: 27; pl. XIX: 11.

65. U. Franke, personal communication.

66. For the motif, see HARGREAVES, 1929: pl. XVII: 25; pl. XVIII: 15; FAIRSERVIS, 1975: fig. 41: 22.

Fig. 6 – *Shahr-i Sokhta, Nal-related ceramics.* ***1****, MAI 8692, fragment of a cylinder-conical bowl with a frieze of zig-zag lines and semicircles, Period I, phases 8-9;* ***2****, MAI 2302, carinated beaker with frieze of triangles on the exterior. Period I, found on surface (see also fig. 4.2).*

Finally, MAI 6127 (fig. 4: 8) a quite similar cylinder-shaped jar with saddle-like shoulder comes from Grave 10 (10/2). It bears on the outer wall a continuous frieze of repeated stepped lines, a design of radiating lines on the shoulder. It was used as a lid for a large globular jar with a short cylindrical neck.[67] As stated before, such stepped lines are a very common expedient in the Nal graphic tradition. Finally, the inventory of Nal-related wares in the Helmand civilization should also include some probable fragments found in the test trench at Said Qala

67. PIPERNO and SALVATORI, 2007: fig. 46.

Tepe[68] and other finds at Mundigak,[69] ascribed to Periods III to IV of the general site sequence.

The first seven specimens date back to Period I (about 3200-2800 BCE). Although information on the provenience is not currently available, the earliest specimens, found together with nose-lug painted jars, Southern Turkmenian geometric pottery, Jemdet Nasr style seal impressions and a single inscribed proto-elamite tablet, might be actual imports. Later Nal-like vessels could have been made at Shahr-i Sokhta or other sites imitating the Nal style. The latest specimens on record are datable to late Period II (phase 5a of the Central Quarters ceramic sequence, mid 3rd millennium BCE). As far as cylindrical jars—the most common type—are concerned, earlier cylindrical specimens with a flat/sloping shoulder and a clearly distinguished neck (such as MAI 7041 and probably MAI 8724) might have transformed into smaller versions with a concave profile, a limited concave shoulder, marked on the upper corner point by a sharp ledge in Period II. In these later small jars, the neck disappears, turned into a simple vertical projection of the shoulder itself. At Mehrgarh, versions that might be intermediate between the two forms appear starting from Period V.[70] In the test trenches excavated at Anjira and Siah Damb, Surab (Pakistan) the first type was encountered in Period III[71] while the second type appears in Period IV.[72]

ANOTHER VESSEL AND MORE POTSHERDS

Franke mentions a brown-slipped Kot-Dijian jar found in the Central Quarters, originally described by the excavators as "definitely extraneous to the local production of Shahr-i Sokhta for its shape, ceramic texture and surface treatment"[73] (fig. 7: 1). This jar, MAI 8383, was generically compared to flat sub-globular vessels commonly found in Kot-Dijian complexes.[74] The vessel was found in room CDLXX of the "House of the Jars" in the Central Quarters. The upper part of the vessel was found in a secondary position in the dump of late Period II, phase 5b, but the base was recovered in its primary context of deposition, a basal layer of phase 7 (around 2900-2800 BCE). Although Franke remarks that "Similarities are confined to the

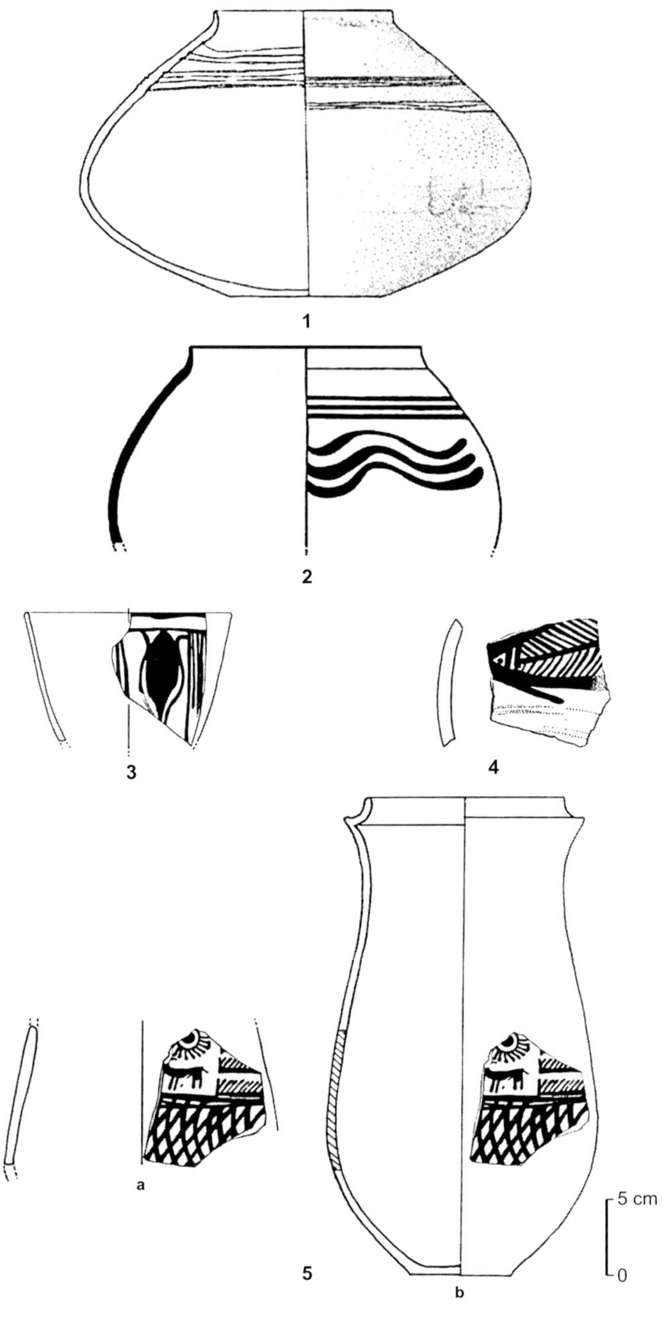

Fig. 7 – *Other Indus related vessels from Shahr-i Sokhta and Mundigak.* ***1****, Shahr-i Sokhta. MAI 8383 Kot-Dijan jar, Period II, Phase 7;* ***2****, Mundigak. Possible Kot-Dijian jar (CASAL, 1961: fig. 253);* ***3****, Shahr-i Sokhta. MAI 8666, high truncated-cone shaped bowl fragment with a pipal leaf decoration, found on surface (see also fig. 8);* ***4****, Shahr-i Sokhta. MAI 8734 Globular jar potsherd with "X-rayed" fish design, found on surface (see also fig. 9);* ***5a****, Shahr-i Sokhta. MAI 8675, fragment of a black-on-red painted vessel, found on surface (see also fig. 5a and fig. 10);* ***5b****, Shahr-i Sokhta. MAI 8675, black-on-red painted jar (conjectural reconstruction).*

68. SHAFFER, 1978: 63-65.
69. CASAL, 1961: fig. 58: 113; fig. 87: 361; fig. 93: 413.
70. JARRIGE *et al.*, 1995: 152, figs. 2-14 n-o.
71. See DE CARDI, 1965: fig. 16:7.
72. *Ibid.*: fig. 17: A7.
73. SALVATORI and VIDALE, 1997: 78.
74. KHAN, 1965; HALIM, 1971.

Fig. 8 – *Shahr-i Sokhta. High truncated-cone shaped bowl fragment with a pipal leaf decoration found on surface (photograph by F. Aghadian).*

groove decoration, but the shape is different",[75] in our view the short, vertical sharp neck is a feature of Kot-Dijian ceramic traditions; the lower body was obtained by paddle and anvil, a technique recurring in the Indus valley, but otherwise totally absent from Shahr-i Sokhta. Similarly, the series of horizontal parallel grooves made by combing inside and outside the shoulder finds no matches in the technical inventory of the Shahr-i Sokhta production.[76] As remarked by Franke, another possible Kot-Dijian jar (fig. 7: 2) is reported from Mundigak.[77]

These likely imports from the Indus core area are not isolated. A fragment of a high truncated-cone shaped bowl with S-shaped profile, a simple raised rim and a pipal leaf decoration (MAI 8666, fig. 7: 3 and fig. 8), found on its surface, is equally extraneous to local wares, and might have been imported from the south-east or from the east. This small and tall bowl, 13 cm wide at the mouth, and originally 12-13 cm high, was manufactured on a potter's wheel and carefully turned on the same device along the base. It was baked in a strongly oxidizing atmosphere: the colour of the paste, yellowish red (Munsell 5YR 5/8), contrasts with the yellowish-buff hue dominating the common buff wares of Shahr-i Sokhta. The compact and homogeneous paste is devoid of macroscopic inclusions and is covered by a reddish yellow slip (5YR 6/6). The pigment, poorly preserved, is reddish brown (2.5 YR 5/6). Two parallel lines run below the rim and limit a frieze divided in a series of long and narrow metopes, separated by bands of five or more parallel lines. The only surviving metopa shows a vertical elongated leaf, with two symmetrical swirling lines rising from the base on both sides of the leaf. *Pipal* leaves are a distinctive motif of the pre-Indus ceramic complexes across wide regions of the Subcontinent; they become very common in the Kot-Dijian phase (approximately 2800-2600 BCE), and often appear on the famous *verres ballon* of the Helmand civilization.[78] The variant with the symmetrical swirling elements is well-known at Kalibangan, in Haryana, where it occurs in different versions, painted or incised in a plastic state.[79] Besides such a possible link with Kalibangan and Haryana, similar sherds are reported from an enormous region, stretching from Kech Makran, Period IIIb[80] (about 2800-2600 BCE) to Mundigak.[81] Although this potsherd demonstrates another connection with the Subcontinent in the first half of the 3rd millennium BCE, it seems impossible to further detail its geographical context of provenience.

Another potsherd[82] (MAI 8734, fig. 7: 4 and fig. 9), found on the surface of Shahr-i Sokhta and presently on exhibit at the Museo Nazionale d'Arte Orientale "G. Tucci" in Rome, provides a link with northern Baluchistan during the Kot-Dijian phase. It belonged to a globular buff ware jar painted on the exterior with an "X-rayed" fish design. Such a motif is well known at Rehman-Dheri.[83] The details of the eye, placed in the extremity of the head, and its proximal fins are identical; nonetheless, the orientation of the fish-bone, anatomically correct in the Rehman-Dheri specimens, in the fragment of fig. 4: 4 is inverted.

A fragment of a black-on-red painted vessel (MAI 8675, fig. 7: 5 and fig. 10), equally collected from the surface of Shahr-i Sokhta, may be related to the Indus valley both on stylistic and technical grounds. The fine reddish yellow paste (7.5 YR 7/6) shows a limited amount of pores and rare white

75. FRANKE, 2002: 349.
76. SALVATORI and VIDALE, 1997: fig. 95: 2.
77. CASAL, 1961: fig. 253.

78. For Mundigak, see CASAL, 1961: fig. 56: 84; fig. 59: 116; figs. 64-65; fig. 88: 372; fig. 93: 412; fig. 102: 485; fig. 104: 502; fig. 128: 711. For Shahr-i Sokhta, in the settlement area, see SALVATORI and VIDALE, 1997: fig. 185: 11; in the graveyard, see PIPERNO and SALVATORI, 1983: pl. VIIc; SAJJADI, 2003: fig. 23e; PIPERNO and SALVATORI, 2007: figs. 674-675.
79. SANKALIA, 1974: 346.
80. DIDIER, 2007: 4-169/P12.
81. CASAL, 1961: fig. 65: 179-182; fig. 88: 372; fig. 93: 412; fig. 102: 485; fig. 128: 711.
82. TOSI, 1973: fig. 10c.
83. DURRANI, 1981; SHAH and PARPOLA, 1991: 376-377, Rhd 255-260.

Fig. 9 – *Shahr-i Sokhta. MNAOR 8734 (TOSI, 1973: fig. 10c). Globular jar potsherd with "X-rayed" fish design, found on surface. No scale.*

Fig. 10 – *Shahr-i Sokhta. MAI 8675 Fragment of an Indus black-on-red painted jar, found on surface (photograph by F. Aghadian).*

inclusions, and is covered inside and outside by a yellowish red slip (5 YR 5/6), with irregular horizontal-oblique burnishing marks. Technical features suggest a coiling and wheel-throwing process, each coil of the wall being about 3.5 cm high. This potsherd, too, finds no comparison with the local buff ware production. It belonged to an elongated, probably pear-shaped jar, about 17 cm wide at its maximum diameter, originally located at the lower edge of the preserved sherd portion, where a coil joint is still visible. The outer surface of the vessel was painted with a dark reddish grey pigment (5 YR 4/2) in superimposed registers or friezes. The upper frieze retains part of a quadruped, most probably a male goat or an antelope, below a sun-like radiating circle filled with a dot; to the right one sees what is left of three superimposed "combs" with oblique teeth, turned upside down. The lower register, separated by the upper one by a double line, is covered by a cross-hatched pattern. The vessel is easily distinguished from the local product because of its form, technical features and painted figuration. The only possible match is an Indus pear-shaped jar (as reconstructed in fig. 7: 5b), the most intensively decorated vessel of the Indus valley ceramic repertory. These jars may have a rounded or flat base,[84] and are lavishly painted with dark pigments on bright red slips on the greatest part of the outer surface. Figures, as a rule, are organized in superimposed registers and often metope-like subdivisions. The four motifs visible on the Shahr-i Sokhta potsherd, namely the radiating "sun" filled with a dot, the animal, the upturned "combs" (interchangeable with branches of tree-like patterns), and cross-hatched horizontal bands figure among the most common designs on this peculiar ceramic type.[85] An example of "net" motif associated with the radiating "sun" design also appears on a vessel ascribed to the Nal tradition.[86] The designs, in our case, are carefully traced. As painted figuration, during the long time-span of the Integration Era, is affected by a gradual process of smoothing, we would tentatively date our sherd to a Harappa 3A horizon (approximately 2600-2450 BCE).

OTHER CLASSES OF ARTEFACTS

Besides ceramics, the artefact record of similarities or imports from the Indo-Pakistani Subcontinent found at Shahr-i Sokhta and Mundigak is both substantial and heterogeneous. The most relevant classes of artefacts and their implications

84. See MARSHALL, 1931: pl. LXXXV: 6; pl. LXXXVII: 4-5; MACKAY, 1938: pl. LIV: 5, LVIII: 7; DALES and KENOYER, 1986: 74-76, fig. 1.
85. See, among many possible examples, the tables in MARSHALL, 1931: pl. XC-XCIII; MACKAY, 1938: pl. LXVIII- LXX; MACKAY, 1943: pl. XXX-XXXVII; DALES and KENOYER, 1986: 75.
86. PARPOLA, 1981: fig. 23: 7.

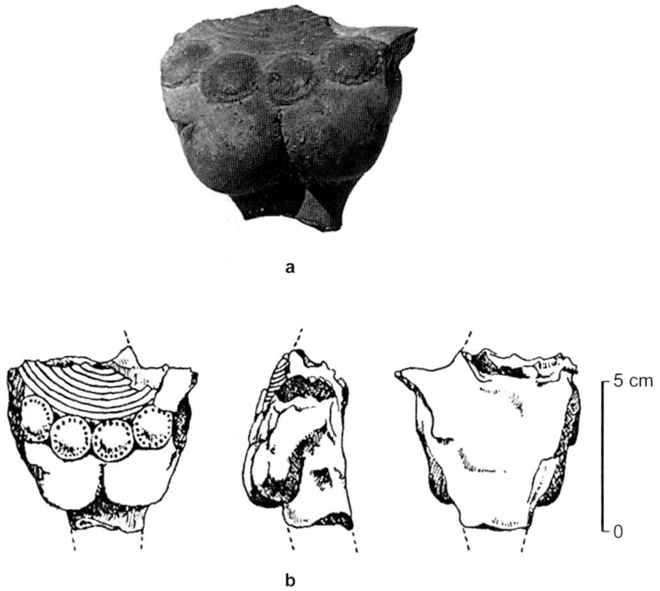

Fig. 11 – *Shahr-i Sokhta. MAI 2307, Terracotta female figurine in "Zhob" style, Period II, phase 6 (**a**, photograph by F. Bonardi; **b**, drawing by V. La Bianca).*

are briefly discussed in the following pages. They include (in chronological order) a famous fragmentary figurine in unbaked clay with Zhob-like features, a substantial series of terracotta cakes and fragments of the same objects, at least three types of stamp seals and three types of beads, the same type of terracotta mouse traps found at Mundigak, Mohenjo-Daro and Bampur, different types of gaming pieces and dice, two types of shell ornaments, a fragmentary cylinder seal in steatite with an Indus inscription probably found in Sistan, three well-known fragments of small-to-medium sized male stone statues.

ZHOB-LIKE FRAGMENTARY CLAY FIGURINE

This fragment[87] (fig. 11) (MAI 2307) was found at Shahr-i Sokhta in room XIX, 8, and ascribed to Period II, phase 6: it should be dated around 2650-2570 BCE, according to Salvatori and Tosi.[88] The female torso, 5.5 cm high and 6.6 cm wide, shows a series of large disk-like beads embellished by dots at the edge, hanging below a multiple necklace. The circulation of similar figurines in the centres of the Helmand civilization is also demonstrated by the discovery of two other fragmentary specimens at Mundigak.[89] The Mundigak figurines were ascribed by the excavator to Period IV,1 and IV,2, *i.e.* to the first half of the 3rd millennium BCE, a date consistent with the evidence from Shahr-i Sokhta. These figurines, like the potsherd with the "X-rayed fish" point to connections with sites in northern Baluchistan occupied during the Kot-Dijian phase.[90]

TERRACOTTA CAKES

Variously called "terracotta tablets",[91] "triangular plaques"[92] or "triangular terracotta cakes"[93] these artefacts (fig. 12, tables 2 and 3), made of coarse chaff-tempered clay, are a very common find in several protohistoric sites of the Subcontinent from the late Regionalization Era (2800-2600 BCE) to the Localization Era (1900-1700 BCE). In this latter time-span they frequently assume irregular rounded shapes, to finally retain the form of a lump of clay squeezed in the hand. Despite abundant and often unnecessary speculation, archaeological evidence demonstrates that they were used in pyrotechnological activities, both in domestic and industrial contexts. The most likely hypothesis is that these objects, in the common kitchen areas, were heated to boil water, and used as kiln setters in other contexts. Shahr-i Sokhta is the only site in the eastern Iranian plateau where such terracotta cakes, triangular or more rarely rectangular, are found in great quantity. Their use, perhaps by families or individuals having special ties with the Indus region, might have been part of simple domestic activities, but this conclusion is questioned by the fact that several terracotta cakes, at Shahr-i Sokhta, bear stamp seal impressions or other graphic signs (in more than 30% of the total cases). In many cases the actual impressions are poorly preserved, and require detailed study. Perhaps these objects used in some form of administrative practice.[94] Although many specimens are fired or burnt, a small percentage of the "cakes" found at Shahr-i Sokhta is unfired (see table 2). On the other hand, their modification in the frame of one or more unknown semantic contexts is not unknown in the Indus valley. At Kalibangan (Haryana, India), for example, two terracotta cake fragments respectively bear a cluster of signs of the Indus writing system and a

87. TOSI, 1983: pl. LXIX: 61.
88. SALVATORI and TOSI, 2005: 285.
89. CASAL, 1961: 253, pl. XLI: 8-9.
90. For Damb Sadaat II-III, see FAIRSERVIS, 1956: 224, fig. 16: a-f; for Periano Ghundai in the Zhob valley, FAIRSERVIS, 1959: 330-331, fig. 5: c-d; for Mehrgarh, Periods VI-VII, see JARRIGE *et al.*, 1995: fig. 2: 36-37.
91. MARSHALL, 1931: 480.
92. MACKAY, 1938: 429.
93. RAO, 1985: 517.
94. E. Cortesi, ongoing research.

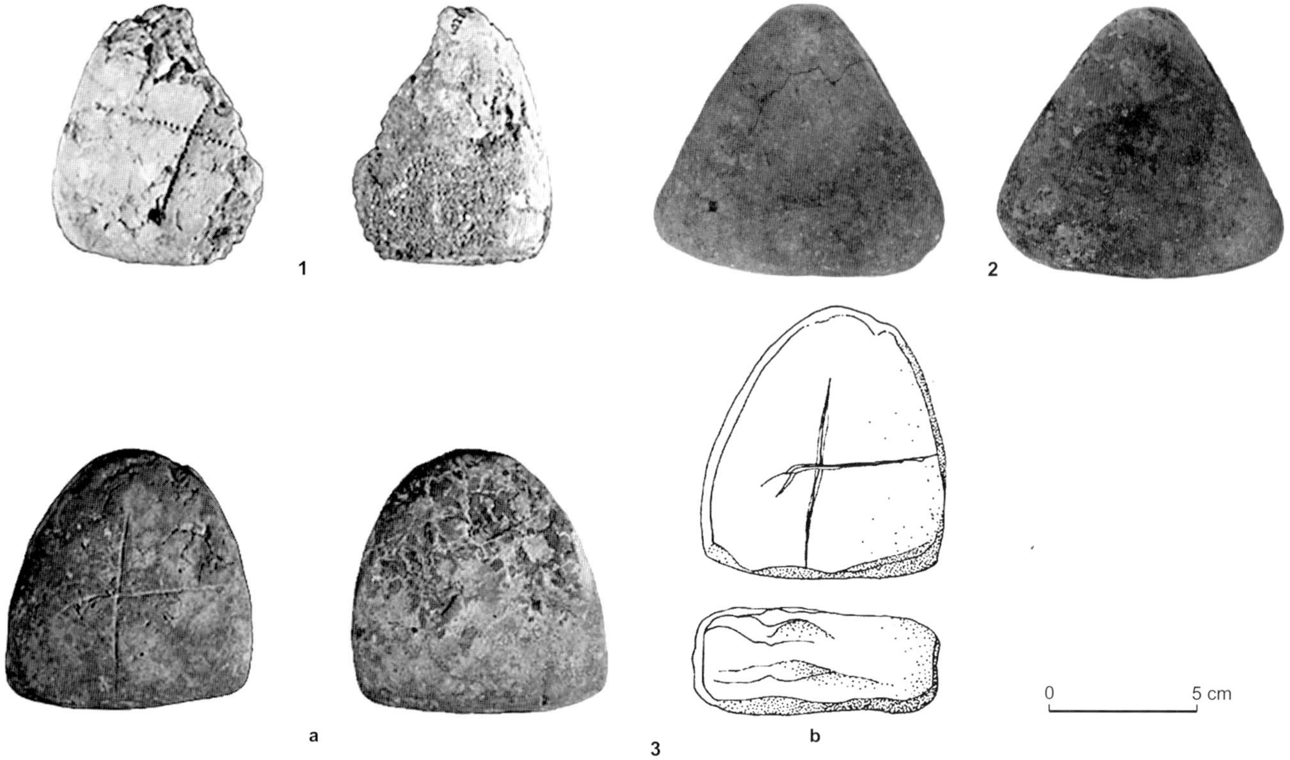

Fig. 12 – *Shahr-i Sokhta, terracotta cakes, Periods II and III.* ***1****, MAI 1026 (front and rear);* ***2****, MAI 376 (front and rear);* ***3****, MAI 9794 (**3a**, photograph of front and rear;* ***3b****, drawing).*

possible scene of animal sacrifice in front of a possible divinity.[95] While a terracotta cake found at Chanhu-Daro (Sindh, Pakistan) bears a star-like design, another has three central depressions.[96] The most important group of incised terracotta cakes comes from Lothal, where the record includes specimens with vertical strokes, central depressions, a V-shaped sign, a triangle, and a cross-like sign identical to those found at Shahr-i Sokhta.[97]

Tables 2 and 3 show a complete inventory of these objects (most so far unpublished), their provenience and proposed dating, and finally summarize their frequencies across the Shahr-i Sokhta sequence.

The data suggest that terracotta cakes are absent from Period I. This might be due to the very small amount of excavated deposits in the earliest settlement layers, but the almost total absence of terracotta cakes in layers datable to phases 8-7, exposed in some extension both in the Eastern Residential Area and in the Central Quarter, is remarkable. The majority of the finds belong to Period II, phases 6 and 5 (amounting together to about 60% of the cases). As the amount of sediments investigated for Period III in the settlement areas, for various reasons, is much less than what was done for Period II, the percentage of about 40% obtained for Period III (which, we believe, dates to the second half of the 3rd millennium BCE) actually demonstrates that the use of terracotta cakes at Shahr-i Sokhta continued to increase.

STAMP SEALS

Three types of stamp seals known at Shahr-i Sokhta might imply links with the Subcontinent. The first one (MAI 115) is attested by a fragment of a rectangular stamp seal found on surface and described in the records as "white limestone" (but it could also be in fired steatite). Split in half, the seal originally bore a grid of incised lines, three of which are

95. JOSHI and PARPOLA, 1987: K-79A and K-79B, K-119A and K-119B: 324.
96. MACKAY, 1934: pls. XXXVIII: 1 and XCII: 27.
97. See RAO, 1985: 517-518, pls. CXXXI-CXXXIV.

Table 2 – *Inventory of terracotta cakes so far found at Shahr-i Sokhta (mostly so far unpublished), with context of provenience and dating. After* SALVATORI *and* VIDALE, *1997: 79.*

MAI	Description	Size (in cm.)	Square	Room	Cut	Period	Phase
4087	Fragment of small triangular cake	7,3 x 8,00		CDLXVIII	7	II	7
4758	Triangular cake	4,4 x 5,8 x 7,4		XII	8	II	6
5195	Corner fragment of triangular cake	4 x 3,6		XII	8	II	6
5196	Corner fragment of triangular cake, burnt	3,6 x 7,1		XII	8	II	6
1200	Triangular cake, bearing a square stamp seal impression, unbaked	2,1 x 9,5 x 10,2	XIH	XXV	6	II	6
5199	Triangular cake, bearing two rectangular stamp seal impressions	11,2 x 11,4		LXXVII	7	II	6
1026	Fragment of triangular cake bearing a cross-like sign made with dots	4 x 6,45 x 8,65	XIG/XIH	XX	8	II	6
6864	Corner fragment of triangular cake, burnt	2,4 x 6,30	SCP, S		2	II	5
6876	Corner fragment of triangular cake, burnt	2,9 x 5	XHJ		13b	II	5
525	Fragment of triangular cake	3,2 x 5,5 x 10,5	XIH		2	II	5
637c	Rectangular cake, unbaked	2,5 x 6,4	RYL		4	II	5a
376	Triangular cake	3 x 9,5 x 9,2	RYL		5-6	II	5a
726	Fragment of triangular cake, burnt	3,2 x 5 x 6,2	RYL	Under wall	7	II	5a
5197	Rectangular cake, burnt, unbaked	4,6 x 7,3		CC	6	II	5a
6880	Fragment of triangular cake, with square stamp seal impression	3,4 x 6,7		CDII	3	II	5a
582	Triangular fragment of cake	3,9 x 5,7 x 6,2	XIB	XVIII	2-5	II	5a
493	Fragment of triangular cake	3,5 x 4,5 x 5,3	XIB	XVIII	3	II	5a
662	Fragment of cake	4,5 x 5,7	RYM		2	II	5b
6836	Triangular cake, bearing a trapeze-shaped stamp seal impression	4,3 x 9,6		DIII	5	III	4
6837	Triangular cake, bearing a copper square stamp seal impression	4,1 x 7,2		DII	2	III	4
6838	Triangular cake, bearing a stamp seal impression with a cross design, burnt	3,3 x 9,5	EWP		1	III	4
9794	Triangular cake, bearing a linear cross-like sign	3,5 x 8,4 x 8,9		DIII	3	III	4
nd	Fragment of triangular cake	nd	SCP			III	4*
nd	Fragment of cake, indeterminable shape bearing a seal impression	nd	NXK			III	4*
5081	Fragment of triangular cake, bearing a stamp seal impression	6,1 x 10,6	Surface			III	4-3
2243	Fragment of cake, indeterminable shape, bearing a square seal impression, unbaked	2 x 6,5 x 8,3	XHI		2-3	III	4-3
3368	Fragment of triangular cake, unbaked	3,5 x 8,1	XIG	XXI-XXII	4	III	3
2533	Fragment of triangular cake, probably with round stamp seal impression, unbaked	3,2 x 4,5 x 9,7		CXX Corridor	3	IV	1
3106	Round flat cake, unbaked	nd		CXX Corridor	2	IV	0
5198	Round cake, two pierced ends, unbaked	4,8 x 8,7	XJA	Section	6	nd	nd
430	Triangular cake	3,5 x 9 x 9	RYM	Bulk		nd	nd

Table 3 – *Number and percentages of terracotta cakes so far found at Shahr-i Sokhta, by Periods and Phases. nd = 2; tot. 31 (100%).*

Period	Phase	Nr.	%
II	7	1	3.4
II	6	6	20.7
II	5	11	38
III	4-3	9	31
IV	1-0	2	6.9

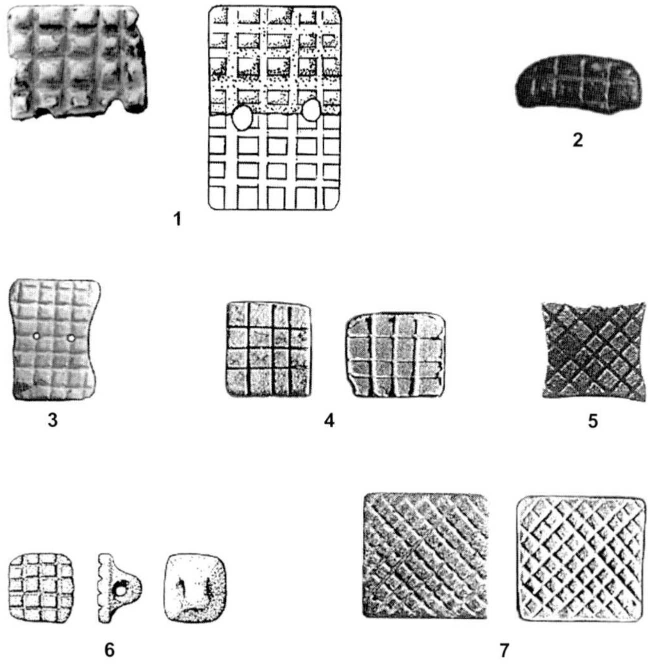

Fig. 13 – *Stamp seals bearing grid-like motives from Shahr-i Sokhta and other sites. Scale not available.* ***1****, Shahr-i Sokhta, MAI 115, found on surface (photograph by F. Bonardi and drawing by M. Vidale);* ***2****, Rehman Deri (DURRANI, 1994-1995: 204, fig. 2);* ***3****, Rehman Deri (ibid.: 202, fig. 2);* ***4****, Banawali (JOSHI and PARPOLA, 1987: 346, B-20 A);* ***5****, Tarakai Qila (SHAH and PARPOLA, 1991: 414, Trq-4 A);* ***6****, Ras' al-Jinz (CLEUZIOU and TOSI, 2000: fig. 16,7);* ***7****, Mohenjo-Daro (JOSHI and PARPOLA, 1987: 86, M-351 A,a).*

orthogonal to the short side, and perhaps seven are in the other direction (fig. 13: 1). Grid-like designs, although admittedly rather simple, are not common in the Helmand seal repertory, but frequent in the Indus region. Figure 13 compares this specimen with similar stamp seals from Rehman-Dheri[98] (northern Baluchistan, Pakistan) (here fig. 13: 2-3), with a square steatite seal (fig. 13: 4) found at Banawali[99] (Haryana, India), and with another example (fig. 13: 5) from Tarakai Qila[100] (northern Baluchistan, Pakistan). While this type of seal was definitely used in the Kot-Dijian phase, its use in the Integration Era as well is suggested by the find of similar specimens (fig. 13: 6) at Mohenjo-Daro,[101] Ras' al-Jinz[102] (fig. 13: 7) and Lothal.[103] The popularity of similar simple grid patterns is also demonstrated by a series of cubic terracotta tokens from Mohenjo-Daro[104] and similar seal-like objects from Harappa.[105] The best match for the Shahr-i Sokhta grid seal is the specimen from Rehman-Dheri, and it is possible (if not likely) that the former find dates back to the first half of the 3rd millennium BCE.

To Rehman-Dheri, and northern Baluchistan in general, is related another fragmentary seal (MAI 6032) from the surface of the protohistoric city (fig. 14: 1). It is a fragment of a fired steatite seal originally having a distinctive lobate contour. Each lobe hosted a double concentric circle, with a central dot. In this specimen survive two lobes, part of a third, and two holes asymmetrically placed in the centre for the string. Originally the object had five protruding lobes. Almost identical seals (fig. 14: 3-5), with 4 to 6 lobes were found at Rehman-Dheri.[106] The identity of the Shahr-i Sokhta seal fragment to those found at Rehman-Dheri places the contact within the Kot-Dijian time horizon (*ca* 2800-2600 BCE). Other fourfold lobate seals come from the nearby site of Tarakai Qila[107] (fig. 14: 6), from Bhirrana,[108] India, and from Nausharo[109] (fig. 14: 7). A rectangular seal in bone, bearing six concentric circles (MAI 7649, fig. 14: 2) was found in Grave 801/3.[110] The grave has only three artefacts, but one is a copper/bronze pin with the head formed by spirals, comparable to a specimen found in the Central Quarters, sq. NXK, cut 2,[111] would ascribe the grave to phase 5B, *i.e.* between 2600 and 2500 BCE. Steatite stamp seals with concentric circles, but included in square or round frames,

98. DURRANI, 1994-1995: 202-204, fig. 2.
99. JOSHI and PARPOLA, 1987: 346, B-20A-D.
100. SHAH and PARPOLA, 1991: 414, Trq-4 A.
101. JOSHI and PARPOLA, 1987: 86, M-351.
102. CLEUZIOU and TOSI, 2000: fig. 16: 7.
103. RAO, 1985: pl. CCLXXXII: A 7-8, 11-12; pl. CCXCIII: 4.
104. JOSHI and PARPOLA, 1987: 123.
105. *Ibid.*: 195, H-126A.
106. DURRANI, 1994-1995: 202, 204, figs. 1-2.
107. SHAH and PARPOLA, 1991: 414, Trq-3 A.
108. V. Shinde, personal communication; SHINDE *et al.*, 2007.
109. SHAH and PARPOLA 1991: 407, Ns-2 A, dated to the Integration Era.
110. PIPERNO and SALVATORI, 2007: 331.
111. SALVATORI and VIDALE, 1997: fig. 241: 1.

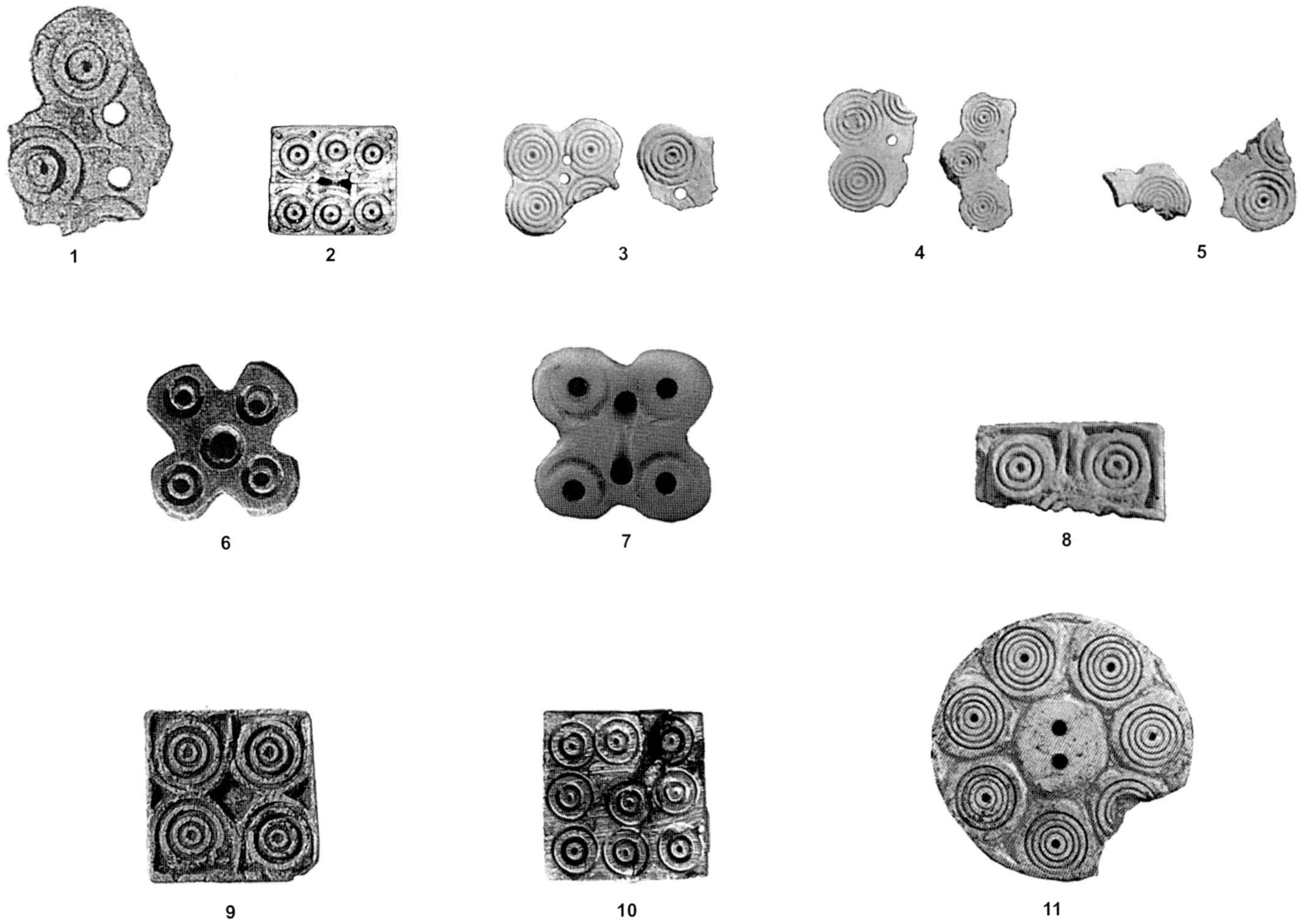

Fig. 14 – *Stamp seals with concentric circles from Shahr-i Sokhta and other sites. Scale not available.* ***1****, Shahr-i Sokhta, MAI 6032, found on surface;* ***2****, Shahr-i Sokhta MAI 7649, bone, from Grave G801/3, Period II, phase 5b;* ***3****, Rehman Dheri (DURRANI, 1994-1995: 204, fig. 1);* ***4****, Rehman Dheri (ibid., fig. 2);* ***5****, Rehman Dheri (ibid.: 204, fig. 1);* ***6****, Tarakai Qila (SHAH and PARPOLA, 1991: 414, Trq-3 A);* ***7****, Nausharo (ibid.: 407, Ns-2 A);* ***8****, Rehman Dheri (DURRANI, 1994-1995: 203, fig. 2);* ***9****, Tarakai Qila (SHAH and PARPOLA, 1991: 414, Trq-2 A);* ***10****, Harappa (JOSHI and PARPOLA, 1987: 196, H 128-A);* ***11****, Nausharo (SHAH and PARPOLA, 1991: 407, Ns-3 A).*

are also reported from Rehman-Dheri[112] (fig. 14: 8), Tarakai Qila[113] (fig. 14: 9), Nausharo[114] (fig. 14: 11), Mehrgarh,[115] Area MR 1, Periods V/VI, Harappa[116] (fig. 14: 10), and Mohenjo-Daro.[117]

The craftsmen of Shahr-i Sokhta also made a series of stamp seals in various stones, shell and copper, where the main motif was a central cross (fig. 15). The cross divides the field in four quadrants, further decorated with concentric angular patterns, and in one case, with round drilled dots (fig. 15: 8). The motif of the cross expanding with geometric angular traits, or limited by holes drilled within the arms, is well known in a class of steatite stamp seals found at Harappa,[118] Mohenjo-Daro[119] and Lothal[120] (fig. 15: 10); the same motif, as suggested by some impressed terracotta tablets from Mohenjo-Daro, was

112. DURRANI, 1994-1995: 203, fig. 2.
113. SHAH and PARPOLA, 1991: 414, Trq-2 A, Kot-Dijian phase.
114. *Ibid.*: 407, Ns-3 A, Integration Era.
115. JARRIGE *et al.*, 1995: fig. 10: 27.
116. JOSHI and PARPOLA, 1987: 196, H-128 A.
117. SHAH and PARPOLA, 1991: 158, M-1259 A-1260 A.
118. *Ibid.*: 1991: 304, H-630 A-H-635 A.
119. SHAH and PARPOLA, 1991: 158, M-1254 A-1258 A; see our fig. 9: 11 and 14, respectively from JOSHI and PARPOLA, 1987: 87, M-352 A and 86, M-349 A.
120. JOSHI and PARPOLA, 1987: 256, L-75 A.

Fig. 15 – *Stamp seals bearing a central cross motif from Shahr-i Sokhta and other sites. Scale not available.* ***1****, Shahr-i Sokhta, 1505.16 , Grave IUL 1505 (*SAJJADI*, 2003: fig. 36,a), Period I (?);* ***2****, Shahr-i Sokhta. MAI 8019, Grave 725 sup., late Period II or early Period III (*PIPERNO *and* SALVATORI*, 2007: fig. 626);* ***3****, Shahr-i Sokhta, MAI 8020, from the same grave, late Period II or early Period III (ibid.);* ***4****, Shahr-i Sokhta, MAI 7684, Grave 311, Period II, phases 6-5a (ibid.: fig. 461);* ***5****, Shahr-i Sokhta, MAI 7685, from the same grave, phases 6-5a (ibid.);* ***6****, Shahr-i Sokhta, MAI 6025, found on surface;* ***7****, Shahr-i Sokhta, MAI 6047, found on surface;* ***8****, Shahr-i Sokhta, MAI 6027, found on surface;* ***9****, Harappa (*SHAH *and* PARPOLA*, 1991: 304, H-635 A);* ***10****, Lothal (*JOSHI *and* PARPOLA*, 1987: 256, L-75 A);* ***11****, Mohenjo-Daro (ibid.: 87, M-352 A);* ***12****, Mohenjo-Daro (ibid.: 112, M-464 A);* ***13****, Harappa (*SHAH *and* PARPOLA*, 1991: 304, H-630 A);* ***14****, Mohenjo-Daro (*JOSHI *and* PARPOLA*, 1987: 86, M-349 A).*

carved on cylinder seals[121] (fig. 15: 12). There is a basic affinity between the geometry of the Shahr-i Sokhta seals and the Indus examples. Although the earliest specimens at Shahr-i Sokhta date back to Period I (like the lapis lazuli seal in our fig. 15: 1), other seals of this type come from middle Period II (fig. 15: 4-5) or late Period II-early Period III contexts (fig. 15: 2-3). The use of this geometric pattern at Shahr-i Sokhta thus seems to spread between 2600 BCE and the following centuries, in correspondence with the main urban peak of the Integration Era in the Indus valley.

BEADS

The manufacture of disk beads in steatite, at Mehrgarh, is attested from the beginning of the 5th millennium BCE.[122] Thin, fired steatite disk beads, according to excavation data from Harappa, were serially produced in the Indus valley from the last centuries of the 4th millennium BCE.[123] At Mundigak, some specimens were found in a Period I, 3-4 context, *i.e.* around the middle of the 4th millennium, thus showing the early adoption of this type of ornament in the repertories of the Helmand civilization.[124] White steatite disk beads of variable size may be found on the surface of Shahr-i Sokhta as well. Their chronological context is detailed by the discovery of two complete necklaces in well controlled stratified contexts.

The first necklace (MAI 3571, fig. 16), comprising 271 beads, was stuck by salt crusts onto the plaster of a Period III house, where it had been lost. This surface was partially exposed below the foundations of the Burnt Building. The second necklace was found in room CCXV, cut 3, a context also dated to Period III. A sample of seven beads (MAI 2757, fig. 17) is presently stored in the facilities of IsIAO at Rome. The maximum diameter of Indus steatite disk beads usually ranges between 0.7 and 0.9 cm, the thickness is about 0.2 cm and the hole wavers around 0.1-0.3 cm. In the Indus cities, they were mass produced with a technology involving the use of copper saws for cutting steatite blocks in flat tablets, grooving and snapping the tablets in rectangular sticks, sawing the sticks into parallelepipeds, perforation of these latter pre-forms, probably followed by rounding, slicing the disk beads and a final firing stage.[125]

121. *Ibid.*: 112, M-464 A.
122. VIDALE, 1995; VANZETTI and VIDALE, 1995.
123. KENOYER, 1997: 262 and 266.
124. CASAL, 1961: fig. 138: 1.
125. VIDALE, 1989, 2000 and 2004; KENOYER, 1997.

Fig. 16 – *Shahr-i Sokhta, necklace MAI 3571, with 271 fired steatite disk-beads of Indus fashion. It was found soldered by salt onto the surface of a house dating to Period III, phase 4. No scale.*

In 2007, Vidale carried out a preliminary investigation of steatite bead making at Tepe Hissar (Damghan, Iran)[126] based on samples of bead making debris originally recovered on the site by R.H. Dyson and presently kept in Rome. The aim was to compare the steatite bead manufacturing sequence at Tepe Hissar with that previously reconstructed at Mohenjo-Daro and Harappa, and to ascertain if the beads found at Shahr-i Sokhta were locally produced or imported from the Indus valley.[127]

126. DYSON and HOWARD, 1989.
127. VIDALE *et al.*, 2007.

The steatite processed at Tepe Hissar, being lower in silicon, and higher in calcium, iron and aluminium, was easily distinguished even by a preliminary quantitative microprobe analysis from the stone worked at Harappa and Mohenjo-Daro. Moreover, while in the analyzed samples there was no evidence of thin disk bead production, the manufacturing sequence involved the preliminary use of coarse blades in a kind of metalworking glassy slag, followed by sawing with metallic saws, and the application of not less than 5-6 stages of firing for whitening and rendering steatite hard.

While such technical choices might have been entirely dependent upon the mineralogical and chemical features of locally available raw material, the overall evidence suggested an independent technical tradition not affected by Indus models. The beads found at Shahr-i Sokhta, in contrast (taken from the collection in CCXV, cut 3), were morphologically identical to the Indus specimens; the chemical characterization showed minor variations from the Indus beads (slightly higher amounts of aluminium, magnesium, potassium and calcium). The initial inference was that the steatite disk beads found at Shahr-i Sokhta, Period III, were imports. However, after this paper had been completed[128] we identified in the surface collection from the site a piece of a fired steatite perforated cylinder with slicing marks (see Addendum and fig. 23: 1). This piece alone might suggest, contrary to previous assumptions, that fired steatite disk beads could have been locally produced at Shahr-i Sokhta with a distinctively Indus technique.

Two other types of beads might be relevant to the present discussion. Etched carnelian beads, another indicator of Indus trade and exchange activities, are reported from Mundigak.[129] A possible ivory bead was reported at Shahr-i Sokhta, in the Central Quarters, in a Period III, phase 4 context[130] but still awaits study and publication.

MOUSE TRAPS FOUND AT MUNDIGAK AND MOHENJO-DARO

Two pottery mouse traps found at Mohenjo-Daro, in relatively recent occupation layers,[131] can only be compared with two similar devices found at Mundigak, Period IV,1[132] and one from Bampur.[133] The technical principles of the traps found in

128. M. Vidale and F. Egloff, ongoing research.
129. CASAL, 1961: 138: 23.
130. SALVATORI and TOSI, 2005: 287.
131. MACKAY, 1938: DK-12299, pl. LIV: 20-22, 427-428.
132. CASAL, 1961: 145-146, fig. 84: 314-314a.
133. DE CARDI, 1970: 329.

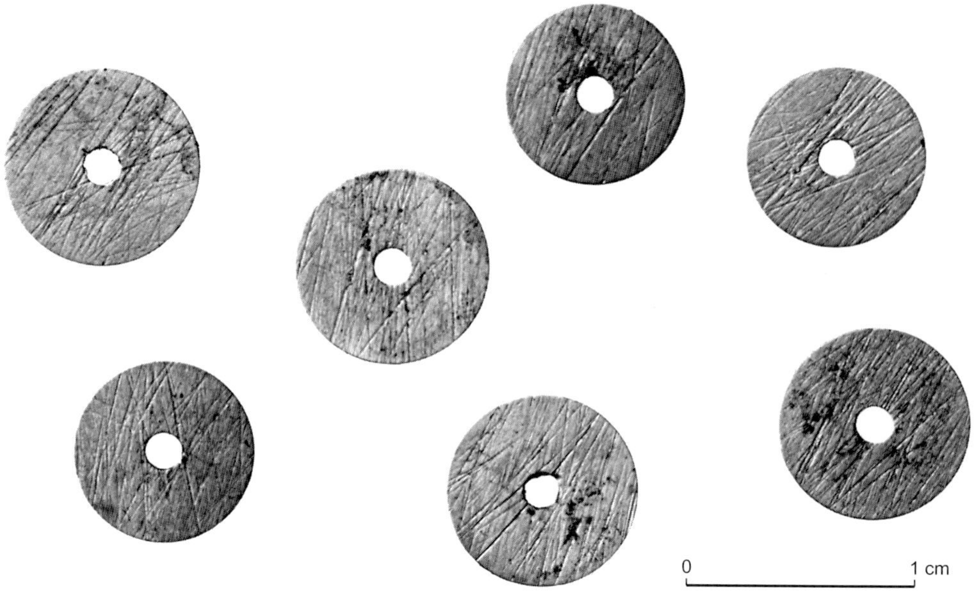

Fig. 17 – *Shahr-i Sokhta, necklace MAI 2757, detail of 7 beads (Period III, phase 4) (Photograph F. Aghadian).*

the two protohistoric cities (fig. 18) might have been different, perhaps involving the use of a knot at Mohenjo-Daro against a downward sliding pottery lid at Mundigak. Nonetheless, the overall similarity of the ceramic containers suggests a parallel adaptation, based upon shared know-how, for coping with common problems of rodent infestations in the "domestic universes" of the two civilizations.[134] The specimens from Mundigak might be several centuries older than the Mohenjo-Daro ones, suggesting that such an adaptation was as widespread in time as in space.

GAMING PIECES AND DICE

A type of gaming piece in a bluish green stone (MAI 6190 a-b), found in Grave 12 of the Shahr-i Sokhta graveyard, dated to the beginning of Period III[135] (fig. 19: 1-2), may be compared to ivory pieces, having presumably the same function, found at Altyn Depe[136] and at Gonur North[137] in Turkmenistan. Another similar specimen was found at Mohenjo-Daro[138] (fig. 19: 3). In spite of the different base material, the shape of the gaming pieces and their decoration are very similar.

The idea that the shape of dice might reflect early contacts between Mesopotamia and the Indus civilization goes back to a suggestion first proposed by E.J.H. Mackay[139] and was later developed in a paper by G.F. Dales.[140] Dales considered cubic dice a relatively late innovation in the history of material culture of the Near East, and noticed their scarcity in the west, opposing this evidence to the apparent high frequency of the same type at Mohenjo-Daro and Harappa. Actually, at Mohenjo-Daro dice in the form of cubes, tablets, long parallelepipeds, and prism-like pieces with triangular section[141] were unearthed. Although cubic dice (seven specimens known) are less frequent than the other types, they were found at variable depths. The standards envisaged 1 opposed to 2, 3 to 4 and 5 to 6, although there are variations.[142]

At Shahr-i Sokhta, while a cubic die was found in a layer datable to Period IV,[143] parallelepiped-like pieces (or rectangular dice) are more common (fig. 20). The inventory includes three specimens from the settlement area and four from Grave 731, where they accompanied the famous gaming board in

134. CALOI and COMPAGNONI, 1977: 188-190.
135. PIPERNO and SALVATORI, 2007: fig. 60.
136. SHETENKO, 1968: fig. 14: 8-9 and 1970: 59; KIRCHO, 2000: fig. 2.I.5.
137. SARIANIDI, 1998: fig. 21.16-17.
138. MARSHALL, 1931: pl. CXXXIV: 3.
139. MACKAY, 1938: 559.
140. DALES, 1968.
141. MACKAY, 1938: 559-562.
142. *Ibid.*: 559.
143. TOSI, 1983: 174, fig. 10.

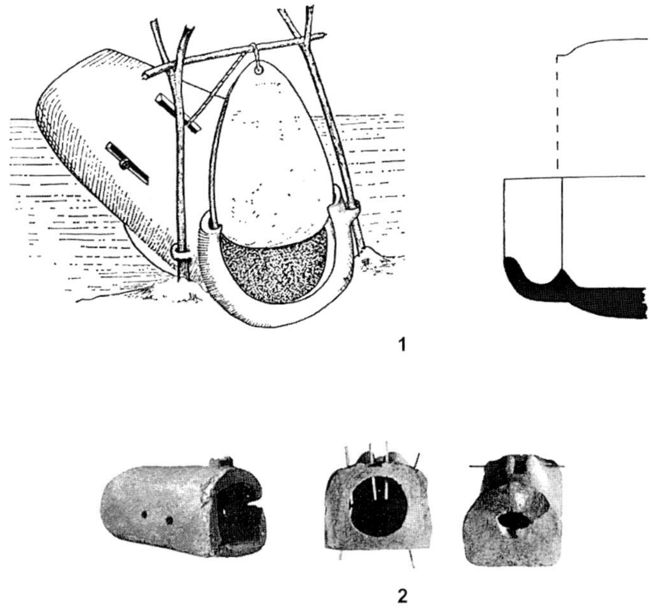

Fig. 18 – *Pottery mouse trap. Scale not available.* ***1****, Mundigak (DE CARDI, 1967: fig. 4);* ***2****, Mohenjo-Daro (MACKAY, 1938: pl. LIV, 15-17, 20-22).*

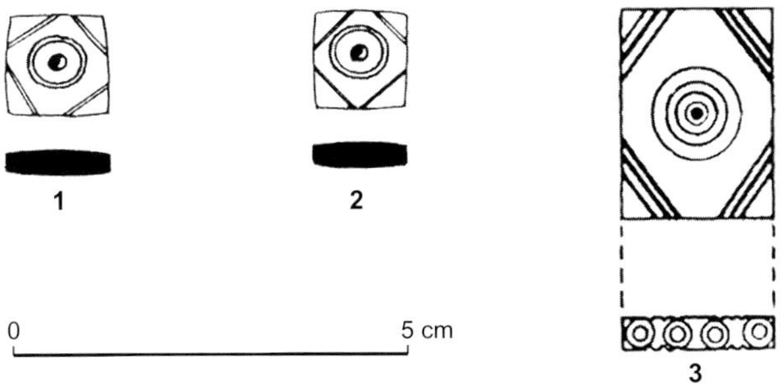

Fig. 19 – *Gaming pieces.* ***1****, Shahr-i Sokhta. MAI 6190a, Grave 12 (PIPERNO and SALVATORI, 2007: fig. 60), Period III, phase 4;* ***2****, Shahr-i Sokhta. MAI 6190b, from the same grave (ibid.);* ***3****, ivory gaming piece found at Mohenjo-Daro (MARSHALL, 1931: pl. CXXXIV,3).*

pipal (*Dalbergia sissoo*) wood carved with a snake motif and a complete set of gaming pieces[144] (Period III, phase 4). These dice belong to a single type. In three specimens, the faces are marked by drilled holes filled with bone, ivory or shell inlays, possibly set with bitumen or another glue. Numbers are opposed in the following way: 1 to 4, 2 to 3 in one case, 1 to 3, 4 to 2 in another case. One die is split in half, and only the face bearing number 2 is preserved. The fourth specimen is different: numbers are made by a system of triangles and lozenges. Thus, we have a central lozenge for 1, two opposed triangles for 2, a lozenge and two opposed triangles for 3, two central lozenges and two opposed triangles for 4. The opposition pattern is 1 to 3 and 4 to 2. Note that in this die the numbers are expressed by plane geometrical figures, or by enclosed areas, and not by the number of linear incisions.

MAI 781 (fig. 20: 2) is another limestone die measuring 3.8 x 1.5 x 1.5 cm. It was found on the surface, in an area covered by ceramics of Periods III and IV. Numbers, in this case, are indicated with drilled dots: 1 opposed to 4, 2 opposed to 3. MAI 3446 (fig. 20: 3), like the one just described, it is in limestone and comes from a surface context of Period III. Broken in half, it should have originally measured 4.2 x 1.1 x 1.1 cm. The numbers are expressed by a drilled circle with a central dot: in this case, 1 was opposed to 2, 4 to 3.

Die MAI 2377[145] (fig. 20: 1a interpreted in fig. 20: 1b) came from square XHI, 2-3, Period III. It is in a slightly translucent calcite, and measures about 3 x 1.1 x 1 cm. Like the last die from Grave 731, numbers are expressed by means of linear incisions, but in this case the interpretation of the numeric notation is less easy. On one side we can see three "X" signs, the central one narrower than the others; two horizontal lines run on the base of the proximal and distal crosses. The overall design is strangely asymmetric. The opposed face is plain. Another face bears two "V" signs, while two convergent oblique lines appear on the opposed face. Interpreting the numeric values on the basis of the number of individual incised traits we would have 0 opposed to 2 and 4 opposed to 8, but we might instead count the areal divisions or sections, as in the last die in Grave 731. In this case, we might have 1 (the plain face) opposed to 4 (the face with the asymmetric pattern: the vertical traits would close the geometric partitions to be counted) and 2 (the face marked by the two convergent traits) opposed to 3 (the face with the two "V" signs).

At Mohenjo-Daro and Harappa, besides cubic dice various types of elongated dice, "throwing sticks" or gaming pieces, were in use. They are said to be most frequently made of ivory. The form, section and notation are quite variable, and sometimes very complex. The closest match with the rectangular dice from Shahr-i Sokhta are rectangular sticks, square in section, whose sides are marked with 1, 2 and 3 concentric circles, the fourth side bearing two parallel central lines, probably

144. PIPERNO and SALVATORI, 1983 and 2007: figs. 691-693.

145. TOSI, 1983: 174, fig. 11.

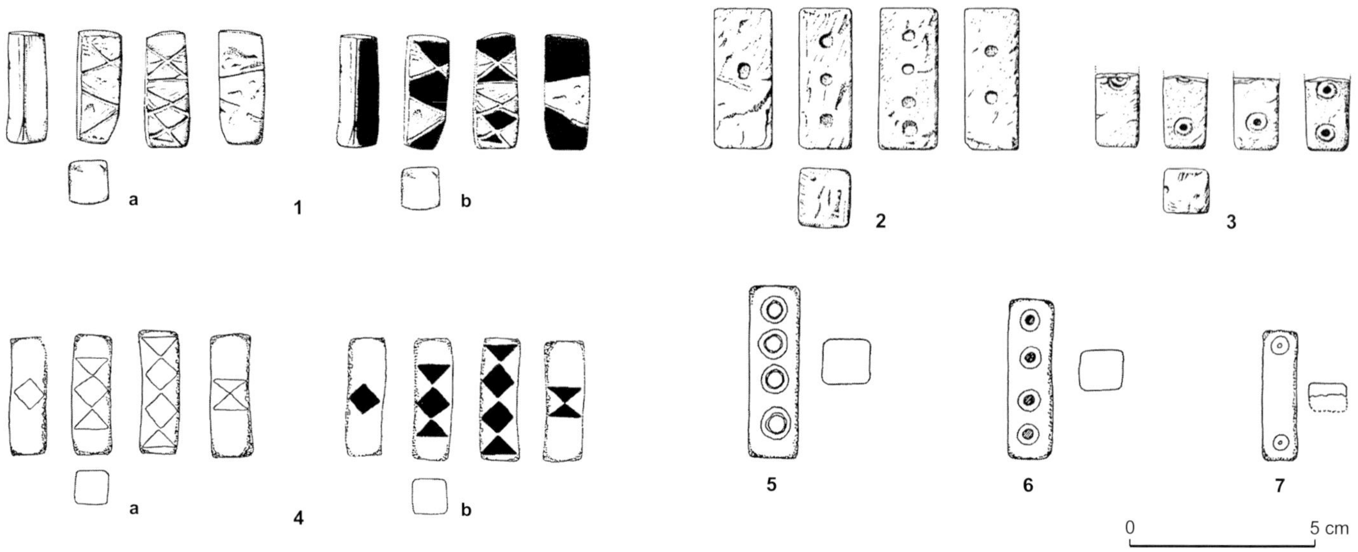

Fig. 20 – *Shahr-i Sokhta, rectangular dice.* ***1a****, MAI 2377 (*TOSI*, 1983: 174, fig. 11), Period III, phase 4;* ***1b****, MAI 2377, black areas show a possible numeral interpretation;* ***2****, MAI 781, found on surface;* ***3****, MAI 3446, found on surface in a Period III context (as associated to Period III ceramics);* ***4a****, MAI 8099, Grave 731, Period III, phase 4 (*PIPERNO *and* SALVATORI*, 2007: fig. 692,4);* ***4b****, MAI 8099, possible numeral interpretation;* ***5****, MAI 8099, from the same Grave (ibid.: fig. 692,1);* ***6****, MAI 8099, from the same Grave (ibid.: fig. 692,2);* ***7****, MAI 8099, from the same Grave (ibid.: fig. 692,3).*

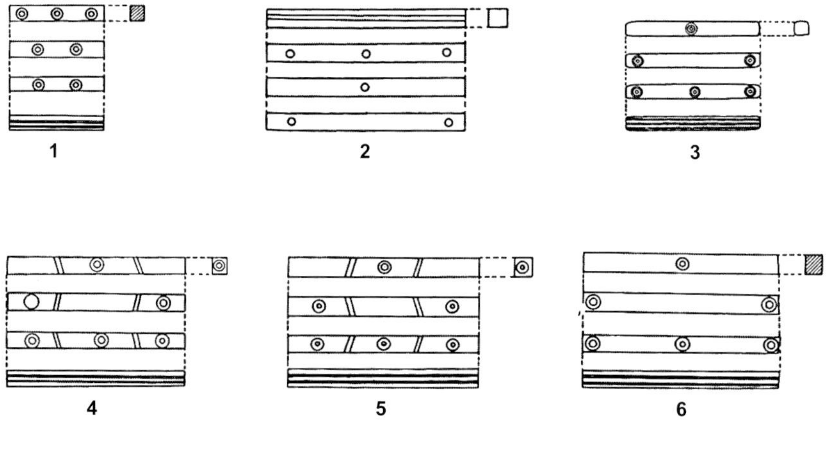

Fig. 21 – *Mohenjo-Daro, rectangular dice. No scale.* ***1****, (*MARSHALL*, 1931: pl. CXLIII, 41);* ***2****, (*MACKAY*, 1938: pl. CXXXVIII, 48);* ***3****, (ibid.: pl. CXXXVIII, 41);* ***4****, (*MARSHALL*, 1931: pl. CXLIII, 49);* ***5****, (ibid.: pl. CXLIII, 47);* ***6****, (ibid.: pl. CXLIII, 51).*

denoting 4 (fig. 21). A recurrent opposition pattern is 1 to 3, 2 to "4", but here, again, there are variations. Some of these ivory rectangular dice bear linear incisions like orthogonal or oblique strokes, crosses and zig-zag designs, and have a superficial resemblance with the two Shahr-i Sokhta specimens described above.[146] Interestingly, in the ivory dice from Mohenjo-Daro (see fig. 21: 4-5) in plates CXLIII, 47 and 49 the faces bearing numbers 1-3 are subdivided by means of oblique strokes, and concentric circles mark the segments that actually express the numeric value. While at Shahr-i Sokhta rectangular dice may bear drilled dots or, in alternative, incised lines, the dice or throwing sticks found at Mohenjo-Daro often associate these two different systems of notation in quite complex patterns, thus reflecting a noticeably wider range of games and game rules.

In sum, besides the use of cubic dice, the rectangular dice abundantly used in the Indus valley cities and in Central Asia had local versions at Shahr-i Sokhta, apparently limited to Period III. While the snake-board is structurally identical to the inlaid boards from the Royal Cemetery of Ur, this similarity of the dice points to a first-hand knowledge of games and gaming rules.

SHELL ARTEFACTS

The exploitation of the shell of the mollusc *Turbinella pyrum* by the craft communities of the Indus civilization has been the subject of intensive research.[147] The shell was gathered by specialized fishing communities along the Karachi coast, the coast of Gujarat and Kutch and traded inland, as a raw unmodified material or after being transformed into bangles, shell ladles and other finished commodities. Twenty fragments of *Turbinella pyrum*, identified from the morphology of the columella, were collected on the surface of Shahr-i Sokhta.[148] Out of this group, twelve are fragments of bangle manufacturing waste. The rest belong to stamp seals or are fragments of other objects. The evidence shows that complete shells were brought inland from the coasts to Sistan, in the frame of a sporadic long-distance trade. As all the fragments come from surface deposits, their preferential association with Period III, before the urban crisis of Period IV, is quite likely. The bangles

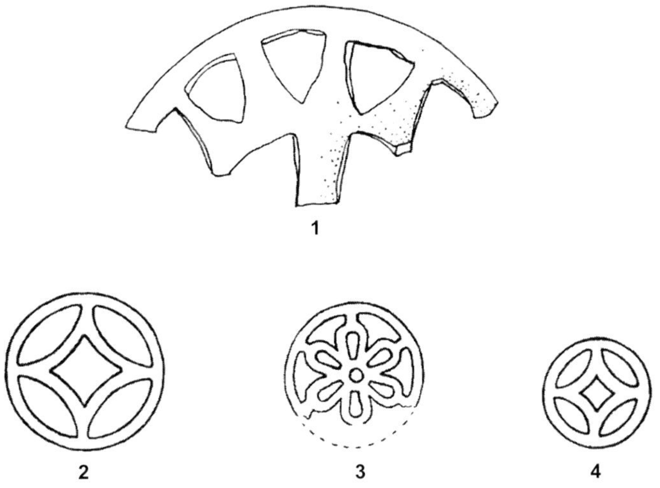

Fig. 22 – *Shell inlay pieces from Shahr-i Sokhta (**1**, found on surface) and Mohenjo-Daro (**2-4**). No scale.*

found at Shahr-i Sokhta are rarely standardized, having variable sections and different size.[149] The modification of the suture joint of the end of the spiral with a "V"-shaped cut pattern, a typically Indus expedient, is not on record, and the manufacturing techniques seem to be a local variation. From the surface of Shahr-i Sokhta comes also a fragment of a complex "perforated disk"[150] (fig. 22: 1), an inlay piece made from shell of the same species. The disk (Dep. CS 8392/1) shows on the outer edge a series of triangles the vertex of which points to the centre. Similar disks, sometimes with very complex geometry, are well known at Harappa and Mohenjo-Daro (fig. 22: 2-4); most probably, they were applied as inlay pieces to wooden boxes, musical instruments, and decorative panels or to other costly wooden artefacts. As some precious wood varieties, such as *pipal*, reached Sistan from south-east, on this basis we may hypothesize a trade of shell-inlaid wood artefacts along the same trade route.

A STEATITE CYLINDER SEAL WITH AN INDUS INSCRIPTION

The only Indus inscription so far reportedly found in Sistan appears on a fragmentary cylinder seal belonging to the antiques collection of General H. MacMahon,[151] donated to the

146. See MARSHALL, 1931: pl. CXXII: 22-45; MACKAY, 1938: pl. CXXXVIII: 41, 48; pl. CXLIII: 18-54, and particularly 41, 47, 49, 51; VATS, 1940: some artefacts in pl. CXIX.

147. DURANTE, 1979a and 1979b; KENOYER, 1984a, 1984b, 1985, 1991c and 1991d; BHAN and GOWDA, 2003; BHAN and KENOYER, 1984; SONAWANE *et al.*, 2003; VIDALE, 2000.

148. DURANTE, 1979a.

149. *Ibid.*: fig. 3.

150. TOSI, 1983: pl. LXXVIII: 92.

151. KNOX, 1994.

British Museum in the 1960s. The artefact was probably collected from the surface of a site in Sistan, during a visit in 1906. The information cannot be confirmed, but this provenience appears more probable than Swat, the other source area of the artefacts gathered by this military officer. The seal fragment (1.9 cm high, with a diameter of 1.4 cm) corresponds to the lower part of a solid cylinder provided, in the missing end, with a perforated projection. Similar seals were found in Margiana (Turkmenistan) at Gonur South, Taip and Togolok[152] as well as in the graveyard of Sibri near Mehrgarh[153] (Baluchistan, Pakistan). It retains a sequence of six signs running at the base of the original imprint, while another possible register with other signs run on the missing portion. The preserved base of the cylinder bears a damaged sign apparently formed by a cluster of triangles, of uncertain interpretation. The archaeological contexts of these finds, both in the Murghab delta and in northern Baluchistan, indicate a dating to the first centuries of the 2nd millennium BCE.

From Tepe Yahya comes a seal impression on pottery with few Indus signs.[154] From the same context (belonging to Period IV-A, though to be contemporary with the Indus Civilization) another potsherd shows an anthropomorphic figure in low relief. The figure sits with crossed legs, a position very similar to the "yogic" posture common in Indus seals.[155]

STONE STATUES OF KNEELING PERSONAGES

The series of fragmentary stone statues of kneeling male figures found in the upper levels of Mohenjo-Daro is often mentioned as the main example of the Indus statuary in the round; it has been described by the excavators and later in greater detail by Ardeleanu-Jansen.[156] The most famous sculpture is the so-called "Priest king", a steatite fragmentary bust found in DK-B area. Although such sculptures have not been found at Harappa, another find of this type was recently unearthed at Dholavira[157] in Kutch, India. The recurrent traits are a typical posture, with one knee touching the floor, the other leg bent higher, the arms bent on the knees; elaborate head-dressings, with fillets (or ribbond) departing in two bands on the neck or hair knots or single braids on the rear; strongly stylized of face and head traits, with ears reduced to abstract curvilinear patterns. Three examples of limestone heads comparable to the statues found in the Indus valley were found in the Helmand area, between Sistan and Kandahar. The first and most famous comes from Mundigak;[158] the second (MAI 8379) was picked up from the surface of a little hummock named Tepe Chah-i Torog 2,[159] 15 km south of Shahr-i Sokhta; the third and last is a limestone head from a private collection[160] from the surface of an unknown site in north-western Sistan.[161] The Mundigak head was found while digging a ceramic dump datable to Period IV, 3, *i.e.* late Period III or Period IV of the Shahr-i Sokhta sequence. The Tepe Chah-i Torog 2 head was dated to Period IV on the basis of the associated surface ceramics.[162] The chronology of the Helmand examples thus fits with the latest time span of Harappa IIIc (*ca* 2250-1900 BCE) and places these artefacts, and the complex of ideas that accompanied them, at the end of the 3rd millennium BCE. From an iconographic viewpoint, such male figures have little to do with the representative and plastic tradition of the Indus valley, but are faithfully reproduced in their basic features in the late 3rd millennium BCE art of central Asia (see for example the figures on a silver vessel from a looted grave in Bactria).[163] The contemporary occurrence of these sculptures in Sistan, Sindh and Kutch, in this light, rather than pointing to a more limited Helmand-Indus connection, might be related to the general evidence of strong links between the urban poles of south-western Central Asia, Baluchistan and the Indus valley at the close of the millennium.[164] The MacMahon seal fragment could have similar implications, but might reflect the activities of Indus traders settled in Margiana and travelling across Sistan and Baluchistan towards the Gulf.

CONCLUSIONS

In summary, the evidence points to a continuous contact between the Helmand the Indus civilizations throughout the 3rd millennium BCE. The links can hardly have been limited to the first half of the millennium, when few but meaningful finds firmly link Period II with the Kot-Dijan horizons. As far as Shahr-i Sokhta Period III is concerned, although the record

152. SALVATORI, 2000: 133.
153. SANTONI, 1984; SHAH and PARPOLA, 1991: 412-413.
154. LAMBERG-KARLOVSKY, 1976: 42ff.
155. PARPOLA, 1981: 181, fig. 23: 14.
156. ARDELEANU-JANSEN, 1984.
157. J.M. Kenoyer, personal communication.
158. CASAL, 1961: pl. XLII-XLIV.
159. JARRIGE and TOSI, 1981: fig. 5b.
160. DALES, 1985: pl. 1.
161. *Ibid.*: 219.
162. JARRIGE and TOSI, 1981: 132.
163. WINKELMANN, 1994; fig. 693: 3d; SARIANIDI, 1986: 168-169.
164. PARPOLA, 2006.

Table 4 – *Material links of the Shahr-i Soktha sequence. It includes the two artefacts discussed in between the Helmand Civilization, Baluchistan and the Indus valley in the 3rd millennium BCE, in the different Periods (see the Addendum below).*

Period	Probable imported goods	Formal-functional convergence, cases of technical adaptation
I	Nal vessels	
II	Lobate steatite stamp seals Grid-like stamp seals Wet wares, grey wares (??) Kot-dijan vessels Zhob-like figurine	Locally made Nal-like vessels Wet wares, grey wares (??) Pipal leaf motifs on pottery Use of triangular-rectangular terracotta cakes Terracotta mouse-traps
II – III	Early Indus (?) pear-shaped jar	Stamp seals with cross-like patterns
III	*Turbynella pyrum* (raw material) *Turbynella pyrum* (elaborate inlays) Pipal wood (raw material or finished goods) Etched carnelian beads Faience bangles "Ivory bead" (perhaps fired steatite)	Pipal leaf motifs on pottery *Turbynella pyrum* (local production of bangles) Local production of steatite disk beads (?) Use of triangular-rectangular terracotta cakes Dice and gaming pieces similar to Indus specimens
IV		Cubic dice Round terracotta cakes

from reliably stratified contexts is admittedly limited, we would preliminarily establish a correlation with Period IIIb at Harappa, while Shahr-i Sokhta Period IV might partially overlap with Harappa IIIc (around 2100-1900 BCE). Although the above-mentioned chronological disagreement cannot be considered solved, and more evidence is expected after the current excavations in Makran and at Sohr-Damb, the archaeological picture is presently wider and more coherent, and can be discussed in greater detail. The "interaction sphere", on the basis of the evidence so far reviewed, can thus be articulated in time and somehow detailed in scale and function.

Table 4 summarizes the general picture of the material links. The Murghab-related seal fragment and the stone statues of male kneeling personages were omitted, because in the present context we deem them more related to the role played in South Asia by Central Asian Bronze Age urban civilizations in the later centuries of the 3rd millennium BCE than to the issue of the Helmand-Indus contacts. Periods I and IV were excavated to a lesser extent, and are therefore underrepresented. Thus, they can hardly be meaningfully compared to the record of Periods II and III (second half of the 3rd millennium BCE) when the scarcity of excavated lots is somehow tempered by the extensive collection of surface small finds, carried out for years at the time of the excavation.

The evidence does not add much to the image of a limited, trickle-like interregional exchange or trade, well fit with the proposed reconstruction of diffuse trade-exchange networks at the opposite pole of the Iranian plateau.[165] In this picture, nothing suggests large-scale trade, centralization of trading network or centrally managed entrepreneurship. The import from the Indus valley and northern Baluchistan to Sistan of seals and ceramics (perhaps used as containers of perishable foodstuffs), of few classes of finished ornaments, as well as of a limited range of valuable base materials used for making costly luxury objects seems to reflect sporadic trade contacts, individual trips or marriages rather than systematic, specialized forms of long-distance trade. The evidence also fits with the preliminary analytical results of R. Law,[166] who carried out an INAA[167] analysis on a series of carnelian rough-outs from the surface of Shahr-i Sokhta and compared them with samples of the same stone from the Chagai areas and from several 3rd millennium BCE sites in the Indo-Pakistani Subcontinent. Shahr-i Sokhta evidently exploited a local carnelian source easily distinguished from that collected in southern and in the

165. POTTS, 1994.
166. LAW, 2005.
167. INAA means Instrumental Neutron Activation Analysis.

abundant south-eastern source areas. It seems to have been a market consuming a local product, more than an intermediate transhipment centre within a wider trade network.[168] In a similar light, a comprehensive study of lapis lazuli production at Shahr-i Sokhta suggests that bead production in the EWK-EWP craft areas was carried out by independent part-time specialists for local demand rather than being monitored by the urban elites for long-distance western trade.[169]

The word "interaction sphere" initially figured prominently in the headings of this research, but eventually we removed it from the title of the paper. In general, the cultural relationships between the Helmand centres and the Subcontinent appear to have been ephemeral and sporadic, had a secondary economic impact, and probably do not justify the use of the term. The links, anyhow, are well-established archaeological facts. Some classes of ceramics, tools and ornaments, throughout the whole sequence, seem to be the local versions of artefacts and habits more firmly rooted beyond the south-eastern frontier. Interestingly, at Shahr-i Sokhta we can recognize at least three technologies that may have been adopted from Indus ones: the possible evidence of local steatite disk bead production, the still unknown technology involving the making and use of terracotta cakes, and the import of *Turbinella pyrum* for a local, scarcely standardized bangle industry; the habit of impressing stamp seals onto terracotta cakes or marking them with incised signs points to a local adaptation or transformation of the original functions of these cheap objects. The carnelian abundantly present on the surface of Shahr-i Sokhta would appear to have been fired in order to enhance its colour, according to the traditional Indian technology, although at present this cannot be demonstrated (the same technique, anyhow, has been recognized at Mundigak).[170] But if this was the case, firing would have enhanced the contrast between red and transparent bands, instead of simply deepening its red hue. Although these activities cannot be considered as economically very relevant, and possibly involved just a minority of the population of Shahr-i Sokhta, the city, particularly in Period III, appears quite permeable to the influence of Indus habits and fashions, definitely more than any other early urban centre of eastern Iran so far excavated.

In Period II, both for the items presumably imported and those locally produced with stylistic or technical features comparable to the Subcontinent, the links include basic domestic equipment (seals, pottery, terracotta figurines, mouse-traps, terracotta cakes). In contrast, when one moves to Period III, the connections shift to a different functional realm, ornamentation and status display: we encounter steatite, ivory and carnelian beads, shell inlays and bracelets, gaming pieces. The use of terracotta cakes, as we have seen, continued and might even have increased. It is hard to say whether, or to what extent, this peculiar pattern reflects an actual cultural process (a growing cultural interaction between the elites of the two cultures) or this is simply a by-product of the conservation of the site and excavation itself. Unfortunately the Period II deposits excavated at Shahr-i Sokhta are mainly formed by architectural fillings and well preserved dumping areas within a dense network of private households, meanwhile Period III is represented by few large residential compounds including relatively large houses, and by hectares of deposits eroded on the surface from similar elite architectural contexts.

168. TOSI and PIPERNO, 1973.

169. VIDALE, forthcoming.

170. BARTHÉLEMY DE SAIZIEU and CASANOVA, 1993.

ADDENDUM

In December 2007, after this paper had been completed, Vidale and Egloff[171] identified in the surface collection from Shahr-i Sokhta two other artefacts that, without contradicting the picture so far outlined, widen the repertory of the material links between the site and the Indus world. One is the above-mentioned perforated cylinder in fired steatite (fig. 23: 1, MAI 8691). It bears on its preserved face a series of transverse cutting marks, and on the outer surface a series of spiral-like traces of abrasion. Both marks are consistent with the manufacturing traces commonly visible on Indus-like disk beads in fired steatite (see figs. 16 and 17), and for this reason we are inclined to interpret this piece as possible evidence of a local production at Shahr-i Sokhta of similar ornaments. The second object (fig. 23: 2, MAI 8682), an unexpected surprise, is most probably a fragment of an Indus faience bangle. The fragment is light greenish blue, and shows distinctive glass-like fractures. The section is rounded, slightly pointed (as if had been pinched while being formed) on the interior. Many Indus bracelets, both in terracotta and in faience, bear the same feature. The most immediate and reasonable hypothesis, given the present absence of evidence of faience-based technologies at Shahr-i Sokhta, is that this piece was imported from the Indus valley. This might be ascertained in future by means of chemical tests. In both cases we are dealing with personal ornaments, and although both artefacts lack a reliable

171. M. Vidale, ongoing research.

archaeological context, they would fit very well in the archaeological picture we reconstructed for Period III.

Fig. 23 – *Shahr-i Sokhta. Two Indus-related artefacts found on surface.* ***1****, MAI 8691, perforated tube in fired steatite with sawing marks, possibly a residue of the local manufacture of steatite disk beads like those visible in figs. 16 and 17;* ***2****, MAI 8682, fragment of an Indus bangle in light greenish blue faience.*

Elisa CORTESI – Maurizio TOSI
Dipartimento di Archeologia
Facoltà di Conservazione dei Beni Culturali
Università di Bologna, Via S. Vitale, 28 – 48100 Ravenna
ITALY
elisa.cortesi@gmail.com
maurizio.tosi@tiscali.it

Alessandra LAZZARI
CNR - Istituto per le Tecnologie Applicate ai Beni Culturali,
Area della Ricerca di Roma 1 –
Via Salaria Km. 29,300
Monterotondo Scalo – Roma
ITALY
Alessandra.Lazzari@itaBCE.cnr.it

Massimo VIDALE
Istituto Superiore per la Conservazione e il Restauro
Piazza S. Francesco di Paola, 9-00184 Roma
ITALY
massimo.vidale@beniculturali.it

BIBLIOGRAPHY

ALGAZE G.

1993 *The Uruk World System: The Dynamics of Expansion of Early Mesopotamian Civilization*. Chicago and London: University of Chicago Press.

AMIET P. and TOSI M.

1978 Phase 10 at Shahr-i Sokhta: Excavations in Square XDV and the Late 4th Millennium B.C. Assemblage of Sistan. *East and West* 28: 9-31.

ARDELEANU-JANSEN A.

1984 Stone Sculptures from Mohenjo-daro. *In*: JANSEN M. and URBAN G. (eds.), *Interim Reports* 1: 139-157. Aachen: IsMEO/RWTH.

BARTHÉLEMY DE SAIZIEU B. and CASANOVA M.

1993 Semi-Precious Stones Working at Mundigak: Carnelian and Lapis Lazuli. *In*: GAIL A.J. and MEVISSEN G.J.R. (eds.), *South Asian Archaeology* 1991: 17-30. Stuttgart: Franz Steiner Verlag.

BESENVAL R.

1994 The 1992-1993 field-seasons at Miri Qalat: new contributions to the chronology of protohistoric settlement in Pakistani Makran. *In*: PARPOLA A. and KOSKIKALLIO P. (eds.), *South Asian Archaeology* 1993: 81-91. Helsinki: Suomalainen Tiedeakatemia.

1997a The Chronology of the Ancient Occupation in Makran: Results of the 1994 Season at Miri Qalat, Pakistan Makran. *In*: ALLCHIN R. and ALLCHIN B. (eds.), *South Asian Archaeology* 1995: 199-235. Cambridge: Oxford IBH.

1997b Entre le Sud-Est Iranien et la plaine de l'Indus : le Kech-Makran. Recherches archéologiques sur le peuplement ancien d'une marche des confins indo-iraniens. *Arts Asiatiques* 52 : 5-36.

2000 New data for the chronology of the Protohistory of Kech-Makran (Pakistan) from Miri Qalat 1996 and Shahi-Tump 1997 Field-Seasons. *In*: TADDEI M. and DE MARCO G. (eds.), *South Asian Archaeology* 1997: 161-187. Roma: IsIAO.

BESENVAL R. et DIDIER A.

2004 Peuplement protohistorique du Kech-Makran au 3e millénaire av. J.-C. : l'assemblage céramique de la période IIIc sur le site de Miri Qalat. *Paléorient* 30,2 : 159-178.

2005 Plundering in the Indo-Iranian Borderlands (Afghanistan, Pakistan and Iran). A case study: the Dasht Plain (Kech-Makran). Paper given at the 18th international conference of *South Asian Archaeology*. London, 4-8 July 2005.

BHAN K.K. and GOWDA D.

2003 Shell Working at Nagwada (North Gujarat) with Special Reference to Shell Industries of the Harappan Tradition in Gujarat. *Man and Environment* XXVIII, 2: 51-80.

BHAN K.K. and KENOYER J.M.

1984 Nageshwara: A Mature Harappan Shell Working Site on the Gulf of Kutch, Gujarat. *Journal of the Oriental Institute* 34: 67-80.

BINFORD L.R.

1965 Archaeological systematics and the study of culture process. *American Antiquity* 31: 203-210.

BISCIONE R.

1984 Baluchistan Presence in the Ceramic Assemblage of Period I at Shahr-i Sokhta. *In*: ALLCHIN B. (ed.), *South Asian Archaeology* 1981: 69-80. Cambridge: Cambridge University Press.

1990 The Elusive Phase 2 of Shahr-i Sokhta Sequence. *In*: TADDEI M. (ed.), *South Asian Archaeology* 1987: 391-409. Roma: IsMEO.

BISCIONE R., BULGARELLI G.M., COSTANTINI L., PIPERNO M. and TOSI M.

1974 Archaeological Discoveries and Methodological Problems in the Excavations of Shahr-i Sokhta, Sistan. *In*: VAN LOHUIZEN-DE LEEUW J.E. and UBAGHS J.J.M. (eds.), *South Asian Archaeology* 1973: 12-52. Leiden: E.J. Brill.

CALDWELL J.R.

1964 Interaction Spheres in Prehistory. *In*: CALDWELL J.R. and ROBERT L.H. (eds.), *Hopewellian Studies. Illinois State Museum Scientific Papers* XII: 135-143. Springfield: Illinois State Museum.

CALOI L. and COMPAGNONI B.

1977 I mammiferi. *In*: TUCCI G. (ed.), *La Città Bruciata del Deserto Salato*: 183-213. Venice: Erizzo.

CASAL J.M.

1961 *Fouilles de Mundigak*. Paris: Librairie C. Klincksieck.

CHANG K.C.

1986 *The Archaeology of Ancient China*. New Haven: Yale University Press.

CLEUZIOU S. and TOSI M.

2000 Ra's al-Jinz and the Prehistoric Coastal Cultures of the Ja'alan. *The Journal of Oman Studies* 11: 19-74.

D'AMORE P.

1999 *L'argilla e il tornio. Tecniche e tipologie vascolari iraniche dal periodo del ferro all'età dell'impero sasanide*. Roma: Museo Nazionale d'Arte Orientale.

DALES G.F.

1962a Harappan Outposts on the Makran Coast. *Antiquity* 36: 86-92.

1962b A search for ancient seaports. *Expedition* 4: 3-10.

1968 Of Dice and Men. *In*: HALLO W.W. (ed.), *Essays in Memory of E.A. Speiser*: 14-22. (*Journal of the American Oriental Society* 1).

1985 Stone Sculptures from the Protohistoric Helmand Civilization, Afghanistan. *In*: GNOLI G. and LANCIOTTI L. (eds.), *Orientalia Josephi Tucci Memoriae Dicata*: 279-300. Roma: IsMEO.

DALES G.F. and KENOYER M.

1986 *Excavations at Mohenjo-Daro: the Pottery*. Philadelphia: University Museum Monographs.

DALES G.F. and LIPO C.P.

1992 *Explorations on the Makran Coast, Pakistan: A Search for Paradise*. Berkeley: University of California Press (*Contributions of the Archaeological Research Facility* 50).

DE CARDI B.

1965 Excavations and Reconnaissance in Kalat, West Pakistan. *Pakistan Archaeology* 2: 86-182.

1967 The Bampur Sequence in the 3rd Millennium BCE. *Antiquity* XLI, 161: 33-41.

1968 Excavations at Bampur, S.E. Iran: a Brief Report. *Iran* VI: 135-155.

1970 *Excavations at Bampur, a Third Millennium Settlement in Persian Baluchistan, 1966*. New York: The American Museum of Natural History (*Anthropological Papers of the American Museum of Natural History* 51,3).

1983 *Archaeological Surveys in Baluchistan, 1948 and 1957*. London: Institute of Archaeology (*Occasional Publication* 8).

DIDIER A.

2007 *Archéologie des confins indo-iraniens: Étude de la production céramique du Kech-Makran (Pakistan) dans la première moitié du 3e millénaire av. J.-C. (2800-2500 av. J.-C.)*. Thèse de doctorat non publiée. Paris: Université de Paris 1 – Panthéon-Sorbonne.

DITTMANN R.

2002 Iran als Mittler zwischen Ost und West. *In*: HAUSLEITER A., KERNER S. and MÜLLER-NEUHOF B. (eds.), *Material Culture and Mental Sphere: Rezeption Archäologischer Denkrichtungen in der Vorderasiatischen Altertumskunde*: 329-344. Munich: Ugarit-Verlag.

DURANTE S.

1979a The utilization of Xancus pyrum (L.) at Shahr-i Sokhta. Further evidence for cultural relations between India and Iran in the 4th-3rd millennium B.C. *In*: VAN LOHUIZEN-DE LEEUW J.E. (ed.), *South Asian Archaeology* 1975: 27-42. Leiden: E.J. Brill.

1979b Marine shells from Balakot, Shahr-i Sokhta and Tepe Yahya: their significance for trade and technology for ancient Indo-Iran. *In*: TADDEI M. (ed.), *South Asian Archaeology* 1977: 317-344. Naples: Instituto Universitario Orientale.

DURRANI F.A.

1981 Rehman Dheri and the Birth of the Civilization in Pakistan. *Bulletin of the Institute of Archaeology* 18: 191-207.

DURRANI F.A. (ed.)

1994-1995 *Excavations in the Gomal Valley*. Peshawar: University of Peshawar (*Ancient Pakistan* 10).

DYSON R.H. and HOWARD S.M. (eds.)

1989 *Tappeh Hesar. Reports of the Restudy Project, 1976*. Florence: Le Lettere.

FAIRSERVIS W.A. Jr.

1956 *Excavations in the Quetta Valley, West Pakistan*. New York: The American Museum of Natural History (*Anthropological Papers of the American Museum of Natural History* 45).

1959 *Archaeological survey in the Zhob and Loralai Districts, West Pakistan*. New York: The American Museum of Natural History (*Anthropological Papers of the American Museum of Natural History* 47).

1975 *The Roots of Ancient India*. Chicago: University of Chicago Press.

FLUSCH L.

2003 *Futur antérieur. Trésors archéologiques du 21e siècle après J.-C.* Lausanne-Vidy: Infolio éditions.

FRANKE U.

2002 Iran Between East and West – Widening Horizons: South-eastern Iran and Its "Neighbours" during the 3rd Millennium BCE. Response to Reinhard Dittmann's Paper. *In*: HAUSLEITER A., KERNER S. and MÜLLER-NEUHOF B. (eds.), *Material Culture and Mental Spheres: Rezeption Archäologischer Denkrichtungen in der vorderasiatischen Altertumskunde*: 345-362. Munich: Ugarit-Verlag.

2003-2004 Sohr Damb/Nal (Balochistan, Pakistan): Ergebnisse der Grabungen 2001, 2002 und 2004. *Archäologische Mitteilungen aus Iran und Turan* 35-36: 83-141.

2005 Excavations at Sohr Damb/Nal (Pakistan): Results of the 2002 and 2004 seasons. *In:* FRANKE U. and WIESHAAR J. (eds.), *South Asian Archaeology* 2003: 63-76. Bonn: Linden Soft.

2007 Borderland Chronologies as seen from Sohr Damb/Nal (3200-2400 BCE). Paper presented on the 4th of July at the *19th International Conference on South Asian Archaeology*. Ravenna 2-7 July 2007.

2008 Baluchistan and the Borderlands. *In*: PEARSALL D. (ed.), *Encyclopedia of Archaeology* 1: 651-670. New York: Elsevier.

FRANKE U. and IBRAHIM A.

2005 A New Perspective of an Old Site: Reopening Excavations at Sohr Damb/Nal (Balochistan). *In*: JARRIGE C. and LEFÉVRE V. (eds.), *South Asian Archaeology* 2001: 105-115. Paris: Éditions Recherche sur les Civilisations.

GAMBLE C.

1999 *The Palaeolithic Societies of Europe*. Cambridge: Cambridge University Press (*Cambridge World Archaeology*).

HALIM M.A.

1971 Excavations at Sarai Khola. Part I. *Pakistan Archaeology* 7: 23-89.

HARGREAVES H.

1929 *Excavations in Baluchistan 1925, Sampur Mound, Mastung and Sohr Damb, Nal*. New Delhi (*Memories of the Archaeological Survey of India* 35).

JARRIGE C., JARRIGE J.-F., MEADOW R.H. and QUIVRON G. (eds.)

1995 *Mehrgarh. Field Reports 1974-1985. From Neolithic Times to the Indus Civilization*. Karachi: Department of Culture and Tourism, Government of Sindh.

JARRIGE C. and TOSI M.

1981 The natural resources of Mundigak. *In*: HÄRTEL H. (ed.), *South Asian Archaeology* 1979: 115-142. Berlin: Dietrich Reimer Verlag.

JARRIGE J.-F.

1993 The Question of the Beginning of the Mature Harappan Civilization as seen from Nausharo excavations. *In*: GAIL A. and MEVISSEN G. (eds.), *South Asian Archaeology* 1991: 149-164. Stuttgart: Franz Steiner.

1996 Les fouilles de Nausharo au Balochistan pakistanais et leur contribution à l'étude de la civilisation de l'Indus. *Comptes-rendus des séances de l'Académie des Inscriptions et Belles Lettres*: 821-878.

JARRIGE J.-F. and QUIVRON G.

2008 The Indus Valley and the Indo-Iranian Borderlands at the end of the 3rd Millennium and the Beginning of the 2nd Millennium BC. *In*: RAVEN E.M. (ed.), *South Asian Archaeology* 1999: 61-83. Leiden: Forsten.

JOSHI J.P. and PARPOLA A. (eds.)

1987 *Corpus of Indus Seals and Inscriptions. 1. Collections in India*. Helsinki: Suomalainen Tiedeakatemia.

KENOYER J.M.

1984a Shell Industries at Moenjodaro, Pakistan. *In*: JANSEN M. and URBAN G. (eds.), *Interim Reports* 1: 99-115. Aachen: IsMEO/RWTH.

1984b Shell working Industries of the Indus Civilization: a Summary. *Paléorient* 10,1: 49-63.

1985 Shell Working at Moenjo-Daro, Pakistan. *In*: SCHOTSMANS J. and TADDEI M. (eds.), *South Asian Archaeology* 1983: 297-344. Naples: Istituto Universitario Orientale, Dipartimento di Studi Asiatici.

1991a The Indus Valley Tradition of Pakistan and Western India. *Journal of World Prehistory* 5,4: 331-385.

1991b Urban Process in the Indus Tradition: A Preliminary Model from Harappa. *In*: MEADOW R.H. (ed.), *Harappa Excavations 1986-1990. A Multidisciplinary Approach to Third Millennium Urbanism*: 29-60. Madison: Prehistory Press.

1991c Ornament Styles of the Indus Valley Tradition: Evidence from Recent Excavations at Harappa, Pakistan. *Paléorient* 17,2: 79-98.

1991d Shell Working in the Indus Civilization. *In*: JANSEN M., MULLOY M. and URBAN G. (eds.), *Forgotten cities on the Indus*: 216-219. Mainz: Phillip von Zabern.

1997 Trade and Technology of the Indus Valley: New Insights from Harappa, Pakistan. *World Archaeology* 29,2: 262-280.

KENOYER J.M. and MEADOW R.H.

2000 The Ravi Phase: A New Cultural Manifestation at Harappa. *In*: TADDEI M. and DE MARCO G. (eds.), *South Asian Archaeology* 1997: 55-76. Roma: Istituto Italiano per l'Africa e l'Oriente.

KHAN F.A.

1965 Excavations at Kot Diji. *Pakistan Archaeology* II: 11-85.

KIRCHO L.B.

2000 Bogatoe pogrebenie epokhi srednej bronzy na Altyn-depe [A Rich Grave of Middle Bronze Age at Altyn-Depe]. *Arkheologicheskie Vesti* 7: 70-76.

KNOX R.

1994 A New Indus Valley Cylinder Seal. *In*: PARPOLA A. and KOSKIKALLIO P. (eds.), *South Asian Archaeology* 1993: 375-378. Helsinki: Suomalainen Tiedeakatemia.

KOHL P.L.

1975 Carved Chlorite Vessels: a Trade in Finished Commodities in the Mid-Third Millennium. *Expedition* 18, I: 18-31.

1978 The Balance of Trade in Southwestern Asia in the Mid-Third Millennium. *Current Anthropology* 19,3: 463-492.

1979 The "World Economy" of West Asia in the Third Millennium BCE. *In*: TADDEI M. (ed.), *South Asian Archaeology* 1977: 55-85. Naples: Instituto Universitario Orientale.

LAMBERG-KARLOVSKY C.C.

1970 *Excavations at Tepe Yahya, Iran 1967-1969*. Cambridge MA: Peabody Museum of Archaeology and Ethnology, Harvard University (*American School of Prehistoric Research Bulletin* 27).

1972 Tepe Yahya 1971: Mesopotamia and the Indo-Iranian Borderlands. *Iran* 10: 89-100.

1976 Foreign Relations in the third Millennium at Tepe Yahya. *In*: DESHAYES J. (éd.), *Le plateau iranien et l'Asie centrale des origines à la conquête islamique. Leurs relations à la lumière des documents archéologiques*: 23-31. Paris: Éditions du CNRS (*Colloques internationaux du CNRS* 567).

2004 New Centers of Complexity in the Iranian Bronze Age. *The Review of Archaeology* 25,1: 5-10.

in press Structure, Agency and Commerce in the Ancient Near East. *Iranica Antiqua.*

LAMBERG-KARLOVSKY C.C. and TOSI M.

1973 Shahr-i Sokhta and Tepe Yahya: Tracks on the Earliest History of the Iranian Plateau. *East and West* 23: 21-58.

LAW R.

2005 Indus Rock and Mineral Trade Network: The View from Harappa. Presented at the *18th conference of South Asian Archaeology.* London, 4-6 July 2005.

MACKAY E.J.H.

1938 *Chanhu-Daro Excavations 1935-1936.* New Delhi: Government of India.

1938 *Further Excavations at Mohenjo-Daro.* New Delhi: Government of India.

1943 Chanhu-Daro Excavations 1935-1936. *In*: NORMAN BROWN W., SHRYOCK J.K. and SPEISER E.A. (eds.), *American Oriental Series* 20. New Haven: Kraus Reprint Corporation.

MADJIDZADEH Y.

2003 *Jiroft: The Earliest Oriental Civilization.* Teheran: Cultural Heritage Organization.

2007 Excavations at Konar Sandal: Jiroft. Campaign 2006-2007. Paper presented to the *Middle Asian Intercultural Space Conference.* Ravenna, 8-9 July 2007.

MARCHESI G.

2007 Inscribed "Intercultural Style" Vessels from 3rd millennium Mesopotamia. Paper presented to the *Middle Asian Intercultural Space Conference.* Ravenna, 8-9 July 2007.

MARSHALL sir J.

1931 *Mohenjo-Daro and the Indus Civilization.* London: Probsthain.

MUGAVERO L. and VIDALE M.

2003 The Use of Polychrome Containers in the Helmand Civilisation: A Female Function? *East and West* 53: 67-94.

2006 The Helmand Polychrome Wares as Skeuomorphs. *East and West* 56: 433-438.

PARPOLA A.

1981 New Correspondences between Harappan and Near Eastern glyptic art. *In*: ALLCHIN B. (ed.), *South Asian Archaeology* 1981: 176-195. Cambridge: Cambridge University Press.

2006 Administrative contact and acculturation between Harappans and Bactrians: Evidence of sealings and seals. *In*: JARRIGE C. and LEFÈVRE V. (eds.), *South Asian Archaeology* 2001: 281-288. Paris: ERC.

PIPERNO M.

1977 La necropoli. *In*: TUCCI G. (ed.), *La Città Bruciata nel Deserto Salato*: 113-148. Venice: Erizzo.

PIPERNO M. and SALVATORI S.

1983 Recent Results and New Perspectives from the Research at the Graveyard of Shahr-i Sokhta, Sistan, Iran. *Annuali dell'Istituto Universitario Orientale* 43,2: 173-191.

2007 *The Shahr-i Sokhta Graveyard (Sistan, Iran). Excavation Campaigns 1972-1978.* Roma: *IsIAO* (*IsIAO Reports and Memoirs New Series* VI).

PITTMAN H.

2001 Glyptic Art of Period IV. *In*: LAMBERG-KARLOVSKY C.C. and POTTS D.T. (eds.), *Excavations at Tepe Yahya, Iran 1967-1975. The Third Millennium*: 231-268. Cambridge, MA: Harvard University (*American School of Prehistoric Research* 45).

POTTS D.T.

1994 *Mesopotamia and the East.* Oxford: Oxford University Committee for Archaeology (*Oxford University Committee for Archaeology Monograph* 37).

RAIKES R.

1968 Archaeological Explorations in Southern Jhalawan and Las Bela (Pakistan). *Origini, Preistoria e Protostoria delle Civiltà Antiche.* Roma: Università degli Studi di Roma. Istituto di Paletnologia-Museo delle Origini.

RAO S.R.

1985 Lothal. A Harappan Port Town 1955-62. *Memories of the Archaeological Survey of India* 78, Vol. I-II. Naba Mudran.

SAJJADI S.M.S.

2003 Excavations at Shahr-i Sokhta. First Preliminary Report on the Excavations of the Graveyard 1997-2000. *Iran* 41: 21-97.

2004 Archaeological Report. Sistan and Baluchestan Project. *Iran* 42: 247-250.

SAJJADI S.M.S., COSTANTINI L. and BIASIMI COSTANTINI L.

2004 Nata j-e akhern barresiha-ye giahbastanshenasi dar Shahr-e Sokhta [The Results of the Latest Archeobotanical Researchs at Shahr-e Sokhta]. *Gozareshha-ye Bahstanshenasi* 2: 67-86.

SALVATORI S.

2000 Bactria and Margiana Seals. A New Assessment of Their Chronological Position and a Typological Survey. *East and West* 50: 97-145.

SALVATORI S. and TOSI M.

2005 Shahr-i Sokhta Revised Sequence. *In*: JARRIGE C. and LEFÈVRE V. (eds.), *South Asian Archaeology* 2001: 281-292. Paris: ERC.

SALVATORI S. and VIDALE M.

1997 Shahr-i Sokhta 1975-1978: Central Quarters Excavations. Preliminary Report. Roma: IsIAO (*IsIAO Reports and Memoirs Series Minor* I).

SANKALIA H.D.

1974 *Prehistory and Protohistory of India and Pakistan.* Poona: Deccan College.

SANTONI M.

1984 Sibri and the South Cemetery of Mehrgarh. Third Millennium Connections between the Northern Kachi Plain (Pakistan) and Central Asia. *In*: ALLCHIN B. (ed.), *South Asian Archaeology* 1981: 52-60. Cambridge: Cambridge University Press.

SARIANIDI V.I.

1983 The Pottery of Shahr-i Sokhta I and its Southern Turkmenian Connections. *In*: TOSI M. (ed.), *Prehistoric Sistan I*: 183-198. Roma: IsMEO (*IsMEO Reports and Memories* XIX, 1).

1986 *Die Kunst Des Alten Afghanistan.* Leipzig: Seemann.

1998 *Margiana and Protozoroastrism.* Athens: Kapon Editions.

SHAFFER J.G.

1978 *Prehistoric Baluchistan: with Excavation Report on Said Qala Tepe.* New Delhi: B.R. Publishing Corp.

1992 The Indus Valley, Baluchistan and Helmand Traditions: Neolithic through Bronze Age. *In*: EHRICH R. (ed.), *Chronologies in Old World Archaeology*. Vol. I. Chicago: The University of Chicago Press.

SHAH S.G.M. and PARPOLA A. (eds.)

1991 *Corpus of Indus Seals and Inscriptions. 2. Collections in Pakistan*. Helsinki: Suomalainen Tiedeakatemia.

SHETENKO

1968 Raskopki na Altyn-depe v Juzhnoy Turkmenii (Excavations at Altyn-depe in Southern Turkmenistan). KSIA - *Kratkie soobschenija Instituta Arkheologii* [*Short Communications of the Institute of Archaeology*] 114: 39-43.

1970 O torgovykh putjakh epokhi Bronzy po materialam turkmenistano-kharappskikh paralleley [Bronze Age Trade Routes on the bases of Turkmenistan-Harappa parallels]. KSIA - *Kratkie soobschenija Instituta Arkheologii* [*Short Communications of the Institute of Archaeology*] 122: 59-63.

SHINDE V., OSADA T. and SHARMA M.M.

2007 Harappan Archaeology in the Ghaggar Basin: Results of the First Season's (2006-07) Work. Paper given at the 19th international conference on *South Asian Archaeology* 2007. Ravenna, 2-6 July.

SONAWANE V.H., AJIITPRASAD P., BHAN K.K., KRISHNAN K., PRATHAPACHANDRAN S., MAJUMDAR A., AJITA K.P. and JAVA M.

2003 Excavations at Bagasra - 1996-2003: A Preliminary Report. *Man and Environment* XXVII, 2: 21-50.

TOSI M.

1973 The Cultural Sequence of Shahr-i Sokhta. *Bulletin of the Asia Institute of Pahlavi University* 3: 64-80

1983 Development, Continuity and Cultural Change in the Stratigraphic Sequence of Shahr-i Sokhta. *In*: TOSI M. (ed.), *Prehistoric Sistan 1*: 127-180. Roma: IsMEO (*IsMEO Reports and Memoirs* XIX,1).

TOSI M. (ed.)

1983 *Prehistoric Sistan 1*. Roma: IsMEO (*IsMEO Reports and Memoirs* XIX,1).

TOSI M. and PIPERNO M.

1973 Lithic Technology Behind the Ancient Lapis Lazuli Trade. *Expedition* 16,1: 15-23.

TUCCI G. (ed.)

1977 *La Città Bruciata nel Deserto Salato*. Venice: Erizzo.

VANZETTI A. and VIDALE M.

1995 Formation Processes of Beads: Defining Different Levels of Craft Skills among the Early Beadmakers of Mehrgarh. *In*: PARPOLA A. and KOSKIKALLIO P. (eds.), *South Asian Archaeology* 1993: 763-776. Helsinki: Suomalainen Tiedeakatemia.

VATS M.S.

1940 *Excavations at Harappa*. New-Delhi: Government of India.

VIDALE M.

1989 A Steatite-cutting Atelier on the Surface of Moenjodaro. *Annali dell'Istituto Universitario Orientale* 49,1: 29-51. Naples: Istituto universitario orientale di Napoli.

1995 Early Beadmakers of the Indus Tradition: the Manufacturing Sequence of Talc Beads at Mehrgarh in the 5th Millennium B.C. *East and West* 4: 45-80.

2000 *The archaeology of Indus crafts. Indus Craftpeople, and why we study them*. Roma: Istituto Italiano per l'Africa e l'Oriente (*IsIAO Reports and Memoirs*, Series Minor IV).

2004 Pirotecnologie Euroasiatiche della Steatite. *In*: GIANNICHEDDA E. (ed.), *Metodi e Pratica della Cultura Materiale. Produzione e Consumo dei Manufatti*: 69-93. Bordighera: Istituto Internazionale di Studi Liguri.

2007 The Cemetery of Mathoutabad. Paper presented to the *Middle Asian Intercultural Space Conference*. Ravenna, 8-9 July 2007.

forthcoming Lapis Lazuli Working at Shahr-i Sokhta. Roma: Istituto Italiano per l'Africa e l'Oriente (*IsIAO Reports and Memoirs Series Minor*).

VIDALE M., PETRUCCI G., PRIORI G. and GUIDA G.

2007 Serial Production of Disk Beads in Soft Stones between Iran and the Indus Valley in the 3rd Millennium BCE. Paper given at 19th the international conference of *South Asian Archaeology* 2007. Ravenna 2-6 July 2007.

WALLERSTEIN I.

1974 *The Modern World System: Capitalist Agriculture and the Origins of the European World in the Sixteenth Century*. New York: Academic Press.

WINKELMANN S.

1994 Intercultural Relations between Iran, the Murgabo-bactrian Archaeological Complex (BMAC) Northwest India and Faylaka in the Field of Seals. *In*: PARPOLA A. and KOSKIKALLIO P. (eds.), *South Asian Archaeology* 1993: 815-831. Helsinki: Suomalainen Tiedeakatemia.

WRIGHT R.P.

1987 The Frontiers of Prehistoric Baluchistan and the development of the Indus Civilization. *In*: TRINKAUS K.M. (ed.), *Polities and Partitions: Human Boundaries and the Growth of Complex Societies*: 61-82. Tucson: Arizona State University (*Arizona State Anthropological Research Papers* 37).

2002 Revisiting Interaction Spheres-Social Boundaries and Technologies on Inner and Outermost Frontiers. *Iranica Antiqua* XXXVII: 403-418.

THE PLASTERED SKULLS AND OTHER PPNB FINDS FROM YIFTAHEL, LOWER GALILEE (ISRAEL)

I. MILEVSKI, H. KHALAILY, N. GETZOV and I. HERSHKOVITZ

Abstract: *Two extended seasons of excavations were recently conducted at the site of Yiftahel (Khalet Khalladyiah) in the fringe of Beit Netofa valley, Lower Galilee. Four new areas (F, G, H and I) were opened in addition to the five areas excavated in the past. This note will present the highlights from the main occupational period at Yiftahel,* i.e., *the Mid-Late PPNB (8,000-7,000 BC). Large PPNB buildings, displaying rectilinear plan with mud-brick and/or stone walls, were discovered in the recent seasons of excavations. All buildings have thick lime-plastered floors. Aside from plentiful lithic assemblages related to the "naviform" industry, rich botanical and faunal remains were found at the site. The most outstanding finds are related to the mortuary practices, including three modeled plastered skulls.*

Résumé : *Récemment deux longues campagnes de fouille ont été conduites sur le site de Yiftahel (Khalet Khalladyiah) en marge de la vallée de Beit Netofa en Basse Galilée. Quatre nouveaux chantiers (F, G, H, I) sont ainsi venus s'ajouter à ceux ouverts dans le passé. Cet article présente les caractères de la principale période d'occupation du site, le PPNB moyen-récent (8000-7000 BC) : grands bâtiments de plan rectangulaire, aux murs en brique crue et/ou en pierre et aux sols recouverts d'un épais enduit de plâtre ; industrie lithique abondante montrant un débitage de type « naviforme » ; grande richesse de vestiges botaniques et fauniques. Les découvertes les plus remarquables touchent aux pratiques funéraires et à la mise au jour de trois crânes surmodelés.*

Keywords: *Yiftahel, Southern Levant, Israel, PPNB, Plant Domestication, "Naviform" technology, Burial customs, Plastered skulls.*
Mots-clés : *Yiftahel, Levant sud, Israël, PPNB, Domestication des plantes, Débitage « naviforme », Pratiques funéraires, Crânes surmodelés.*

Yiftahel (Khalet Kalladyiah) is located on the western fringe of the Beit Netofa valley, Lower Galilee (Israel) *ca* 8 km west of Nazareth (fig. 1). The archaeological remains are embedded within the eastern bank of Nahal Yiftahel (W. Khalladyiah). The site is estimated to extend over approximately 4 ha (40 dunams).

Two extended seasons of excavations were conducted during September 2007-January 2008 and April-November 2008 at Yiftahel. The large-scale archaeological salvage excavations, on behalf of the Israel Antiquities Authority (IAA), were conducted by H. Khalaily, I. Milevski and N. Getzov. Four new areas (F, G, H and I) were opened in addition to the five areas excavated in the 1980s and the 1990s (A, B, C, D and E)[1] (fig. 2). During the present excavations four main chrono-stratigraphical layers were excavated: Pre-Pottery Neolithic B (PPNB), the Pottery Neolithic (PN)—Lodian (Jericho IX) culture, the PN—Wadi Rabah culture and the Early Bronze (EB) IA. This article is devoted to the remains and finds from the PPNB occupation layers and mainly to the plastered skulls found in Area I, dated to the Mid-Late PPNB (8,000-7,000 cal. BC).

1. LAMDAN and DAVIES, 1983; GARFINKEL, 1987; RONEN, BENTUR and SOROKA, 1991; BRAUN, 1997; KHALAILY, MARDER and MILEVSKI, 2000.

Manuscrit reçu le 2 novembre 2008, accepté le 25 novembre 2008

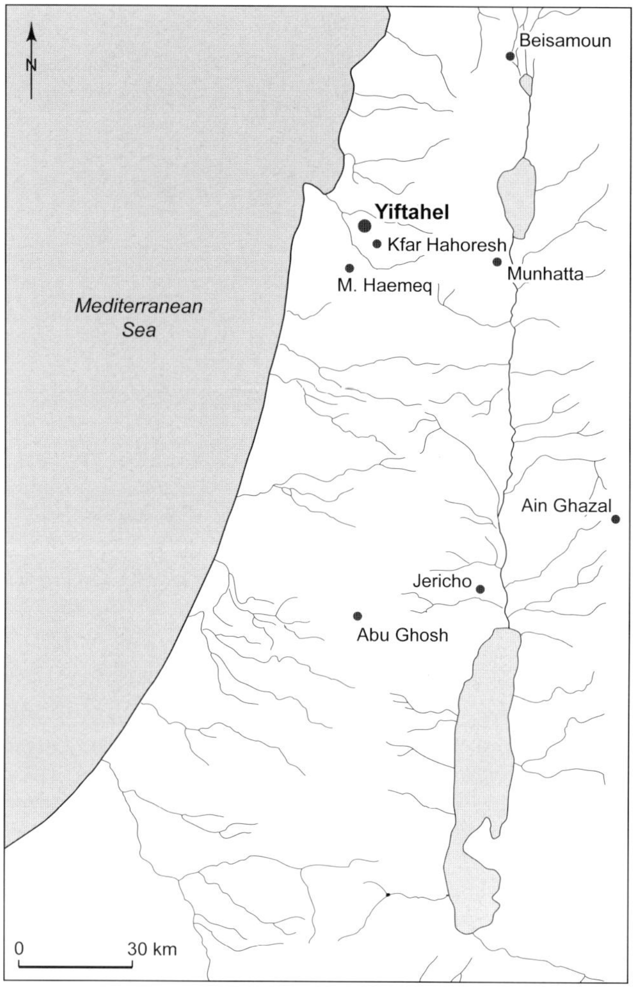

Fig. 1 – *Map of PPNB sites, with location of Yiftahel.*

THE PRE-POTTERY NEOLITHIC B

The main occupational period at Yiftahel is clearly the Mid-Late PPNB of which approximately 2,000 m^2 have been excavated by all expeditions combined.

The internal stratigraphy within the PPNB in each area of excavation is not identical.[2]

The PPNB buildings display a rectilinear plan with mud-brick and/or stone walls. Elements present within the structures include constructed hearths, pits, installations and burials. All of the structures have thick lime-plastered floors.

In Area I, which includes the former Area D, seven construction phases reflected by floor levels were discerned within the PPNB (local Stratum 3), suggesting a very dynamic chapter during this period. The most impressive occupation in Area I is represented by two large rectangular buildings, 501 (14 m x 6 m) and 552 and one smaller building, 502 (8 m x 3 m) (fig. 3). More research is necessary in order to determine whether these structures can be interpreted as being communal buildings, where various activities were carried out. Building 501, constructed of either mud-bricks or *terre pisée*, was well preserved. Entrance to the structure was by either of two narrow entrances in the west. Within, there were postholes, several installations and several burials. The postholes, which penetrated the two thick layers of the plaster floor, indicate that the pillars were 0.20 m in diameter. Area I was rich in lithic, botanic and faunal assemblages.

FINDS

The PPNB of Yiftahel is rich in botanic remains, mainly in Area I. Concentrations of seeds, mainly lentils (*Lens culinaris*) and beans (probably *Vicia faba*)[3] were encountered in that area.[4] In addition, small quantities of emmer, probably wild *Triticum dicoccoides*,[5] were found in the same area, though it is too soon to confirm it.[6]

Within the faunal remains, main species found were gazelles (*Gazella gazella*), wild goat (*Capra aegagrus*), aurochs (*Bos primigenius*) and wild boar (*Sus scrofa*). Cervids were also present although in lesser quantities.

The PPNB lithic industry at the site is outstanding especially in regard to the "naviform" blade technology.[7] Several industrial waste pits were located and several "naviform" workshops were identified[8] indicating the intensive specialized knapping that was performed at the site. It has been suggested elsewhere[9] that large quantities of blades produced at the site were exported to other localities.

The formal flint tools retrieved in the 2007-2008 seasons are inversely retouched sickle blades, arrowheads—"Jericho", "Amuq" and "Byblos" types—(fig. 4) and several axes. Several

2. See KHALAILY *et al.*, 2008.

3. KISLEV, 1985.
4. ZOHARY and HOPF, 1993: 106-107.
5. *Ibid.*: 40-42.
6. U. Weiss, personal communication.
7. QUINTERO and WILKE, 1995; ABBÈS, 2003; BARZILAI, forthcoming.
8. See KHALAILY *et al.*, 2008.
9. KHALAILY, 2006.

Fig. 2 – *Aerial view with the location of the excavated areas.*

caches of flint and bone tools were also associated with Buildings 500 and 501.[10]

MORTUARY PRACTICES AND OTHER CULT PRACTICES

The mortuary practices at the site[11] display the known PPNB pattern of intra-site interments in flexed position, under plaster floors and in pits. Primary and secondary burials were found mainly in Area I, a few in Areas F, G and H. The minimum number of human inhumations during the PPNB in all areas and phases stand today at 32, 23 during the 2007-2008 seasons. Some of the skeletons had their skull intact; others had their skull removed post-mortem (fig. 5). Several infant burials were recovered in Areas F (formerly C) and I.[12] The most emotive case is a triple burial (L 5224) consisting of an adult man (Homo 6), a woman (Homo 7) and a juvenile (Homo 8), all of them embracing each another (fig. 6).

Other aspects of cult practices in Area I could include the interment of *B. primigenius* bones. For example, one pit yielded a horn while a second included the spine, the pelvis and a leg in articulation. At the nearby site of Kfar Hahoresh, similar activities involving the burial of *B. primigenius* and other animal species have been connected with cult activities.[13]

THE PLASTERED SKULLS

The most impressive find related to mortuary and cult practices is a cache of three plastered skulls deposited in a pit (L4187) excavated in Open Area 402, north of Building 501

10. KHALAILY *et al.*, 2008.

11. LAMDAN and DAVIES, 1983; HERSHKOVITZ, GARFINKEL and ARENSBURG, 1986; KHALAILY, MARDER and MILEVSKI, 2000.

12. The preliminary anthropological research reveals the following: three plastered skulls (see below) and one very damaged non-decorated skull in Area I; five primary burials with cranial removal (including one child of *ca* 8 years old), Area I; five primary burials without cranial removal, Areas H and I; nine secondary burials (among them three infants), Areas G and I.

13. HORWITZ and GORING-MORRIS, 2004.

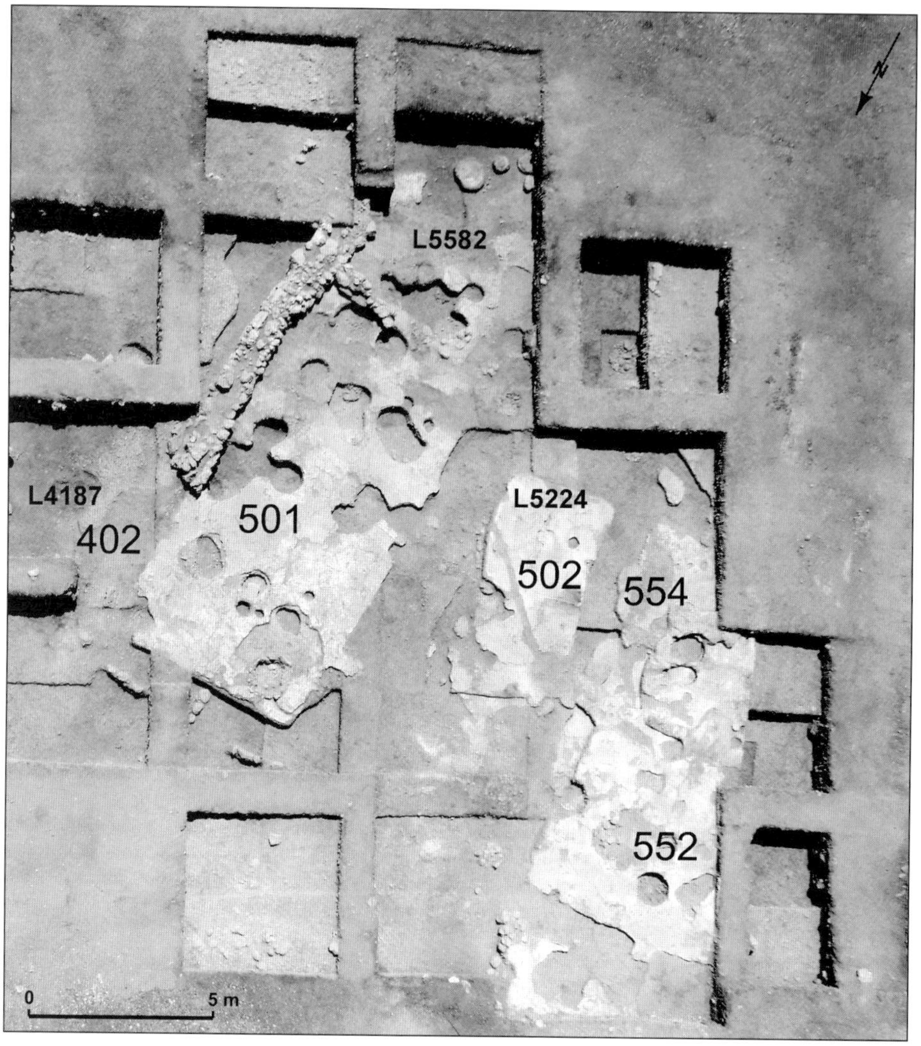

Fig. 3 – *PPNB buildings in Area I, Stratum 3, looking southeast.*

(fig. 7). These skulls can be added to the growing list of modeled skulls from many PPNB sites in the Levant[14] and sites from the PN in Anatolia.[15]

The skulls from Yiftahel, labeled Homo 1 to Homo 3, were placed together in a row. They were all facing the same direction (west), and all belong to adult individuals. Due to restoration difficulties they have not yet been studied thoroughly and therefore age and sex could not be determined with sufficient accuracy (fig. 8). All three skulls show considerable damage, mainly to the calvaria and face.

The skull of Homo 2 bears a well remodeled face, resembling in some aspects the molded skull from the nearby PPNB site of Kfar Hahoresh.[16] The remodeling covers the whole facial skeleton, extending superiorly above the supraorbital arches and laterally into the infratemporal fossae. Special attention had been paid to important facial features, including the nose, the eyes and the mouth. The nose was made of a limestone pebble attached to the central part of the face, it was than carpeted by a thin layer of plaster which stabilized it to the face. There is no doubt that the "eyes" are the most prominent feature. Each eye was made of a single white shell (probably *Donax sp.*) fractured in the middle into two equal

14. *e.g.*, HERSHKOVITZ *et al.*, 1995; BONOGOFSKY, 2003; STORDEUR, 2003; STORDEUR et KHAWAM, 2007; KUIJT, 2008; FLETCHER, PEARSON and AMBERS, 2008.

15. HODDER, 2005; ÖZBEK, 2009.

16. GORING-MORRIS *et al.*, 1995: figs. 8-9.

Fig. 4 – *PPNB formal flint tools.* ***a****, Arrowheads from Area G;* ***b****, Arrowheads from Area I;* ***c****, Sickle blades from Area I.*

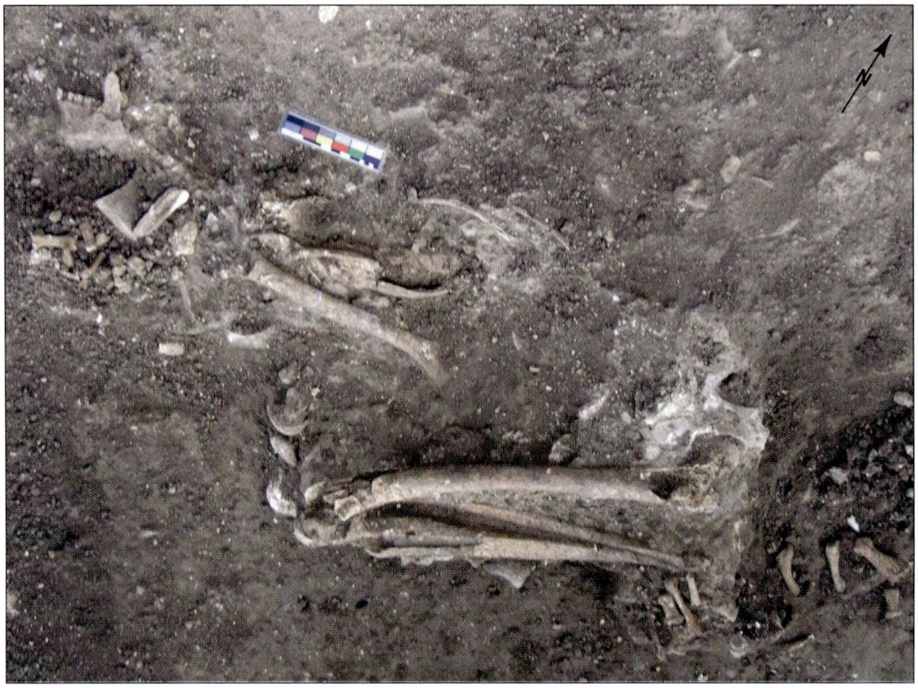

Fig. 5 – *Burial L5582, Area I, looking west.*

Fig. 6 – *Burial L5224, Area I, looking south.*

Fig. 7 – *Plastered skulls* in situ, *L4187, Open area 402, Area I, looking east. From left to right: Homo 1, Homo 2 and Homo 3.*

Fig. 8 – *Plastered skulls from Yiftahel in the laboratory of the Department of Anatomy and Anthropology, Faculty of Medicine, Tel Aviv University. From left to right: Homo 1, Homo 2 and Homo 3.*

segments (to represent the bulbar conjunctiva). The internal (concave) aspect of the shells were pressed toward the plaster, their margins covered by a rounded elongated plaster role to create the impression of an eyelid. Between the two pieces of the shells, a small conical shell was placed (for the iris). The overall image is of a "cat eye". The mouth is represented by a very small, rounded depression 2 cm below the nose. The face in general manifests rounded contours with very prominent and high cheeks.

All these facial characteristics of the remodeled face lend Homo 2 of Yiftahel an expression of contemplation, with open eyes, a half opened mouth and tensed facial muscles. It is surprising that this mask resembles more that of Skull D113 from Jericho[17] rather than that of the nearby Homo 1 from Kfar Hahoresh.

Homo 1 and Homo 3 (on either side of Homo 2) do not posse a full remodeled face, only the eyes were plastered and decorated. To the right (north) is Homo 1. The skull collapsed and had been compressed in the past due to some vertical pressure, calvarial bones were fractured and the facial area pushed towards the vertex. The two orbital cavities were filled with plaster, which was later flattened and carpeted on its upper portion with flat brightly shined shell (probably mother of pearl). At the center, a small piece of black colored flint was inserted (for the iris). The remodeled face of this skull resembles, in some respects, that of Jericho's plastered skull J 5757.[18] Homo 3 (south), has suffered the most severe damage. The calvarial area and the left side of the facial skeleton are for the most part missing, and the remains that survived manifest severe deformation and distortion. The right eye show clear evidence of molding, the orbital cavity was filled with plaster, yet unlike Homo 1 the eye was differently modeled, *i.e.*, the eyelids which are missing in Homo 1 are very thick here, the space between them filled with two pieces of shell, probably also from mother of pearl.

PRELIMINARY REMARKS

The results of the renewed excavations at Yiftahel have considerably modified the understanding of the site during the PPNB period. The combination of large architectural units with small buildings could point out to a social and/or working division within the village.

17. KENYON, 1981: pl. 54a.
18. BONOGOFSKY, 2003: fig. 2.

Craft specialization and a social division inside Yiftahel could be ascertained not just from the architectural units but also from the rich lithic assemblage retrieved, showing the existence of several workshops at the site. Caches of blades and cores, bone tools, and bifacial implements show the existence of tool kits related to the knappers at the site, and associated cult activities.

The excellent preservation of organic remains, especially of pulses, may contribute to the understanding of plant domestication in the Levant at this period, and shed light on the exploitation of the surrounding food resources (faunal and botanical) by the local inhabitants.

Burial practices at Yiftahel follow the known pattern of the PPNB customs, including the cache of three modeled skulls. Cranial removal and modeling is part of a known ritual practice in the PPNB and was interpreted as a means to incorporate the memory of the deceased and a link between life and death.[19]

Additional caches of three skulls found together were found at other PPNB sites,[20] some under plastered floors and some, as at Yiftahel, in extra-mural locations.[21] This pattern of three skulls could symbolize a social arrangement in the worldly PPNB society.[22] In the case of Yiftahel, the differences in modeling between the skulls and their arrangement suggest that the skull in the middle (Homo 2) was the most important one and the two other skulls were guarding it; this explains why only their eyes were modeled. If this was the case, does this indicate social hierarchy within the Yiftahel community?[23]

In summary, following recent discoveries from the 2007-2008 seasons, we can state that Yiftahel was one of the largest PPNB villages in the lower Galilee area, probably interacting with other contemporaneous sites as Kfar Hahoresh. The plan and the structures at the site, the human, botanic and animal remains, the variety of craft activities, both for internal use and external interchange, all suggest that Yiftahel was a central site at the time.

19. BONOGOFSKY, 2003; KUIJT, 2008; FLETCHER, PEARSON and AMBERS, 2008.
20. BONOGOFSKY, 2003.
21. KUIJT, 2008.
22. *Cf.* STORDEUR et KHAWAM, 2007: 21.
23. Certainly, the hypotheses put forward in this paper are preliminary and are not *faits établis*. The future research on the site and a comparative study with near and far sites from the Levantine PPNB will give an answer to this.

ACKNOWLEDGEMENTS

Thanks are due to the other members the staff and workers who participated in the excavations: A. Abu Hamid, E. Amos, Z. Daniel, R. Liran, A. Najjar, E.V. Shlomi, E. Stern, A. Yarosevich, and O. Zidan (field supervisors and architects); V. Essman, T. Kornfeld, R. Moussaief (surveyors); A. Shapiro (registrer and GPS); O. Barzilai (lithic assemblages); G. Le Dosseur (bone tools); H. Kanare (plaster floors); E. Boaretto, L. Regev, and S. Weiner (material analyzes and C^{14} datings); E. Kamaisky (conservator); T. Kuperman, H. May, H. Cohen, B. Madlege, A. Bechar (physical anthropology); L. Weissbrod (microfauna); U. Weiss (archeobotany); Y. Yaakobi, R. Abu Halaf, E. Bakhar and Y. Laban (administrators); workers from Kfar Manda, Nazareth, Tiberiah, Umm el-Fahm; volunteers from Israel and Argentina. Thanks are also due to H. Smithline and S. Wolff for the English editing of a previous version of this note, and to É. Coqueugniot and four anonymous readers who made helpful comments on that version.

Ianir MILEVSKI
Hamoudi KHALAILY
Nimrod GETZOV
Israel Antiquities Authority
POB 586
91104 Jerusalem
ISRAEL
ianir@israntique.org.il
hamudi@israntique.org.il
getzov@israntique.org.il

Israel HERSHKOVITZ
Dept. of Anatomy and Anthropology
Sackler School of Medicine
Tel Aviv University
Tel Aviv
ISRAEL
anatom2@post.tau.ac.il

BIBLIOGRAPHY

ABBÈS F.

2003 *Les outillages néolithiques en Syrie du Nord: Méthode de débitage et gestion laminaire durant le PPNB.* Oxford-Lyon: Archaeopress-MOM (*BAR Int. Ser.* 1150).

BARZILAI O.

forthcoming *Social Complexity the Southern Levantine PPNB as Reflected through Lithic Studies. The Bidirectional Blade Industries.* Jerusalem: Hebrew University of Jerusalem. PhD in progress.

BONOGOFSKY M.

2003 Neolithic Plastered Skulls and Railroading Epistemologies. *Bulletin of the American Schools of Oriental Research* 331: 1-10.

BRAUN E.

1997 *Yiftah'el. Salvage and Rescue Excavations at a Prehistoric Village in Lower Galilee, Israel.* Jerusalem: Israel Antiquities Authority (*IAA Reports* 2).

FLETCHER A., PEARSON J. and AMBERS J.

2008 The Manipulation for Social and Physical Identity in the Pre-Pottery Neolithic. Radiographic Evidence for Cranial Modification at Jericho and its Implications for the Plastering of Skulls. *Cambridge Archaeological Journal* 18: 302-325.

GARFINKEL Y.

1987 Yiftahel: A Neolithic Village from the Seventh Millennium BC in Lower Galilee, Israel. *Journal of Field Archaeology* 14: 199-274.

GORING-MORRIS A.N., GOREN Y., HORWITZ L.K., BAR-YOSEF D. and HERSHKOVITZ I.

1995 Investigations at the Early Neolithic Settlement in the Lower Galilee: Results of the 1991 Season at Kfar Hahoresh. *'Atiqot* 27: 45-62.

HERSHKOVITZ I., GARFINKEL Y. and ARENSBURG B.

1986 Neolithic skeletal remains at Yiftahel, Area C (Israel). *Paléorient* 12,1: 73-81.

HERSKHOVITZ I., ZOHAR M., SPEIRS M.S., SEGAL I., MEIRAV O., SHERTER U., FELDMAN H. and GORING-MORRIS A.N.

1995 Remedy for an 8,500 years Old Plastered Skull from Kfar HaHoresh, Israel. *Journal of Archaeological Science* 22: 779-788.

HODDER I.

2005 New Finds and New Interpretations at Çatal Hüyük. *Anatolian Archaeology* 11: 20-22.

HORWITZ L.K. and GORING-MORRIS A.N.

2004 Animals and ritual during the Levantine PPNB: A Case Study from the site of Kfar Hahoresh, Israel. *Anthropozoologica* 39,1: 165-178.

KENYON K.M.

1981 *Excavations at Jericho. Volume 3. The Architecture and Stratigraphy at the Tell. Plates.* London: British School of Archaeology in Jerusalem.

KHALAILY H.

2006 *Lithic Traditions during the Late Pre-Pottery Neolithic B and the Question of the Pre-Pottery Neolithic C in the Southern Levant.* Beersheva: Ben Gurion University of the Negev. Unpublished PhD.

KHALAILY H., MARDER O. and MILEVSKI I.

2000 New Excavations at the PPNB Site of Yiftahel. *Neo-Lithics* 2-3/00: 18-20.

KHALAILY H., MILEVSKI I, GETZOV N., HERSHKOVITZ I, BARZILAI O., YAROSEVICH A., SHLOMI V., NAJJAR A., ZIDAN O., SMITHLINE H. and LIRAN R.

2008 Recent Excavations at the Neolithic Site of Yiftahel (Khalet Khalladyiah), Lower Galilee. *Neo-Lithics* 2/08: 3-11.

KISLEV M.

1985 Early Horsebean from Yiftah'el, Israel. *Science* 228: 319-320.

KUIJT I.

2008 The Regeneration of Life. Neolithic Structures of Symbolic Remembering and Forgetting. *Current Anthropology* 49,2: 171-197.

LAMDAN M. et DAVIES M.

1983 Le site d'Yiftah'el (Israël). *L'anthropologie* 87: 273-274.

ÖZBEK M.

2009 Remodeled Human Skulls in Köşk Höyük (Neolithic Age, Anatolia): A New Appraisal in View of Recent Discoveries. *Journal of Archaeological Science* 36: 379-386.

QUINTERO L. and WILKE P.J.

1995 Evolution and Economic Significance of Naviform Core-and-Blade Technology in the Southern Levant. *Paléorient* 21,1: 17-33.

RONEN A., BENTUR A. and SOROKA I.

1991 A Plastered Floor from the Neolithic Village, Yiftahel (Israel). *Paléorient* 17,2 : 149-155.

STORDEUR D.

2003 Des crânes surmodelés à Tell Aswad de Damascène (PPNB Syrie). *Paléorient* 29,2 : 109-115.

STORDEUR D. et KHAWAN R.

2007 Les crânes surmodelés de Tell Aswad (PPNB, Syrie). Premier regard sur l'ensemble, premières réflexions. *Syria* 84 : 5-32.

ZOHARY D. and HOPF M.

1993 *Domestication of Plants in the Old World: The Origin and Spread of Cultivated Plants in West Asia, Europe and the Nile Valley*. Oxford: Clarendon Press.

The Middle – Upper Palaeolithic Transition at Yabroud II (Syria). A Re-evaluation of the Lithic Material from the Rust Excavation

A. Pastoors, G.-C. Weniger and J.F. Kegler

Abstract: *The discussion about the Levantine transition from Middle to Upper Palaeolithic is still very intense. Different interpretations of the assemblage from Yabroud II (Syria) make this problem particularly apparent. This article presents the results of our reanalysis, which concentrated on the lithic artefacts from layers 10 to 5. Hence, the updated state of knowledge of Yabroud II allows a comparison to the sequence from Ksar Akil (Lebanon). Acting with all necessary caution that old excavations require, we see evidence for a complete transition from Tabun B-Type via Initial Upper Palaeolithic to Early Ahmarian industries at Yabroud II. Moreover, the cultural change at Yabroud II might be correlated to a climatic event.*

Résumé: *La question de la transition au Levant du Paléolithique Moyen au Paléolithique supérieur est encore aujourd'hui sujet à de vifs débats. Des interprétations différentes des industries de Yabroud II (Syrie) rendent ce problème particulièrement sensible. Un ré-examen des artefacts provenant des couches 10-5 montre que ces assemblages permettent une comparaison avec ceux de la séquence de Ksar Akil (Liban). Prenant en compte toutes les précautions qu'exigent des fouilles anciennes, nous observons à Yabroud II une transition complète qui passerait du type Tabun B à l'Ahmarien ancien via un Paléolithique supérieur initial. Par ailleurs, le changement culturel constaté à Yabroud II serait à corréler à un événement climatique.*

Keywords: *Middle-Upper Palaeolithic transition, Lithic production systems, Old collections, Levant.*
Mots-clés: *Transition Paléolithique moyen-supérieur, Systèmes de production lithique, Anciennes collections, Levant.*

The famous Yabroud rock shelters, which stretch along the northern flank of the Skifta Valley, Syria, have featured in a number of Palaeolithic studies,[1] the main focus of which has been the typological analysis of lithic assemblages; in the case of Yabroud II these led to the identification of several different cultural industries, some of which were described as transitional in character. In this paper we present the results of a revised lithic analysis of material from layers 10 to 5 from this site, undertaken in an effort to clarify a hitherto unsatisfactory picture. As such, at the centre of this paper lies the analysis of lithic production systems and their associated changes in the stratigraphy of Yabroud II. Technological terms and definitions adhere mainly to the work of Boëda, Geneste and Meignen.[2] Our results are then discussed with published comparisons between Yabroud II and Ksar Akil (Lebanon). In spite of the poor stratigraphical documentation of Rust's excavation of Yabroud II, in the course of our re-evaluation it became apparent that the archaeological assemblage displayed features considered typical for the Middle Palaeolithic to Upper Palaeolithic transition in the Levant. Combinations of newly observed technological features now provide

1. *cf.* Bordes, 1955 and 1962; de Sonneville-Bordes, 1956; Solecki and Solecki, 1966; Bakdach, 1982; Ziffer, 1981; El-Kassem, 2001; Frank, 2004.

2. Boëda, 1988 and 1994; Boëda *et al.*, 1991; Delagnes and Meignen, 2006.

 Manuscrit reçu le 20 août 2008, accepté le 25 novembre 2008

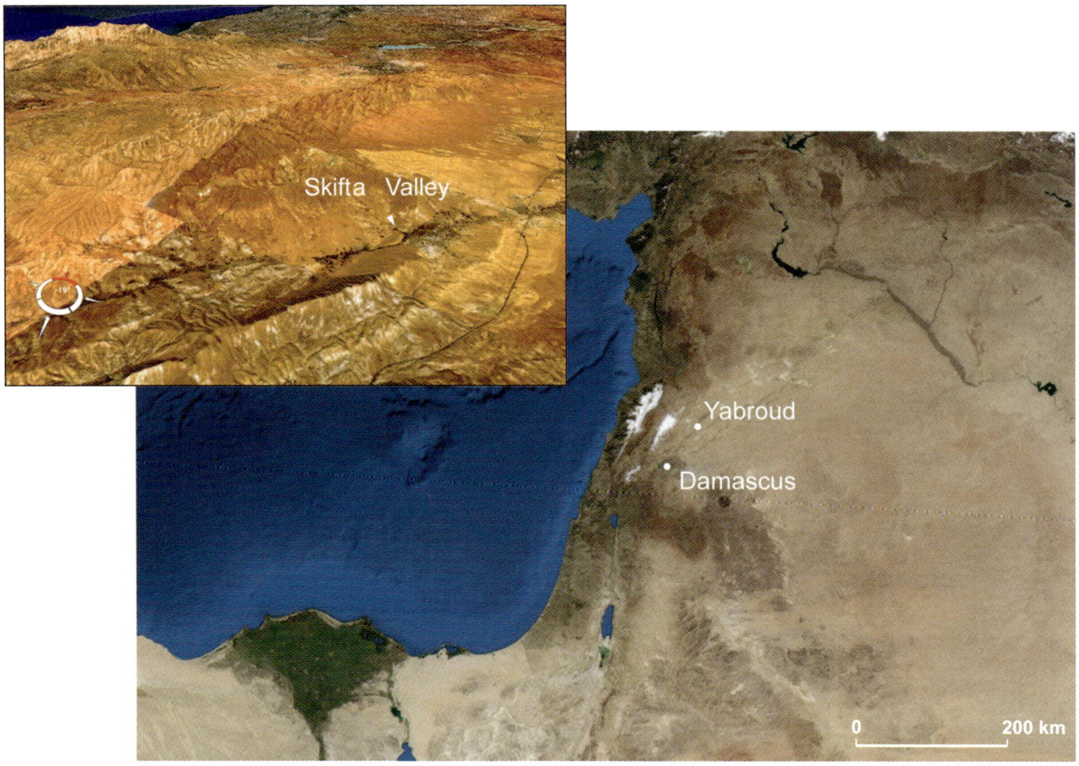

Fig. 1 – *Yabroud II viewed from satellite (source: NASA World Wind).*

new insights into this transitional phase at the Yabroud II rock shelter.

GEOGRAPHIC SETTING AND HISTORY OF RESEARCH IN YABROUD II

Yabroud II, which was discovered by A. Rust in the early 1930s,[3] is situated in the Skifta Valley, about 80 km north-east of Damascus (Syria), on the eastern flank of the Anti-Lebanon Mountains where peaks can reach elevations in excess of 3,000 m above sea level (fig. 1). The rock shelter, which lies at about 1,400 m is located in a steep escarpment of Eocene limestone; it opens to the south-east and occupies the highest elevation of all shelters in the Skifta Valley. A small cavity in the rear wall of the shelter was discovered by Rust. This had been levelled and cleaned, presumably in Roman or Byzantine times, and used for burials (fig. 2); tombs had been cut into the rocky walls. In the western corner of the shelter, a depression of about 4 to 5 m in diameter, with bedrock lying 2.5 m lower than that of the main hall, was excavated down to bedrock by Rust. This turned out to be the main occupation area of the shelter. The main hall, tested by an east-west oriented trench, was found to contain very few remains. Further testing undertaken by Schroeder and Solecki during the Columbia Expedition in 1964 and 1965 confirmed previous observations.[4] Accordingly, only a relatively small part of the rock shelter was actually inhabited during Palaeolithic times, and it would appear that the main chamber was used only from the Mesolithic onwards.

Using the main trench—which provided a vertical stratigraphy spanning some 3 m—Rust defined ten archaeological layers: 1, Micro-Aurignacian (Atlitian); 2-3, Late Aurignacian; 4-5, Middle Aurignacian; 6-7, Early Aurignacian and 8-10, Late Mousterian.[5] Sedimentological observations show that the lower part of the sequence up to layer 7 comprises

3. RUST, 1950.

4. SOLECKI and SOLECKI, 1966.

5. RUST, 1950.

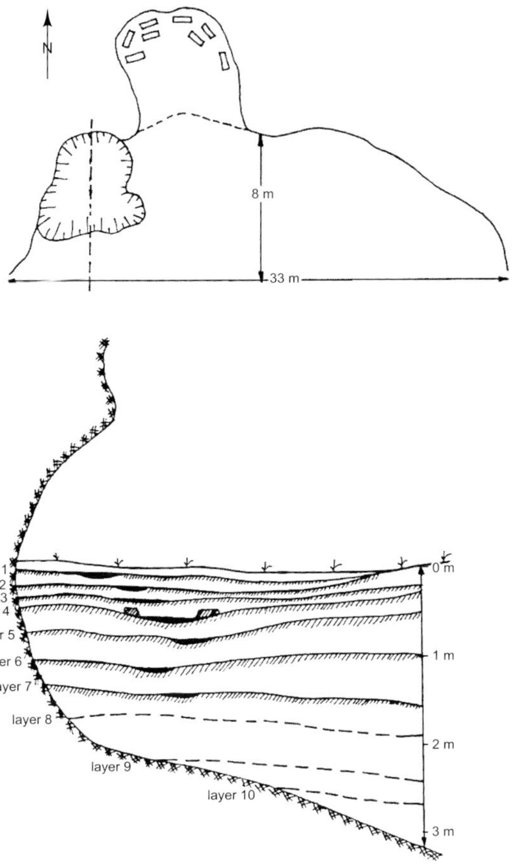

Fig. 2 – *Yabroud II: plan and stratigraphy (after* ZIFFER*, 1981;* RUST*, 1950).*

predominantly coarse-grained sediments. The sediment talus consists of limestone gravel which ranges in diameter from 2 cm to 10 cm. This material stems from the roof of the shelter and became loosely deposited, forming small cavities lacking matrix. From layer 7 onwards the sediment is interspersed by ashy layers,[6] and from layers 6 to 1 the sediment becomes much finer. No breccia was found in this part of the section. These observations, although poorly documented, point to a change between layers 7 and 6, which is also mirrored in the state of preservation of the lithic material. Marks has suggested that this is indicative of a part of the assemblage from layer 7 having been "exposed on the surface for a long period of time."[7] During an inspection of the site by one of us in 1999, lithic fragments and small lenses of ash were observed in the exposed profiles of the 1960s excavations. Therefore, at least some scattered remains of the Palaeolithic occupations are still in place.

LITHIC PRODUCTION SYSTEMS FROM YABROUD II LAYERS 10 TO 5

The collections under re-evaluation in this paper stem from Rust's excavations at Yabroud II, layers 10 to 5. Excavations by Solecki and Schroeder in front the rock shelter in the 1960s yielded only few finds; these still remain unpublished.

The occupation levels and their correlation with the geological layers determined by Rust during his excavation remain ambiguous. Rust's division of cultural layers was based on observations on raw material frequencies made during field work.

> „In beträchtlichem Maße wurde die Möglichkeit einer Abgrenzung der einzelnen Kulturschichten durch den Umstand erleichtert, dass das am Ort vorkommende Silex-Gestein eine ungemein variierende Tönungsskala aufweist. Man kann sich des Eindrucks nicht erwehren, dass die jeweiligen Kulturträger oft ein bestimmtes Rohmaterial, vermutlich einer vorgezogenen Färbung wegen, auswählten. [...] Die dann folgende Schicht [layer 5; *our parenthesis*] ist im Gerätebestand durch braune Töne gekennzeichnet. Der Flint der nächsten Strate [layer 6; *our parenthesis*] ist klar und fast durchsichtig.“[8]

In this passage Rust describes the homogeneous but colourful—from reddish-brown to light beige—lithic material discovered at Yabroud II which he used to define archaeological layers; be this as it may, our attempts to reconstruct Rust's observations proved unsuccessful. Only in layer 6 is a particularly conspicuous variant of chalcedony observed. Indeed, this latter material is very homogeneous and differs clearly from those materials encountered in all other layers. According to Rust, flint found at Yabroud II was procured from a source located upon the plateau above the rock shelter.[9]

Rust did not publish any notes about the thickness of the different layers, and he only mentions their depth below surface (table 1). According to the documented profile, one can assume that the occupation levels were thin layers ("Kulturschichten"). However, we cannot exclude that finds were scattered throughout the sequence and that the thickness of

6. *Ibid.*
7. BERGMAN, 1987b: 151.
8. RUST, 1950: 11.
9. RUST, 1950.

Table 1 – *Yabroud II: amount of studied artefacts.*

Layer	Level	Artefacts
5	- 0,6 m	738
6	- 1 m	304
7	- 1,5 m	124
8	- 2 m	59
9	- 2,5 to - 2,3 m	145
10	- 3 m	7

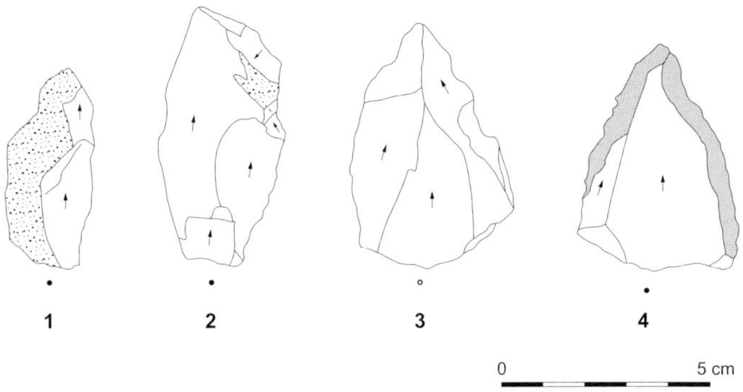

Fig. 3 – *Yabroud II: lithic production system and toolkit of layer 10. **1-4**, Levallois recurrent unidirectional convergent; **4**, convergent sidescraper - (grey shading = retouch).*

levels may have reached up to 50 cm. In their work, Ziffer and Bakdach[10] misunderstood the description given by Rust, and we could find no evidence that Rust's layers correspond directly to defined human occupation units. A detailed study of the taphonomy of the site has not yet been undertaken. In this paper we use the units as defined by Rust, but regard them as artificial units which reflect only a general trend within the chronostratigraphical sequence.

Rust excavated in horizontal layers and sieved all material, except for the Mousterian layers, in an area covering 15 m². According to Rust's description the bottom of the rock shelter was reached in layer 10. Ziffer noticed that Rust did not collect all the waste material.[11] As such, Rust's units comprise a random assortment of artefacts, whereby the original quantity remains obscure. Sample size differs extremely from only seven pieces to more than 700 objects per unit. Not only does this bias make it impossible to study complete operational sequences, but it also renders a comparison of the number of tools within the different lithic inventories impracticable.

Macroscopic analysis of raw material texture has shown that the majority of the material discovered at Yabroud II stems from secondary deposits. As was the case at the rock shelter Yabroud I, rubble-surfaces and weathered cortex were also identified.[12] Secondary flint deposits are known from the immediate vicinity of the site.[13] However, in contrast to Yabroud II, at Yabroud I, Shea noted—based on work by Solecki on the Mousterian material at this site—that people "were evidently more selective, travelling long distances to collect more siliceous flints and bypassing low quality flint nodules eroding from the bedrock in front of the caves."[14] Non-local flint was integrated into the lithic production systems, but neither the source nor the distance is specified by the authors.[15] In retrospect, it is most likely that the different researchers are in fact referring to the very same raw material source, situated about 5 km north-east of Yabroud II.[16]

According to Weniger[17] the radius of local activities at a site may extend up to 20 km. Thus, this would imply that lithic raw material acquisition at Yabroud II was conducted on a more local level and was not dependent on long distance transport. A few pieces of the assemblage show fresh chalky cortex, pointing to the use of primary sources. Indeed, primary flint beds occur in the region around Yabroud II (maximum distance 7 km) in Eocene marl or Upper Cretaceous limestone,[18] though the exact locality of these lithic raw material sources has still not been determined.[19] Consequently, although we cannot rule out that some raw materials were transported over longer distances, the overwhelming majority was certainly collected from the immediate vicinity and is therefore of purely local origin.

LAYER 10

The sample from layer 10 is extremely small (seven pieces) and of low analytical value. In spite of this, four of these seven pieces serve as reliable technological markers, representing the Levallois recurrent unidirectional convergent method. Whereas

10. ZIFFER, 1981; BAKDACH, 1982.
11. ZIFFER, 1981: 73.
12. EL-KASSEM, 2001.
13. RUST, 1950; EL-KASSEM, 2001.
14. SHEA, 2003: 355.
15. SOLECKI and SOLECKI, 1995.
16. EL-KASSEM, 2001.
17. WENIGER, 1991.
18. WOLFART, 1966; DODONOV *et al.*, 2006: fig. 3: 4, 8.
19. EL-KASSEM, 2001: 32.

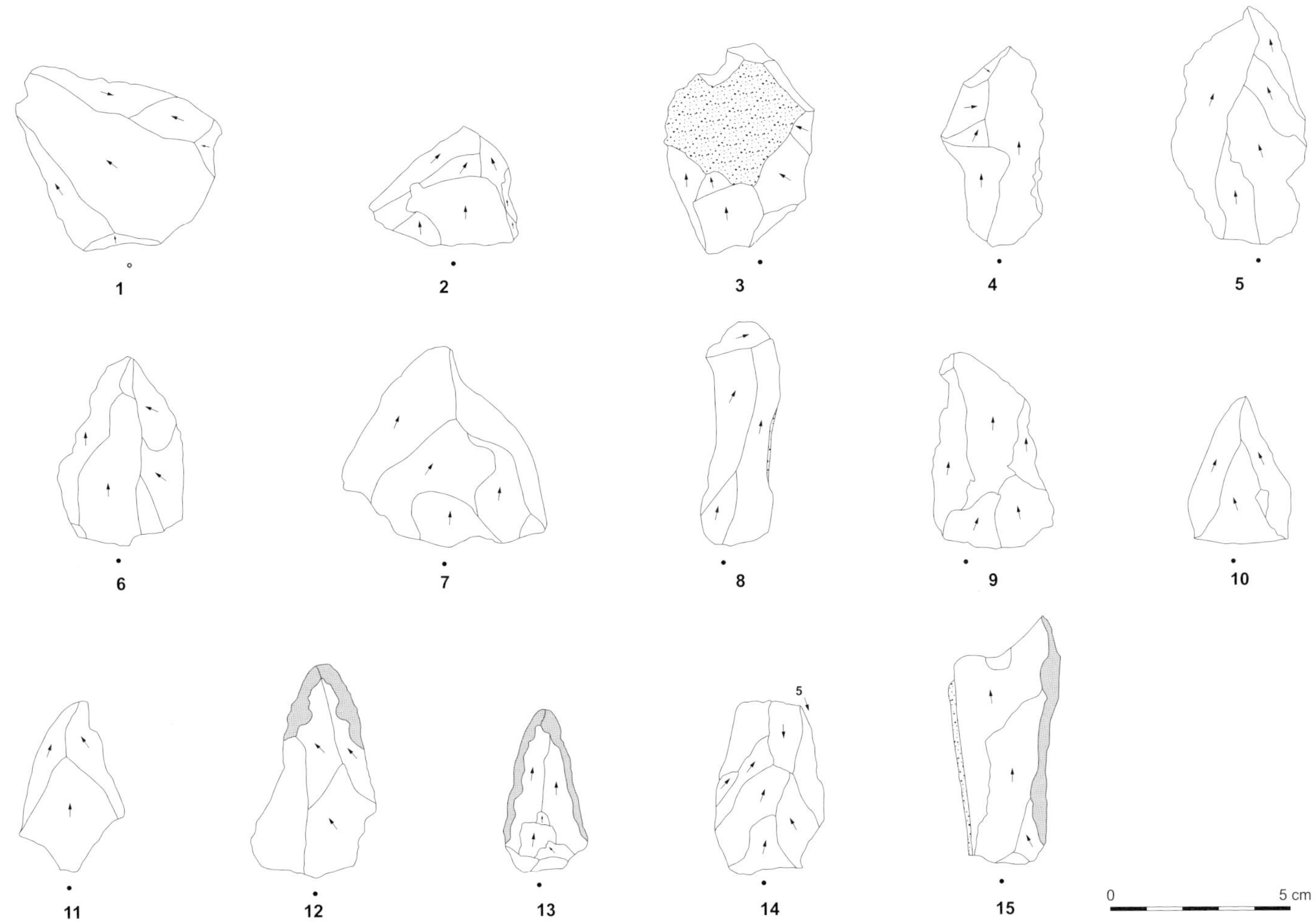

Fig. 4 – *Yabroud II: lithic production system and toolkit of layer 9.* ***1-11*** *and* ***15****, Levallois recurrent unidirectional convergent;* ***12-13****, convergent sidescraper;* ***14****, burin;* ***15****, marginal retouched blade - (grey shading = retouch).*

one elongated *éclat débordant,* bearing more than 50% natural surface, marks the initialisation stage (fig. 3: 1), the three remaining pieces are products of the subsequent exploitation stage: a Levallois blank with negatives resulting from the trimming of the distal convexity of a latero-distal edge (fig. 3: 2), a subtriangular Levallois blank (fig. 3: 3), and a Levallois point with retouched edges (fig. 3: 4). Both triangular pieces show a central triangular and two lateral convergent negative scars.

LAYER 9

Within this larger sample of 145 pieces only the Levallois recurrent unidirectional convergent method could be discerned. Two cores (fig. 4: 1-2) attest the recurrent reduction system with a central triangular and two final lateral convergent negatives. Individual striking platforms are roughly prepared. One core fragment (fig. 4: 1) originated from re-initialisation of a flaking surface. All stages—from initialisation to exploitation and re-initialisation—are present in layer 9. Blanks, either completely or partially covered by natural surfaces, were produced during the decortication of nodules. The dorsal faces display unidirectional (parallel/convergent) removals (fig. 4: 3). There is a wide range of blanks from the exploitation stage which comprise nearly all products from this stage as described by Meignen:[20] *éclats débordants* II (fig. 4: 4-5), Levallois blanks (fig. 4: 6-7), cortical *éclats débordants* (fig. 4: 8 and 15), subtriangular Levallois blanks (fig. 4: 9) and atypical Levallois

20. MEIGNEN, 1995: 365.

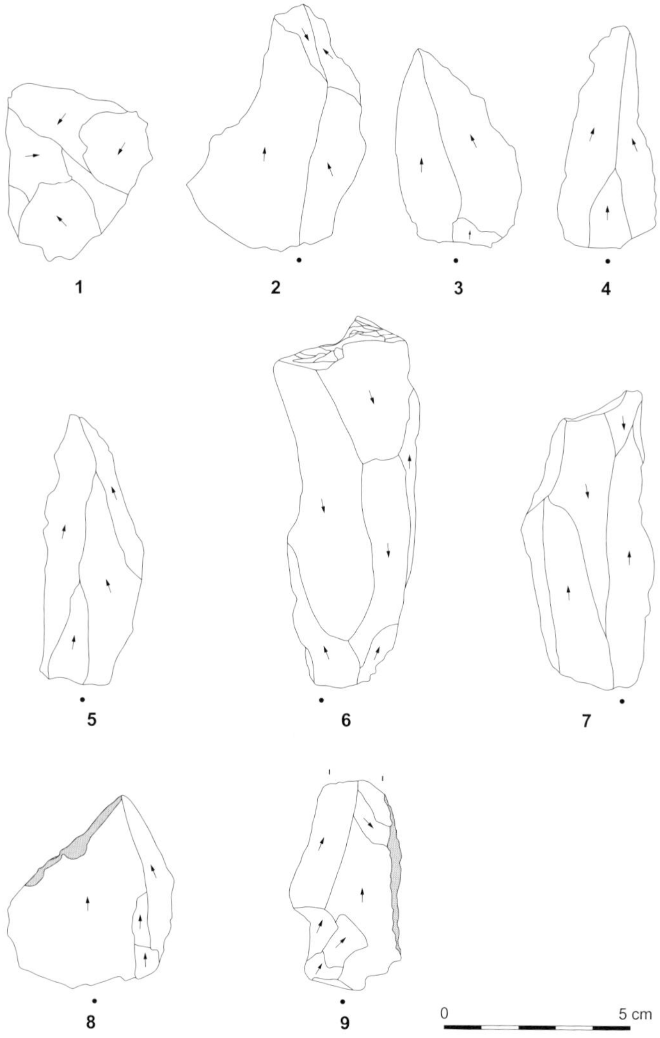

Fig. 5 – *Yabroud II: lithic production system and toolkit of layer 8. **1-7**, Levallois recurrent centripetal; **8-9**, sidescraper - (grey shading = retouch).*

points (fig. 4: 10-11). The convexity of the flaking surface is prepared and continuously maintained by at least two lateral convergent negatives. Only one piece (fig. 4: 4) shows a distal negative which reflects the trimming of the distal convexity from a latero-distal edge. The ensemble of retouched pieces encompasses convergent sidescrapers (fig. 4: 12-13), a burin (fig 4: 14) and a marginally retouched blade (fig. 4: 15).

LAYER 8

The assemblage from layer 8—the uppermost layer attributed to the late Mousterian[21]—comprises a total of 59 pieces. Only one core was recovered from this layer (fig. 5: 1); it has two separate, hierarchically structured surfaces and was reduced by means of the Levallois recurrent centripetal method. The lower surface bears nodule cortex and some isolated removals from preparation. The convexity of the upper surface was maintained by centripetal reduction. The preparation of the striking platform is missing; the cortex on the lower surface was used directly. Only one blank with cortex on its dorsal surface is indicative of the initialisation stage; however, this piece was probably part of a different production system and does not provide evidence for the initial stage of the Levallois recurrent centripetal method. Nevertheless, the exploitation stage of the Levallois recurrent centripetal method is well documented (fig. 5: 2-3), even though some blanks might belong to a Levallois recurrent unidirectional convergent method (fig. 5: 4-5). Two *outrepassés* represent an early stage of the re-initialisation stage during a Levallois recurrent centripetal method reduction (fig. 5: 6-7). The negatives of these two elongated blanks formed the central convexity and preceded the reduction sequence.

According to Rust,[22] layer 8 contains several sidescrapers, one burin, one borer and two *Châtelperronian* points. Our reanalysis could only substantiate the presence of sidescrapers (fig. 5: 8-9).

LAYER 7

As previously mentioned, some of the 124 pieces from this layer display natural edge damage. The assemblage itself provides evidence for two different Levallois methods: recurrent bi-directional and recurrent unidirectional. A single unidirectional bladelet core is present. The Levallois recurrent bidirectional method is well represented by cores and blanks. The hierarchically structured cores show either nodule surface or full preparation on the lower surface (fig. 6: 1-2). To maintain the distal as well as the lateral convexity of the reduction surface, removals were taken from the main striking platform (*contre-bulbe* of the opposite Levallois blank, oblique Levallois blanks or *éclat débordant*). It would appear that the preparation of the lower surface belongs to a previous reduction

21. RUST, 1950; ZIFFER, 1981.
22. RUST, 1950.

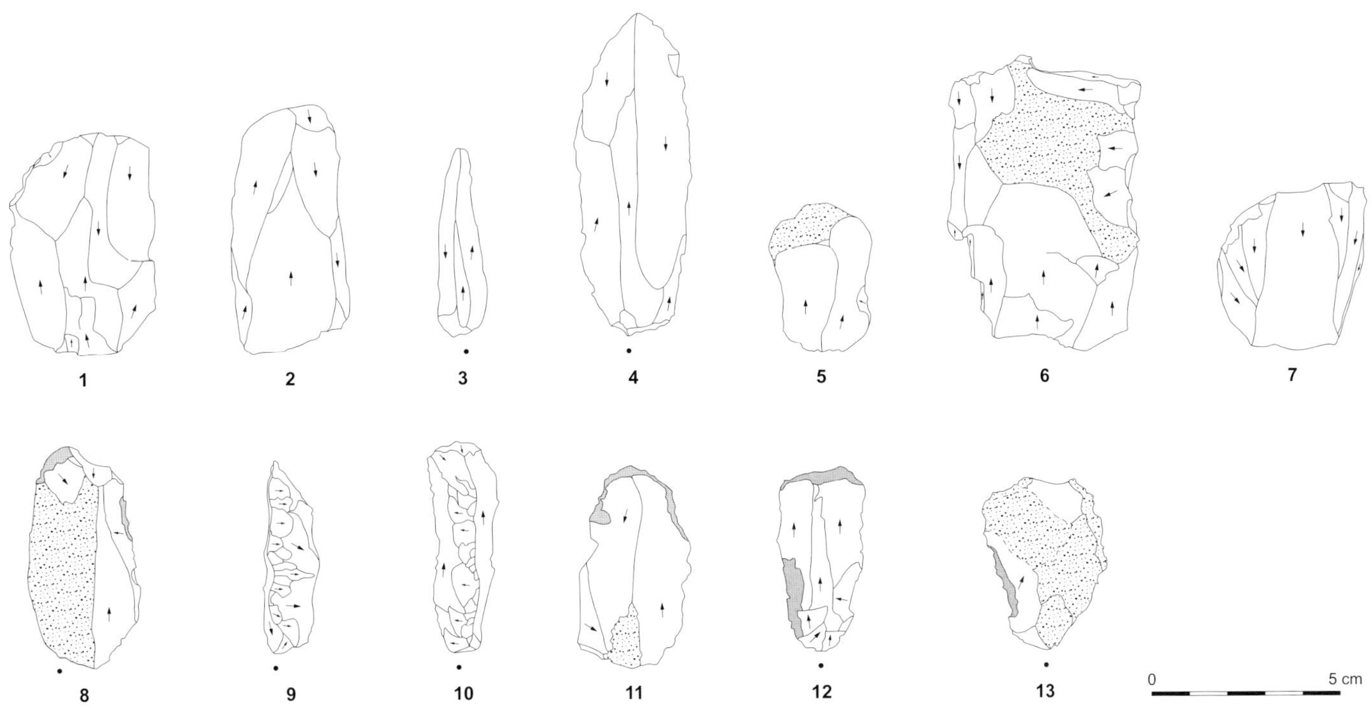

Fig. 6 – *Yabroud II: lithic production system and toolkit of layer 7.* ***1-4****, Levallois recurrent bi-directional;* ***5-6****, Levallois recurrent unidirectional; 7, bladelet unidirectional;* ***8****, cortical blank;* ***9-10****,* éclat débordant*;* ***11-12****, endscraper;* ***13****, sidescraper - (grey shading = retouch).*

stage. The preparation of the striking platform varies from single negative to multiple faceted platforms. The initialisation stage cannot be detected (cortex covering blanks are indifferent). In contrast, the exploitation stage is documented by various elongated blanks or blades with bi-directional negatives on their dorsal surfaces (fig. 6: 3-4).

Even though natural fissures were integrated into the convexity of the flaking surface, two cores represent the Levallois recurrent unidirectional method (fig. 6: 5-6). As was the case with the flaking surface, the lower surface displays cortex and was therefore not prepared. The distal convexity was maintained using the natural surface. The lateral convexity was maintained by *éclats débordants* and oblique Levallois blanks. As already described for the Levallois recurrent bi-directional method, no evidence can be found for the initialisation stage of Levallois recurrent unidirectional method. There are no associated Levallois blanks from the exploitation stage.

The only core to bear witness to the unidirectional bladelet method (fig. 6: 7) also displays a similar system to the Levallois recurrent unidirectional method. Preparation of the striking platform is poorly developed (only one removal). The convexity was simply maintained by the removal of bladelets. Whilst oblique bladelets were removed for the lateral convexity, the distal convexity was achieved by removing curved bladelets using the soft hammer technique. This method of distal convexity preparation marks a change in the concept of core reduction and differs from the Levallois recurrent unidirectional method. Unfortunately, no bladelets from the exploitation stage were found. The isolated appearance of this core without any reduction products leads us to alternative interpretations: either the core is intrusive and belongs to the overlying layer 6, which would mean that there is no evidence for a volumetric concept in layer 7, or the assemblage from layer 7 is incomplete and bladelets are missing due to Rust not sieving all sediments. Thus, to avoid misinterpretation the presence of this core in layer 7 must be treated with caution.

The sample of cortical re-initialisation blanks (fig. 6: 8 and 13) and blades (fig. 6: 9-10) cannot be assigned to a specific concept. These pieces are not crested blades but are *éclats débordants*; as such, they generated lateral convexity. Ziffer describes the technological characteristics of layer 7 as a "non-Levallois industry with only 17.4% Levallois blanks in."[23] This characterisation is in line with his interpretation of the cores: "The cores assemblage is dominated by uni-directional types

23. ZIFFER, 1981: 73.

and by bi-directional prismatic types. The assemblage reflects clearly the Upper Palaeolithic aspect of the whole industry."[24] According to our re-evaluation, Ziffer's prismatic cores do not show a volumetric concept but feature all attributes of a surface concept: Levallois recurrent uni- and bi-directional. According to Rust the dominant components of layer 7 are different types of endscrapers and burins.[25] While the presence of burins cannot be confirmed, endscrapers are common in this inventory (fig. 6: 11-12), with a single side scraper (fig. 6: 13). As in layer 8 neither *Châtelperronian* points nor borers were found. Bakdach mentions a much broader spectrum of tools, including five chamfered pieces.[26] We see no evidence for this.

LAYER 6

The 304 artefacts from layer 6 were made on an unusual raw material and display diverse production systems, comprising both surface and volumetric concepts of core reduction. The surface concept is represented by several methods: recurrent uni- and bi-directional, recurrent centripetal and recurrent unidirectional convergent; the volumetric concept is represented by the unidirectional bladelet method only.

Hierarchically structured cores of the Levallois recurrent unidirectional method show a highly efficient design of the lower surface. Natural nodule or fissure surfaces with some rare preparation removals were used to adapt either the distal or lateral convexity (fig. 7: 1-2). During the exploitation stage the lateral convexity was maintained by *éclats débordants* (fig. 7: 3). The preparation of the striking platform is in line with an economic handling; one single removal or coarse facetting was considered sufficient. Various Levallois blanks, including *éclats débordants*, document the exploitation stage; the different methods employed in the preparation of the convexities are documented on their dorsal surfaces (fig. 7: 4-5).

Cores representing the Levallois recurrent bi-directional method display an identical lower surface as the unidirectional method described previously (fig. 7: 6-7): natural nodule surfaces were prepared by single removals. In contrast, the convexity of the upper surface was adjusted differently. The *contre-bulbes* of the Levallois blanks from the opposite side were handled as distal convexity. The lateral convexity was maintained by oblique Levallois blanks. Striking platforms show either a single negative or fine facetting. From the different stages of the reduction sequence only the exploitation stage is observed (fig. 7: 8-9). Blanks of the re-initialisation stage were not identified. A single core bears witness to the recurrent unidirectional convergent method (fig. 7: 10), but remains close to the system of the described Levallois methods. The appearance of the lower surface is identical to the other two Levallois methods. The convexity of the upper surface was maintained by distal preparation and lateral oblique Levallois blanks. The striking platform is well prepared by fine facetting. Pieces belonging to the exploitation or the re-initialisation stage are not documented. The fourth Levallois method present in layer 6 is the recurrent centripetal method (fig. 7: 11-12). The hierarchically structured cores display either complete natural nodule surface or are characterised by a full preparation of their lower surfaces. The maintenance of the upper surface convexity was assured by slightly oblique, centripetally removed Levallois blanks only. Coarse facetting served to prepare the striking platform. The exploitation stage is documented by Levallois blanks (fig. 7: 13-14). There is no evidence for a re-initialisation stage.

The volumetric concept is specified by the unidirectional bladelet method. A flat fissure or a natural surface of a nodule forms the backside of the cores (fig. 7: 15-16). The initialisation stage is not documented. Later stages show that the distal convexity was maintained by preparation removals from the distal part of the cores or by curved bladelets. The latter was first recognised on a single core from layer 7 (fig. 6: 7). A series of cores and curved bladelets from the exploitation stage were identified in the assemblage (fig. 7: 17-19); these display an efficient solution to maintain the flaking surface, and in layer 6 this becomes a standard practice. The same kind of slightly oblique products were required to maintain the lateral convexity of the core. This was supported by a previous preparation of the core shoulder from the back of the core. The preparation of the striking platform was realised by a single removal. Various bladelet products document the exploitation stage (fig. 7: 17-19); an *outrepassé* (fig. 7: 20) represents the re-initialisation stage. Although the initialisation stage is clearly documented by cortex covered blanks (fig. 7: 21-22), these cannot be assigned to a specific concept or method. The cortex blanks are too unspecific and the different reduction concepts and methods are made upon the same raw-material. The same is true for crested blades (fig. 7: 23-24). As a crested blade they may belong to the initialisation stage of the volumetric concept or as *éclats débordants* to the Levallois recurrent uni- or bi-directional method. Ziffer described the character of the cores as follows: "Among the cores the shapeless type is dominant.

24. ZIFFER,1981: 73.
25. RUST, 1950.
26. BAKDACH, 1982: 53.

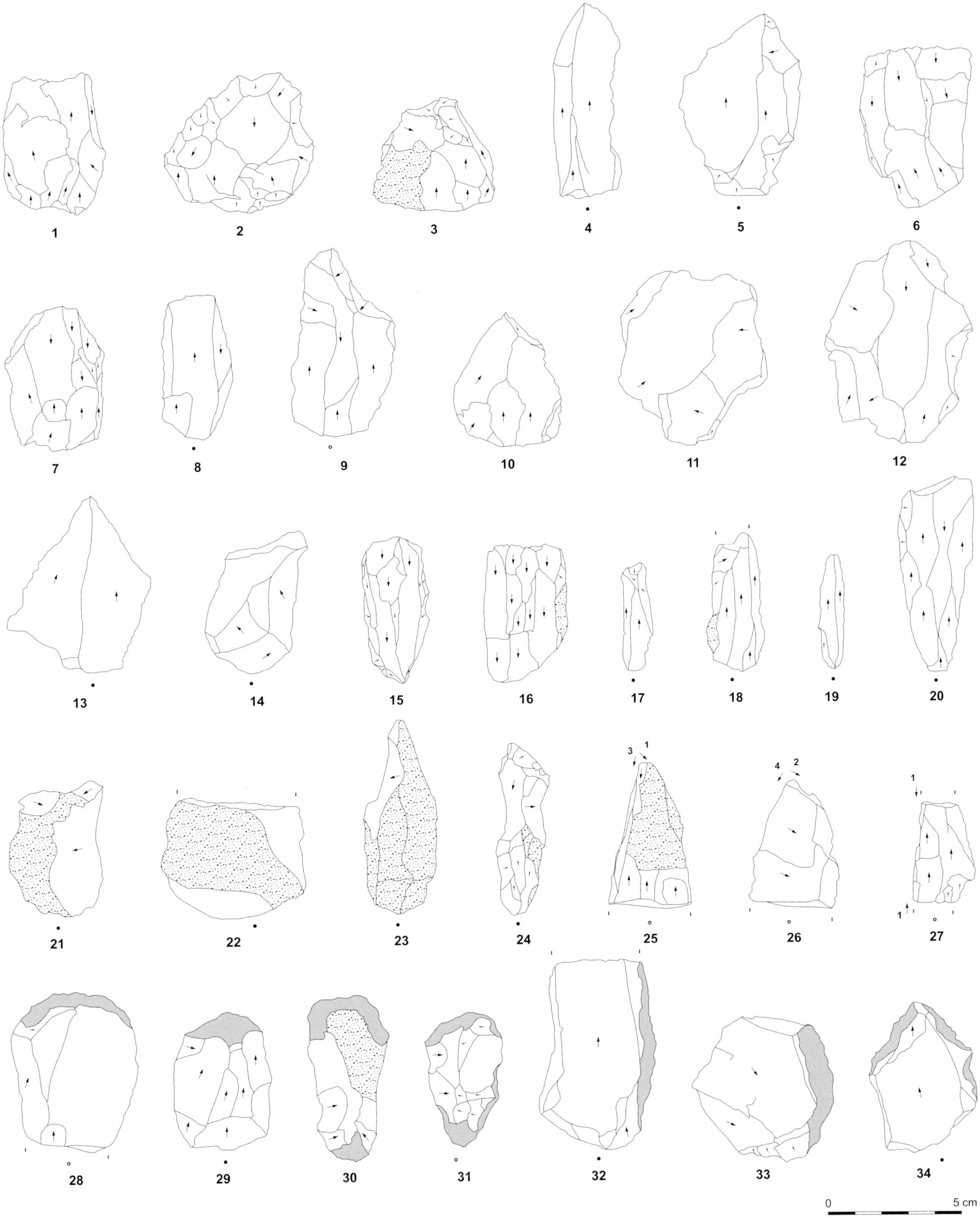

Fig. 7 – *Yabroud II: lithic production system and toolkit of layer 6.* ***1-5****, Levallois recurrent unidirectional;* ***6-9****, Levallois recurrent bi-directional;* ***10****, Levallois recurrent unidirectional convergent;* ***11-14****, Levallois recurrent centripetal;* ***15-20****, bladelet unidirectional;* ***21-22****, cortical blank;* ***23-24****, crested blade or* éclat débordant*;* ***25-27****, burin;* ***28-31****, endscraper;* ***32-33****, sidescraper;* ***34****, convergent scraper - (grey shading = retouch).*

There are also two flat bi-directional cores and some prismatic and pyramidal types too."[27] His description reflects the high technological diversity of the assemblage which results from two technologically different production systems.

The main components of the inventory of retouched pieces are different types of burins and endscrapers (fig. 7: 25-31). Additionally, sidescrapers (fig. 7: 32-33) and a single convergent scraper (fig. 7: 34) are present. Bakdach mentions a much broader spectrum of tools, including two chamfered pieces, borers and other types.[28] This observation could not be substantiated.

LAYER 5

Among the 738 artefacts from this layer no surface concept is documented. Instead, three different (volumetric) methods are observed. The first method, the unidirectional bladelet method, follows a simple strategy: it seeks a continuous reduction of blades or bladelets by low input of preparation (fig. 8: 4-5). Either a ridge was prepared or pebbles with a natural ridge were chosen for exploitation (fig. 8: 1). The back of the cores remained unprepared. The pebble was loped to create a striking platform. The exploitation began with the removal of a crested blade (fig. 8: 2-3). An optional second striking platform on the opposite side allowed the preparation of the distal convexity (fig. 8: 6-8). However, generally speaking, the distal convexity was maintained by curved blades or bladelets (fig. 8: 9-10). The lateral convexity was perpetuated by slightly obliquely removed blanks. The lateral border of the flaking surface adjoins to the back of the core or to a prepared core shoulder (fig. 8: 11-12). Some cores, which are wider than they are long, can be misinterpreted as surface structured unidirectional Levallois cores, though close examination does show that they follow the volumetric concept and mark the final stage of a unidirectional bladelet core (fig. 8: 13). Besides the initialisation and exploitation stage, the *outrepassés* document the re-initialisation stage (fig. 8: 14-15).

The second volumetric method produced twisted bladelets from unidirectional cores (fig. 8: 16-17). The handling of natural pebbles is the same as in the unidirectional bladelet method. A significant difference is that twisted unidirectional cores sometimes carry two flaking surfaces at oblique positions on the striking platform, thus resulting in the torsion of removed bladelets (fig. 8: 18-19). In addition to these exploitation stage products, a triangular preparation blank documents the re-initialisation of the flaking surface (fig. 8: 20). A sample of various cortical blanks results from the initialisation stage; these cannot be attributed to any one of the described bladelet production methods. The third (volumetric) method is represented by a single discoidal core (fig. 8: 21); characteristic are two opposite reduction faces, and the maintenance of convexities via the production of centripetally, slightly obliquely-struck blanks. Evidence is lacking to suggest that the discoidal core marks the end of another reduction process.

"Levallois technology appears in a negligible percentage. Blades dominate again among tools and among waste material."[29] This observation corresponds to the result of our re-evaluation. The surface concept is not documented; instead three different volumetric methods were identified in layer 5. In contrast to Ziffer,[30] we could see no indications for a bi-directional production or a Levallois system.

In addition to a wide assortment of endscrapers (fig. 8: 22-25) and burins (fig. 8: 26-29), Rust also describes different points:[31] two marginally retouched points (El-Wad points) and a backed blade are novelties of layer 5 (fig. 8: 30-32). Bakdach also mentions two chamfered pieces and a borer in this inventory[32] that we were unable to identify.

SUMMARY: LITHIC PRODUCTION SYSTEMS FROM YABROUD II LAYERS 10 TO 5

Prior to presenting an interpretation of the results of our analysis we must first contemplate the incomplete status of the assemblages, sediments having been only partially sieved and waste material not collected during excavation.[33] In particular, most of the small pieces including bladelets are probably missing. It is obvious that a detailed and resilient discussion of a cultural attribution of the layers is very difficult under these conditions. The same is true for the analytical value of the presence or absence of retouched bladelets, Dufour bladelets or El-Wad and Emireh points in Yabroud II. Keeping this unsatisfactory situation in mind, we are nevertheless in the position to recognise at least one major trend in the available material. The observed transition from a Levantine Mousterian surface concept to an Upper Palaeolithic volumetric concept; it follows

27. ZIFFER, 1981: 77.
28. BAKDACH, 1982: 73.
29. ZIFFER, 1981: 77.
30. *Ibid.*
31. RUST, 1950.
32. BAKDACH, 1982: 93.
33. RUST, 1950; ZIFFER, 1981.

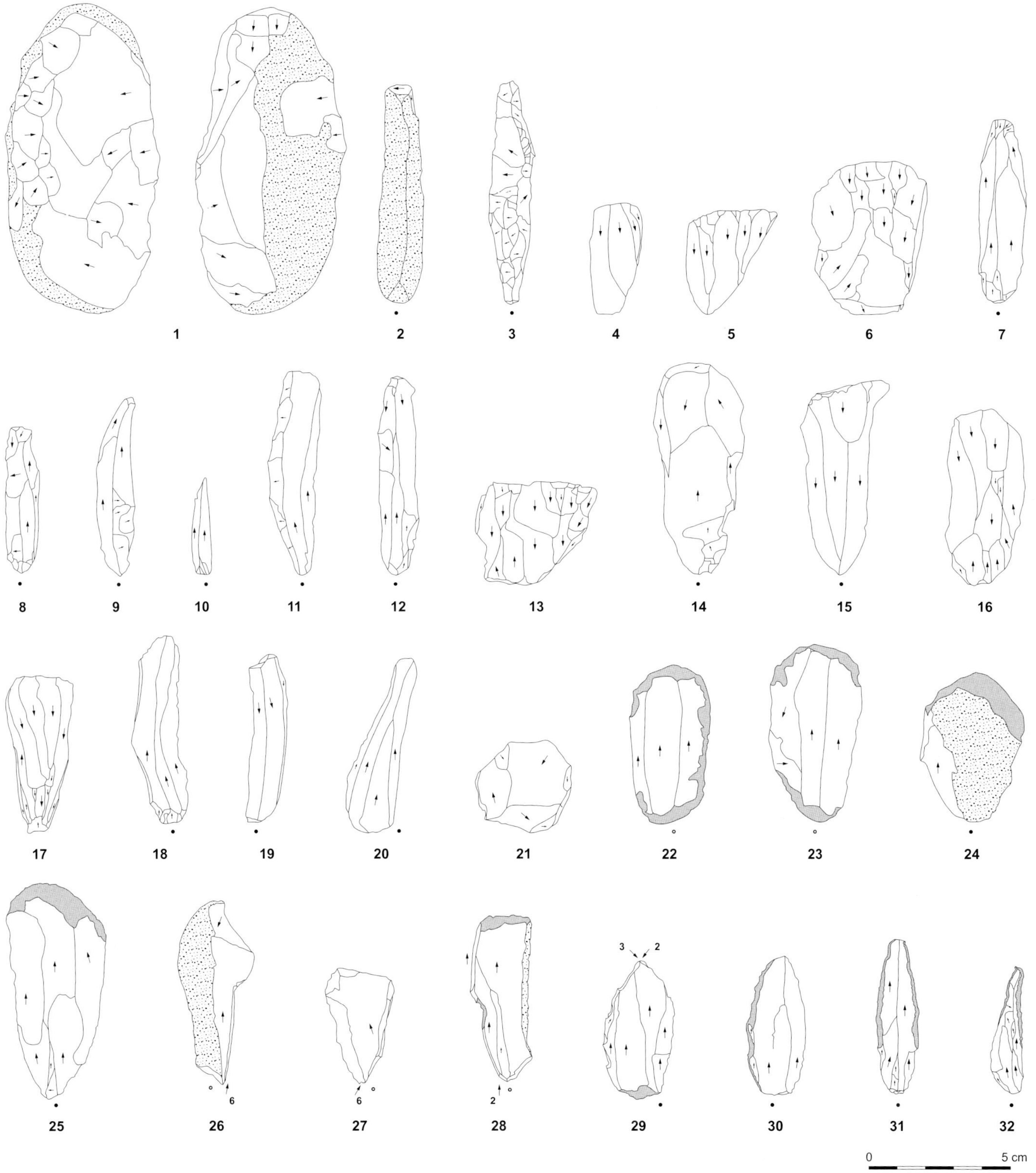

Fig. 8 – *Yabroud II: lithic production system and toolkit of layer 5.* ***1****, prepared pebble;* ***2****, blade with natural ridge;* ***3****, crested blade;* ***4-15****, bladelet unidirectional;* ***16-20****, twisted bladelet unidirectional;* ***21****, discoidal core;* ***22-25****, endscraper;* ***26-29****, burin;* ***30-32****, El-Wad point - (grey shading = retouch).*

Table 2 – *Yabroud II: lithic production systems from layers 10 to 5.*

Layer	Levallois recurrent				Blade/bladelet		Discoidal
	Unidirectional convergent	Centripetal	Unidirectional	Bi-directional	Unidirectional	Unidirectional torse/ twisted	
5					x	x	x
6	x	x	x	x	x		
7			x	x	x ?		
8		x					
9	x						
10	x						

Table 3 – *Yabroud II: relation between blanks and blades in layers 10 to 5.*

Layer	Blank	Blade/ bladelet	n
5	30%	70%	617
6	62%	38%	276
7	29%	71%	114
8	61%	39%	56
9	67%	33%	141
10	67%	33%	6

that this discovery provides a whole new insight into this transitional process at Yabroud II.

From a technological point of view the analysed sequence of Yabroud II (layers 10 to 5) can be divided into three units: whilst the lowermost layers (10 to 7) feature a surface concept, and the upper layer (5) is characterised by a volumetric concept, in the intermediate layer (6) both concepts appear together (table 2). In the case of the former, it is of note that within the spectrum of surface concept only the Levallois recurrent method is recorded. It would appear that throughout the sequence of Yabroud II, the intention—or the cultural choice—of prehistoric knappers was obviously to maintain the convexity of the flaking surface, even during the reduction sequence. However, no indicators for a bi-directional blade/bladelet method were found in layer 5. The focus on a single method in layers 10, 9 and 8 can probably best be explained by the small size of their assemblages; however, there is an obvious change from the unidirectional convergent to centripetal method. The first curved bladelets were produced in layer 6. This marks a change in technology and reflects the tangential gesture in the production systems. Decreasing costs (*i.e.* loss of material) for distal preparation increased the number of blanks produced during exploitation and resulted in a more efficient raw material management. In most cases the assemblages comprise a complete reduction sequence from initialisation, via exploitation, to re-initialisation. Exceptions are the Levallois reduction sequences from layers 10 and 6. The similarity of the reduction sequences in the different assemblages suggests a homogenous economic behaviour with regard to stone tool management.

A closer look at the relative frequencies of blades and blanks within the assemblages proves helpful when attempting to evaluate the development of technological concepts (table 3). The blades from layer 10 to 7 were produced using the surface concept, those in layer 5 using the volumetric concept, and those in layer 6 using both concepts. It is remarkable that the general interest in blades increases suddenly in layer 7; decreases in layer 6, only to recover in layer 5. For the moment we have no conclusive interpretation of this observation, though the

Table 4 – *Yabroud II: main tooltypes from layers 10 to 5.*

Layer	Endscraper	El Wad point	Burin	Sidescraper	Conv. scraper	Ret. blade
5	x	x	x			
6	x		x	x	x	
7	x			x		
8				x		
9			x		x	x
10					x	

Table 5 – *Yabroud II: changes of lithic assemblages within the sequence.*

Layer	Technology	Toolkit	Production	Sediment	Raw material
5	UP	UP	Blade/ bladelet	Fine	
6	MP/UP	MP/UP	Blank	Fine	Special selection
7	MP/UP?	MP/UP?	Blade/ bladelet	Coarse	Damaged edges
8	MP	MP	Blank	Coarse	
9	MP	MP	Blank	Coarse	
10	MP	MP	Blank	Coarse	

pattern independently supports the argument of a change in the lithic production system from layer 7 onwards.

Concerning the cultural development, the typology of retouched pieces is also important (table 4), even though many types are poorly defined and sometimes misinterpreted. Side-scrapers and convergent scrapers are abundant from layers 10 to 6, but they are absent in layer 5. Endscrapers are absent from layers 10 to 8, but appear in the toolkit from layer 7 onwards. The culturally significant El-Wad point appears only in layer 5. Other tools, such as burins and retouched blades prove insignificant in the entire sequence. The presence of chamfered pieces—which Bakdach[34] mentions for layers 7, 6 and 5—could not be confirmed in our re-evaluation. Thus, the available typological data are indicative of a change at layer 7, with a further new input in layer 5. The change within the toolkit predates the change within the technological concept (table 5). This evidence might suggest that the bladelet core in layer 7 is non-intrusive. Generally speaking, it supports a "soft" transitional process.

Three further independent features indicate or support a change of the lithic assemblages from Yabroud II in layers 7 and 6. First, the description of sediments by Rust documents a change in sedimentation from layer 6 onwards; second, in layer 7 there is marked damage to the edges of artefacts; and third, in layer 6 there is clear evidence for the selection of a particular raw material. Thus, altogether five independent features provide evidence for changes in the sequence of Yabroud II in layers 7 and 6 (table 5).

We recall that our results are based on incomplete and not well documented assemblages. In particular, most of the small pieces, including bladelets, are probably missing. With this unsatisfactory situation in mind, our results should be considered as general trends at very best.

34. BAKDACH, 1982.

YABROUD II AND KSAR AKIL: A LONG HISTORY OF COMPARISONS

Interpretation of the sequence of Yabroud II is difficult due to the fragmentary character of the lower assemblages and the lack of radiometric dates. In an attempt to determine the chronocultural interpretation of the sequence we discuss the published comparisons between Yabroud II and Ksar Akil (Lebanon) (table 6). The part of the Ksar Akil sequence of particular significance for this comparison is as follows:[35]

- Aurignacian/Ahmarian with twisted blade/bladelets (Levantine Aurignacian A, layers 13-9),
- Initial Upper Palaeolithic (IUP) - Early Ahmarian (Ksar Akil Phase B, layers 21-15),
- Middle/Upper Palaeolithic transitional entity (Ksar Akil Phase A, layers 25-21).

LAYERS 10 - 8

All researchers have interpreted these layers as Middle Palaeolithic industries. Ziffer[36] defined the material as "Late Levantine Mousterian" and Bakdach[37] termed it "Levallois-Mousterian." The composition of the tool kit and the production system adhere to the description of the "Tabun B-Type Middle Palaeolithic", *e.g.* by Bar-Yosef:

> "It is characterized by blanks removed mainly from unidirectional convergent Levallois cores. Typical products are broad-based Levallois points, commonly with the typical *chapeau de gendarme* striking platform, and often having the special concorde tilted profile when viewed from the side. Blades do occur in this industry, and sometimes form up to 25% of

35. GORING-MORRIS and BELFER-COHEN, 2006.
36. ZIFFER, 1981.
37. BAKDACH, 1982.

Table 6 – *Yabroud II: comparison with Ksar Akil (Lebanon).*

Yabroud II	Ksar Akil	Cultural attribution	Author
5	X-VIII	Aurignacien du Levant B	BESANÇON *et al.*, 1977
	IX-VIII	Aurignacien du Levant B	ZIFFER, 1981
	XIII-XII	Mitteljungpaläolithikum Phase I	BAKDACH, 1982
	X-IX	Undefined tradition	BERGMAN, 1987
	XX	Early Ahmarian	BELFER-COHEN and GORING-MORRIS, 2003
6	XVII-XV	Ksar Akil, Phase B2	ZIFFER, 1981
	XX-XIV	Frühjungpaläolithikum Phase II Ksar Akil, Phase B	BAKDACH, 1982
	XX	Early Ahmarian	BELFER-COHEN and GORING-MORRIS, 2003
7	XX-XVIII	Ksar Akil, Phase B1	BESANÇON *et al.*, 1977
	XVII-XV	Ksar Akil, Phase B2	ZIFFER, 1981
	XX-XIV	Frühjungpaläolithikum Phase I, Ksar Akil, Phase B	BAKDACH, 1982
8		Late Levantine Mousterian	ZIFFER, 1981
		Levallois-Moustérien	BAKDACH, 1982
9		Late Levantine Mousterian	ZIFFER, 1981
		Levallois-Moustérien	BAKDACH, 1982
10		Late Levantine Mousterian	ZIFFER, 1981
		Levallois-Moustérien	BAKDACH, 1982

> the blanks. [...] The common tool types of this industry include side scrapers, Levallois points, rarely burins, some notches, and denticulates."[38]

Even the change to a centripetal preparation from layers 9 to 8 is perfectly matched by the Tabun B-Type configuration: "A later phase within this industry shows a tendency towards a slight increase in centripetal preparation [...]."[39]

LAYER 7

Several researchers have correlated the assemblage from this layer with Ksar Akil layers 20-14. This places the Yabroud II material within the Ksar Akil, Phase B cultural units,[40] which were termed by Bakdach[41] "Early Upper Palaeolithic phase I". Following our re-evaluation, a designation of this material to the early Upper Palaeolithic is difficult. One might search for comparisons in the concept of an Initial Upper Paleolithic (IUP) as it is described by Marks,[42] Kuhn[43] and Shea,[44] but in Yabroud II, layer 7 neither points nor chamfered pieces could be detected. Even if the absence of points were the result of the poor standard of excavation, it remains difficult to assign layer 7 to a particular cultural entity, especially as we see more signs of a Middle Palaeolithic than a Upper Palaeolithic industry. Thus, the precise cultural affinity of this layer must remain unknown.

LAYER 6

As in layer 7, this assemblage was correlated with Ksar Akil layers 20-14, Phase B by Ziffer and Bakdach.[45] This grouping has since been confirmed by a reanalysis undertaken by Sarel and Ronen:[46]

> "Phase B (levels XX-XIV) underlies the Aurignacian layers from which it is separated by the artefact-poor level XIV. [...] Among the 85 cores examined from Phase B, 9.4% are Levallois cores, 85.9% are laminar cores and 4.7% are other flake cores. These levels exhibit both Levallois and blade

38. BAR-YOSEF, 2000: 116.
39. *Ibid.*
40. Phase B1: BESANÇON *et al.*, 1977. Phase B2: ZIFFER, 1981.
41. BAKDACH, 1982.
42. MARKS, 1990.
43. KUHN, 2003.
44. SHEA, 2003.
45. ZIFFER, 1981; BAKDACH, 1982.
46. SAREL and RONEN, 2003: 77.

implements. [...] The blades are detached mostly from opposed platform cores and the bi-directional cores with twisted-axes dominate. Pyramidal or semi-pyramidal cores are not represented. Moreover, soft hammer was used for detaching blades. The main tools are endscrapers and el-Wad points: notches and denticulates occur but chamfered pieces are absent (Azoury, 1986; Ohnuma, 1988)."

In another study Belfer-Cohen and Goring-Morris compare layer 6 from Yabroud II with Ksar Akil layer 20. They ascribe layer 6 to the Early Ahmarian: "More or less standardised blade blanks and tools comprise an important component of all Early Ahmarian industries." The El-Wad (Font Yves) point initially was cited by Garrod[47] as a fossil director of the Levantine Aurignacian industries. Yet, today the primary association of this rather amorphous type is within "Ahmarian" contexts.[48] However, we doubt this observation due to the dominant and diverse surface concepts in the production system of layer 6; El-Wad points are also missing; instead, we see more affinities to the concept of an Initial Upper Paleolithic (IUP).[49]

LAYER 5

In the past, this assemblage has been affiliated to different cultural units within the Ksar Akil sequence layers 13-8; these have included proposed analogies with the Aurignacien du Levant B,[50] with "Middle Upper Palaeolithic Phase I"[51] and with an "undefined tradition."[52] Recently, Belfer-Cohen suggested close relations to Ksar Akil layer 20, and ascribes the assemblage from layer 5 to an Early Ahmarian.[53] Features such as uni- and bi-directional blade-production and twisted unidirectional bladelet-production, discoidal cores and a set comprising endscrapers, burins, El-Wad points and backed blades fit well into the definition of this Early Ahmarian by Gilead,[54] Goring-Morris and Belfer-Cohen,[55] Monigal[56] and Shea.[57]

47. GARROD, 1957.
48. BELFER-COHEN and GORING-MORRIS, 2003: 8.
49. MARKS, 1990; KUHN, 2003; SHEA, 2003.
50. BESANÇON *et al.*, 1977; ZIFFER, 1981.
51. BAKDACH, 1982: 272.
52. BERGMAN, 1987a.
53. BELFER-COHEN and GORING-MORRIS, 2003.
54. GILEAD, 1981.
55. GORING-MORRIS and BELFER-COHEN, 2006.
56. MONIGAL, 2003.
57. SHEA, 2003: 331.

DISCUSSION

The discussion surrounding the Middle to Upper Palaeolithic transition in the Levant has been subject of constant change.[58] Until the mid-1970s all lithic inventories of the Levant chronostratigrafically positioned between the Middle-Upper Palaeolithic transition and the beginning of the Epipalaeolithic were labelled "Aurignacian."[59] Meanwhile, the discussion has become much more sophisticated, but remains controversial, complex and confounding. Today the chronostratigraphy of the Middle-Upper Palaeolithic transition appears to be divided into several cultural units. The so called Initial Upper Palaeolithic still presents some technological features of the Middle Palaeolithic; at least at some sites cultural markers are Emireh Points derived from Levallois blanks and chamfered pieces. The true Early Upper Palaeolithic is the so called Early Ahmarian; this might reflect local developments, or alternatively, it could originate from new immigrants into the Levant, and at some sites it is characterised by El-Wad points. This technocomplex is finally replaced by the Aurignacian. Concerning radiometric dates this Aurignacian appears quite late.[60] The long sequence from Ksar Akil (Lebanon) is still the only continuous chronostratigraphic reference in the Northern Levant.

Our re-evaluation of layers 10 to 5 from Yabroud II provides a new insight into the lithic production systems (table 7) and shows differences with former studies by Rust, Ziffer and Bakdach.[61] Be this as it may, we must keep in mind that our results are based on an unsatisfactory situation concerning the size, completeness and documentation of the different assemblages; we regard our results as indications of general trends. The appearance of tool types such as borers and *Châtelperronian* points in layers 8 and 7 as described in former studies could not be confirmed; neither were chamfered pieces detected. However, the presence of surface concepts for blank production in layers 7 and 6 is a new result. Layer 8 is a clear Middle Palaeolithic assemblage. Layer 7 can be regarded as a Middle Palaeolithic assemblage with some poor Upper Palaeolithic indications. According to Belfer-Cohen and Goring-Morris, layer 5 belongs to the Early Ahmarian.[62] The presence

58. BAR-YOSEF, 2007; BAR-YOSEF and PILBEAM, 2000; BOËDA and MUHESEN, 1993; BOËDA *et al.*, 2001; BOURGUIGNON, 1998; CAUVIN et SANLAVILLE, 1981; GILEAD, 1991; GORING-MORRIS and BELFER-COHEN, 2003; HOVERS, 2006; HOVERS and KUHN, 2006; MARKS, 1990; MEIGNEN *et al.*, 2006; OTTE, 1998; PLOUX et SORIANO, 2003.
59. GORING-MORRIS and BELFER-COHEN, 2006.
60. GORING-MORRIS and BELFER COHEN, 2003.
61. RUST, 1950; ZIFFER, 1981; BAKDACH, 1982.
62. BELFER-COHEN and GORING-MORRIS, 2003: 8.

Table 7 – *Yabroud II: cultural attribution of layers 10 to 5.*

Cultural attribution	Yabroud II	Characterisation
Early Ahmarian	5	Unidirectional blade / bladelet (twisted), endscraper, burin, El-Wad point, backed blade
Initial Upper Palaeolithic	6	Levallois recurrent unidirectional convergent / Levallois recurrent centripetal / Levallois recurrent unidirectional and bi-directional, unidirectional blade, burin, endscraper, sidescraper, convergent scraper
Unspecified Middle Palaeolithic	7	Levallois recurrent unidirectional and bi-directional, endscraper, sidescraper
Tabun B-type (late)	8	Levallois recurrent centripetal, sidescraper
Tabun B-type	9	Levallois recurrent unidirectional convergent, convergent scraper, burin
Levantine Mousterian	10	Levallois recurrent unidirectional convergent, convergent scraper

of diverse surface concepts together with Upper Palaeolithic implements in layer 6 could be related to an Initial Upper Palaeolithic.

Important changes in the lithic assemblage are particularly apparent in layer 7 (table 5). For the first time endscrapers appear in the toolkit. One example of a bladelet core provides evidence of the volumetric production concept, and a bone point originally described by Rust[63]—although no longer part of the collection—documents innovative features. It might be argued that some of these elements are intrusive and belong to layer 6 or even 5, but these are not the only indications for change. Within the lithic production system of layer 7 blades produced by the Middle Palaeolithic surface concept are the main product. These technological changes are accompanied by extensive edge damage that indicates a change of taphonomic conditions. The dynamic of change continues in layer 6. Technology and toolkit show a coexistence of Middle and Upper Palaeolithic lithic technology. The lithic production system is now based on the production of blades. The raw material from layer 6 shows a strong tendency towards a particular local variety, and thus documents a change in behaviour in this phase. From the sedimentological point of view a change from coarse grained to fine grained sediments is reported by Rust which might reflect changes in the environmental system. Due to the uncertain situation at Yabroud II we by no means wish to over-interpret the assemblage from layer 7. With layer 5 the process of change comes to an end.

Considering the poor stratigraphical control and documentation of the archaeological units described by Rust, all our conclusions must be regarded as tentative. However, by mutual control of several independent features our interpretation gains some substantiation. Furthermore, in layers 7-5 Rust mentions fire places with a diameter of 40 cm each, located in the centre of the find concentrations. This increases the credibility of his definition of layers. During our analysis we have observed that the reduction sequences within these assemblages are complete (except the microfraction), which in turn supports the idea of homogenous assemblages.

Sedimentological change between layers 7 and 6 in combination with intense edge damage on artefacts in layer 7 indicate a synchronised shift of environmental conditions.

It appears that subsequent to the deposition of the material comprising layer 7 the environmental circumstances changed: following a hiatus of indefinite time (edge-damaged artefacts) the coarse-grained material was replaced by fine sediments. This observation is possibly linked with a rapid change of climatic conditions. The onset of the Initial Upper Palaeolithic in the Levant is dated between 40,000 and 47,000 BP.[64] Around 42,000 BP water level of Lake Lisan drops very rapidly more than 50 m, which probably corresponds to the Heinrich 4 (H4) event.[65] With great caution we suggest that the sedimentological and cultural change from layers 7 to 6 in Yabroud II marks the transition from Middle Palaeolithic to Initial Upper Palaeolithic, and simultaneously with the H4 event. Nevertheless, this interpretation is requires more research at Yabroud II; however, we cannot exclude local explanations for this observation.

The picture that arises from our re-evaluation of the relevant layers from Yabroud II underlines the importance of the site for the discussion about the Middle to Upper Palaeolithic transition in the Levant. The sequence seems to represent the complete chronostratigraphy from Tabun B-Type via Initial Upper Palaeolithic to Early Ahmarian. In general, the composition of a toolkit underlies functional, ecological and

63. RUST, 1950: 71.

64. SHEA, 2006.
65. BARTOV *et al.*, 2002 and 2003; HAZAN *et al.*, 2005.

social factors, while technological abilities are, at least in the Upper Pleistocene, of minor importance. The spectrum of technological knowledge during the Middle Palaeolithic was well developed; indeed, it would have been quite sufficient to realise the complete Early Upper Palaeolithic lithic production system, including the toolkit.[66] Therefore, the observed changes at Yabroud II reflect deliberate decisions and not changes in cognitive abilities. Within the sequence we observe an increasing efficiency in the lithic production system. Layer 6, the Initial Upper Palaeolithic, provides the first evidence of maintaining distal convexity during the exploitation stage by curved blades/bladelets. This signifies a decrease in the amount of time and material required to prepare the convexity of the distal part of the core. We assume that an increase of environmental stress triggered an increase in efficiency of tool production.

ACKNOWLEDGEMENTS

Ch. Daum, A.-L. Fischer and I. Schmidt helped to organise the material and the study. J. Bakdach accompanied us on our field trip to Yabroud II and gave useful advice. J. Shea provided information on raw material use at Yabroud II. I. Schmidt and L. Clare reviewed the English and Ch. Verna the French text.

Andreas PASTOORS – Gerd-Christian WENIGER
Neanderthal Museum, Talstr. 300
D-40822 Mettmann
GERMANY
pastoors@neanderthal.de
weniger@neanderthal.de

Jan F. KEGLER
Ostfriesische Landschaft
Archäologischer Dienst
Georgswall 1-5
D-26603 Aurich
GERMANY
kegler@ostfriesischelandschaft.de

BIBLIOGRAPHY

AZOURY I.
1986 *Ksar Akil, Lebanon. A technological and typological analysis of the transitional and early upper Palaeolithic levels of Ksar Akil and Abu Halka. I: levels XXV-XII.* Oxford (*BAR Int. Ser.* 289).

BAKDACH J.
1982 *Das Jungpaläolithikum von Jabrud in Syrien.* Unpublished PhD. Köln: Institute of Prehistoric Archaeology, University of Cologne.

BARTOV Y., GOLDSTEIN S.L., STEIN M. and ENZEL Y.
2003 Catastrophic arid episodes in the Eastern Mediterranean linked with the North Atlantic Heinrich events. *Geology* 31,5: 439-442.

BARTOV Y., STEIN M., ENZEL Y., AGNON A. and RECHES Z.
2002 Lake levels and sequence stratigraphy of lake Lisan, the late Pleistocene precursor of the Dead Sea. *Quaternary Research* 57: 9-21.

BAR-YOSEF O.
2000 The Middle and Early Upper Paleolithic in Southwest Asia and Neighbouring Regions. *In*: BAR-YOSEF O. and PILBEAM D. (eds.), *The geography of Neandertals and modern humans in Europe and Greater Mediterranean*: 107-156. Cambridge: Peabody Museum of Archaeology and Ethnology.
2007 Asia, West: Palaeolithic cultures. *In*: PEARSALL D.M. (ed.), *Encyclopedia of Archaeology*: 865-875. Oxford: Academic Press.

BAR-YOSEF O. and PILBEAM D. (eds.)
2000 *The geography of Neandertals and modern humans in Europe and Greater Mediterranean.* Cambridge: Peabody Museum of Archaeology and Ethnology.

BELFER-COHEN A. and GORING-MORRIS A.N.
2003 Current issues in Levantine Upper Palaeolithic research. *In*: GORING-MORRIS A.N. and BELFER-COHEN A. (eds.), *More than meets the eye: Studies on Upper Paleolithic diversity in the Near East*: 1-12. Oxford: Oxbow Books.

BERGMAN C.A.
1987a Hafting and Use of Bone and Antler Points from Ksar Akil, Lebanon. *In*: STORDEUR-YEDID D. (éd.), *La main et l'outil. Manches et emmanchements préhistoriques*: 117-126. Lyon: MOM (*TMO* 15).
1987b *Ksar Akil, Lebanon. A technological and typological analysis of the later Upper Palaeolithic levels of Ksar Akil. II: levels XIII-VI.* Oxford (*BAR Int. Ser.* 329).

BESANÇON J., COPELAND L. et HOURS F.
1975-1977 Tableaux de préhistoire libanaise. *Paléorient* 3: 5-46.

BOËDA É.
1988 De la surface au volume: analyse de concepts des débitages Levallois et laminaire. *In*: FARIZY C. (éd.), *Paléolithique moyen récent et Paléolithique supérieur ancien en Europe. Actes du colloque international de Nemours*: 63-68. Nemours: APRAIF (*Mémoires du musée de Préhistoire d'Île de France* 3).
1994 *Le concept Levallois: variabilité des méthodes.* Paris: CNRS Éditions.

BOËDA É. et MUHESEN S.
1993 Umm El Tlel (El Kowm, Syrie): Étude préliminaire des industries lithiques du Paléolithique moyen et supérieur. *In: Cahiers de l'Euphrate* 7: 47-91. Paris: ERC.

66. PASTOORS, 2009.

BOËDA É., GENESTE J.-M. et MEIGNEN L.

1991 Identification des chaînes opératoires lithiques du Paléolithique ancien et moyen. *Paléo* 2 : 43-80.

BOËDA É., GRIGGO C. et NOËL-SORIANO S.

2001 Différents modes d'occupation du site d'Umm el Tlel au cours du Paléolithique moyen (El Kowm, Syrie centrale). *Paléorient* 27,2 : 13-28.

BORDES F.

1955 Le paléolithique inférieur et moyen de Jabroud (Syrie) et la question de Pré-Aurignacien. *L'Anthropologie* 59 : 486-507.

1962 Sur la chronologie de Paléolithique au Moyen Orient. *Quaternaria* 5 : 57-69.

BOURGUIGNON L.

1998 Les industries du Paléolithique intermédiaire d'Umm el Tlel : nouveaux éléments pour le passage entre Paléolithique moyen et supérieur dans la bassin d'El Kowm. *In* : OTTE M. (éd.), *Préhistoire d'Anatolie. Genèse des deux mondes* : 709-730. Liège : Université de Liège (*ERAUL 85*).

CAUVIN J. et SANLAVILLE P. (éd.)

1981 *Préhistoire du Levant. Chronologie et organisation de l'espace depuis les origines jusqu'au VIe millénaire*. Paris : Éditions du CNRS.

DELAGNES A. and MEIGNEN L.

2006 Diversity of lithic production systems during the Middle Paleolithic in France: are there any chronological trends? *In*: HOVERS E. and KUHN S.L. (eds.), *Transitions before the transition. Evolution and stability in the Middle Paleolithic and Middle Stone Age*: 85-108. New York: Springer (*Interdisciplinary Contributions to Archaeology*).

DODONOV A.E., KANDEL A.W., SIMAKOVA, A.N., MASRI M. and CONARD N.J.

2006 Geomorphology, site distribution and paleolithic settlement dynamics of the Ma'aloula region, Damascus province, Syria. *In*: CONARD N.J. (ed.), *Tübingen – Damascus excavation and survey project 1999-2005*: 257-274. Tübingen: Kerns.

EL-KASSEM M.

2001 *Das Jabrudien von Jabrud (Syrien): Felsdach I, 25*. Unpublished Magister-Thesis. Köln: Institute of Prehistoric Archaeology, University of Cologne.

FRANK T.

2004 *Die Levallois-Spitzen von Jabrud, Felsdach I, Schichten 2-4*. Unpublished Magister-Thesis. Köln: Institute of Prehistoric Archaeology, University of Cologne.

GARROD D.

1957 Notes sur le Paléolithique Supérieur du Moyen Orient. *Bulletin de la Société Préhistorique Française* 54 : 439-445.

GILEAD I.

1981 Upper Palaeolithic tool assemblages from the Negev and Sinai. *In* : CAUVIN J. et SANLAVILLE P. (éd.), *Préhistoire du Levant. Chronologie et organisation de l'espace depuis les origines jusqu'au VIe millénaire* : 331-342. Paris : Éditions du CNRS.

1991 The Upper Paleolithic in the Levant. *Journal of World Prehistory* 5 : 105-154.

GORING-MORRIS A.N. and BELFER-COHEN A.

2006 A hard look at the 'Levantine Aurignacian': how real is the taxon? *In*: BAR-YOSEF O. and ZILHÃO J. (eds.), *Towards a definition of the Aurignacian: Proceedings of the symposium held in Lisbon, Portugal, June 25-30, 2002*: 297-316. Lisboa: Inst. Português de Arqueologia (*Trabalhos de Arqueologia* 45).

GORING-MORRIS A.N. and BELFER-COHEN A. (eds.)

2003 *More than meets the eye: Studies on upper Paleolithic diversity in the Near East*. Oxford: Oxbow Books.

HAZAN N., STEIN M., AGNON A., MARCO S., NADEL D., NEGENDANK J.F.W., SCHWAB M.J. and NEEV D.

2005 The late Quaternary limnological history of Lake Kinneret (Sea of Galilee), Israel. *Quaternary Research* 63: 60-77.

HOVERS E.

2006 Neanderthals and modern humans in the middle palaeolithic of the Levant: What kind of interactions. *In*: CONARD N.J. (ed.), *When Neanderthals and modern humans met*: 65-86. Tübingen: Kerns.

HOVERS E. and KUHN S.L. (eds.)

2006 *Transitions before the transition. Evolution and stability in the Middle Paleolithic and Middle Stone Age*. New York: Springer (*Interdisciplinary Contributions to Archaeology*).

KUHN S.

2003 In what sense is the Levantine initial Upper Paleolithic a 'transitional' industry? *In*: ZILHÃO J. and ERRICO F. de (eds.), *The chronology of the Aurignacian and the transitional technocomplexes. Dating, stratigraphies, cultural implications*: 61-70. Lisboa: Inst. Português de Arqueologia (*Trabalhos de Arqueologia* 33).

MARKS A.E.

1990 The Middle and Upper Paleolithic of the Near East and the Nile Valley: the problem of cultural transformations. *In*: MELLARS P. (ed.), *The emergence of modern humans: an archaeological perspective*: 56-80. New York: Cornell University Press.

MEIGNEN L.

1995 Levallois lithic production systems in the Middle Paleolithic of the Near East: The case of the unidirectional method. *In*: DIBBLE H.L. and BAR-YOSEF O. (eds.), *The Definition and Interpretation of Levallois Technology*: 361-379. Madison: Prehistory Press (*Monographs in World Archaeology* 23).

MEIGNEN L., BAR-YOSEF O., SPETH J.D. and STINER M.C.

2006 Middle Paleolithic settlement patterns in the Levant. *In*: HOVERS E. and KUHN S.L. (eds.), *Transitions before the transition. Evolution and stability in the Middle Paleolithic and Middle Stone Age*: 149-170. New York: Springer (*Interdisciplinary Contributions to Archaeology*).

MONIGAL K.

2003 Technology, economy, and mobility at the beginning of the Levantine Upper Palaeolithic. *In*: GORING-MORRIS A.N. and BELFER-COHEN A. (eds.), *More than meets the eye: Studies on upper Paleolithic diversity in the Near East*: 118-133. Oxford: Oxbow Books.

OHNUMA K.

1988 *Ksar Akil, Lebanon: A technological study of the earlier Upper Palaeolithic levels of Ksar Akil. Vol. III: levels XXV-XIV*. Oxford (*BAR Int. Ser.* 426).

OTTE M. (éd.)

1998 *Préhistoire d'Anatolie. Genèse des deux mondes.* Liège: Université de Liège.

PASTOORS A.

2009 Blades? – Thanks, no interest! Neanderthals in Salzgitter-Lebenstedt. *Quartär* 56: 105-118.

PLOUX S. et SORIANO S.

2003 Umm el Tlel, une séquence du paléolithique supérieur en Syrie centrale: industries lithiques et chronologie culturelle. *Paléorient* 29,2: 5-34.

RUST A.

1950 *Die Höhlenfunde von Jabrud (Syrien).* Neumünster: Karl Wachholtz (*Offa Bücher* 8).

SAREL J. and RONEN A.

2003 The Middle/Upper Palaeolithic transition in northern and southern Israel. A technological comparison. *In*: GORING-MORRIS A.N. and BELFER-COHEN A. (eds.), *More than meets the eye: Studies on upper Paleolithic diversity in the Near East*: 68-79. Oxford: Oxbow Books.

SHEA J.J.

2003 The Middle Paleolithic of the East Mediterranean Levant. *Journal of World Prehistory* 17,4: 313-394.

2006 The Middle Paleolithic of the Levant: recursion and convergence. *In*: HOVERS E. and KUHN S.L. (eds.), *Transitions before the transition. Evolution and stability in the Middle Paleolithic and Middle Stone Age*: 189-212. New York: Springer (*Interdisciplinary Contributions to Archaeology*).

SOLECKI R.S. and SOLECKI R.L.

1966 New data from Yabrud, Syria: Preliminary report of the Columbia University Archaeological Investigations. *Les Annales Archéologiques Arabes Syriennes* 16: 121-153.

1995 The Mousterian Industries of Yabrud Shelter I: A Reconsideration. *In*: DIBBLE H.L. and BAR-YOSEF O. (eds.), *The Definition and Interpretation of Levallois Technology*: 381-397. Madison: Prehistory Press (*Monographs in World Archaeology* 23).

SONNEVILLE-BORDES D. DE

1956 Paléolithique et Mésolithique à Jabroud (Syrie). *L'Anthropologie* 60: 71-83.

WENIGER G.C.

1991 Überlegungen zur Mobilität jägerischer Gruppen im Jungpaläolithikum. *Saeculum* 42,1: 82-103.

WOLFART R.

1966 *Zur Geologie und Hydrogeologie von Syrien unter besonderer Berücksichtigung der süd- und nordwestlichen Landesteile.* Hannover: Bundesanstalt für Bodenforschung (*Beihefte zum Geologischen Jahrbuch* 68).

ZIFFER D.

1981 Yabrud Shelter II: A re-consideration of its cultural composition and of its relevance to the Upper-Paleolithic cultural sequence in the Levant. *Quartär* 31/32: 69-94.

NEW EVIDENCE FOR SIGNIFICANT USE OF BITUMEN IN MIDDLE PALAEOLITHIC TECHNICAL SYSTEMS AT UMM EL TLEL (SYRIA) AROUND 70,000 BP[1]

É. BOËDA, S. BONILAURI, J. CONNAN, D. JARVIE, N. MERCIER, M. TOBEY, H. VALLADAS and H. AL SAKHEL

Abstract: *Mousterian tools with traces of bitumen on their surfaces discovered at Umm el Tlel (Syria) have been dated to 40,000 years (BP). Until now tools with traces of bitumen were known only from the 8th millennium BC, but this new discovery shows usage 30,000 years earlier. Mousterian complex VI 3, dated to 71,000-72,000 BP, has yielded fragments of black cobbles and eleven Levallois artefacts with traces of black material demonstrated to be bituminous. Herein we present the geochemical and microscopy results on materials from artefact and a cobble. Gas chromatography-mass spectrometry analysis of both C_{15+} alkanes and C_{15+} aromatics demonstrates that the black substance is natural asphalt. The source of this asphaltic material or bitumen found on the tools is related to natural outcrops located around 40 km east of Umm el Tlel in the Bichri Massif. For the first time, microscopic analysis of 300 Levallois artefacts with no macro-traces of bitumen visible to the naked eye, has demonstrated that two-thirds of these artefacts have black micro-residues. These micro-traces that were identified on artefacts recovered from a strata in complex VI 3, have been identified as bitumen-bearing following analyses. Thus, rather than being a material used occasionally, bitumen is shown to be of key importance in the technological system of these populations, hypothesized to have been used as an adhesive to attach hafts made of organic matter or bone to tools. These new results extend the oldest known date for the use of natural bitumen by early hominids from 40,000 BP to 71,000 BP and show that older Middle Palaeolithic populations were already acquainted with the use of bitumen as adhesive to attach hafts to flint tools or for other similar uses.*

Résumé: *La première découverte d'outils moustériens avec des traces de bitume sur leurs surfaces, à Umm el Tlel en Syrie, a été datée d'avant 40000 ans BP. Antérieurement à cette découverte, ce type d'outils était connu simplement au VIII*[e] *millénaire BC. Avec ces nouvelles données sur le même site, l'utilisation du bitume se trouve repoussée de 30000 ans. En effet le complexe moustérien VI 3, daté de 71000-72000 BP, a livré des fragments de galets noirs et 11 artefacts Levallois avec des traces de matériau noir, identifiées comme étant bitumineuses. Nous présentons ici les résultats géochimiques et microscopiques sur des matériaux observés dans ou sur des artefacts et des galets. La chromatographie en phase gazeuse couplée à la spectrométrie de masse (CG-SM) des C_{15+} saturés et C_{15+} aromatiques démontre que cette substance noire est de l'asphalte. La source de ce matériau asphaltique ou bitume, présente sur les outils, se trouve dans des affleurements de sables bitumineux, à 40 km de Umm el Tlel dans le massif du Djebel Bichri. Pour la première fois, l'analyse microscopique de 300 artefacts Levallois dépourvus de macro-traces de bitume visibles à l'œil nu, a montré que les deux tiers de ces artefacts avaient des micro-résidus noirs. Ces micro-traces, identifiées sur les artefacts du complexe VI 3, ont été identifiées comme étant à base de bitume après analyse. Ceci montre que ce matériau n'est pas utilisé de façon occasionnelle mais joue un rôle de premier plan parmi les connaissances techniques de ces populations, probablement comme adhésif pour fixer des artefacts sur des manches en matière organique ou en os, ou pour insérer ces artefacts dans des manchons. Ces nouveaux résultats font remonter la plus ancienne date d'utilisation du bitume par des hominidés de 40000 BP à 70000 BP et montrent donc que des populations plus anciennes, du Paléolithique Moyen, savaient déjà utiliser le bitume dans le cadre d'emmanchement ou pour d'autres usages similaires.*

1. A synthesis of this article was previously published in *Antiquity* (BOËDA *et al.*, 2008).

Manuscrit reçu le 19 juin 2008, accepté le 25 novembre 2008

Keywords: *Bitumen, Djebel Bichri, Mousterian, Near East, Micro-traces, Hafting, Microscopic techniques, Geochemical analysis, Stable isotope, Absolute dating.*
Mots-clés: *Bitume, Djebel Bichri, Moustérien, Moyen-Orient, Micro-traces, Emmanchement, Techniques microscopiques, Analyse géochimique, Isotope stable, Datation absolue.*

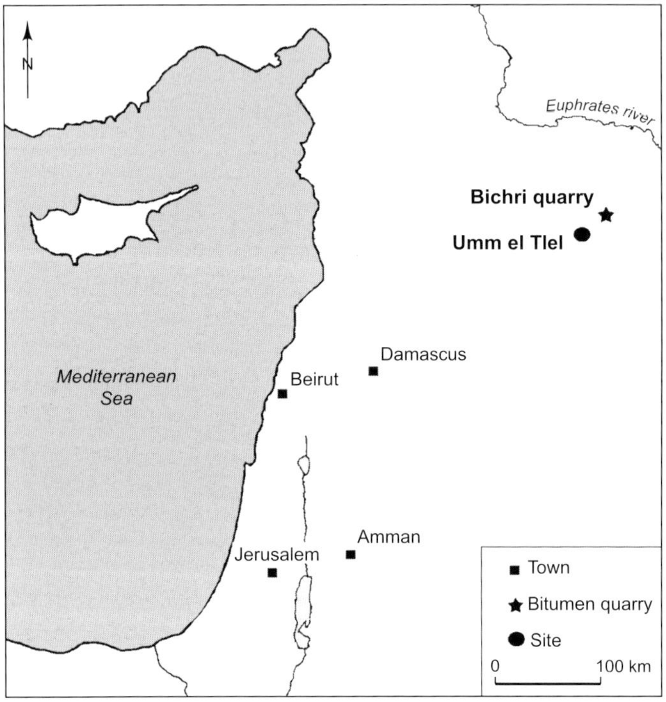

Fig. 1 – *Location of the Umm el Tlel site and bitumen sources (Djebel Bichri).*

In 1992 the discovery of bitumen on convergent side-scrapers and Levallois flakes in Mousterian levels dating to 40,000 BP at Umm el Tlel (el Kowm Basin, Central Syria, fig. 1)[2] led to reconsideration of the commonly accepted idea that bitumen on artefacts first appeared in the Near East near the Dead Sea at Netiv Hagdud (8200-7500 BC) only in the 8th millennium BC.

Since 2000 new discoveries from eight archaeological levels[3] in Mousterian complex VI 3 at Umm el Tlel extended the use of bitumen more than 30,000 years earlier. Seven burnt flint specimens from layer VI 3 dated by thermoluminescence, (a method developed by Aitken)[4] yielded statistically compatible results and a weighted mean of 71 ± 4 ka, which places the deposition of these two layers[5] at the transition between MIS 4 and 3 (table 1).

These discoveries consist of macroscopic and microscopic traces of a black material. Macro-traces are present, both as cobble fragments, as a thin deposit on the surface of certain Levallois artefacts (n = 11) (figs. 2-3) and underlying these artefacts, as an imprint. Microscopic traces were trapped in surface irregularities or quartz veins in flint or adhering to artefact surfaces (fig. 4). While macroscopic residues can be analyzed by geochemical means, microscopic remains cannot, due to the extremely minute amounts of bituminous material. However, these minute residues can be studied using microscopic techniques at various magnifications ranging from 50X to 200X.

ARCHAEOLOGICAL MATERIAL

Macroscopic or microscopic remains are found as small balls and cobble fragments or as fine deposits on the surface of some Levallois artefacts. The small balls and cobble fragments (*e.g.*, n°1782, table 2) were collected from the wadi bed of the Djebel Bichri, some 40 km northeast of Umm el Tlel. They were generated by the disaggregation of *in situ* tar sand layers. Mixed with quartz and gypsum, these cobbles are of highly variable size and diameter ranging from 1-20 cm. Only very small balls, less than 3 cm in diameter, were recovered at the site.

Among more than 1000 artefacts from the eight layers in complex VI 3, 11 lithic artefacts have black macro-residue on

2. BOËDA *et al.*, 1996 and 1998; BOËDA et MUHESEN, 1993.
3. These eight excavated levels over an area of 50 m² have yielded an impressive amount of lithic and faunal material, between 150 and 300 objects per m². Technological analyses have demonstrated the uniformity between these levels, showing clearly the successive occupations of the same group or of groups sharing the same knowledge (BOËDA *et al.*, 2002).
4. AITKEN, 1985.
5. MARTINSON *et al.*, 1987.

Table 1 – *Each flint specimen was treated according to the procedure described by Valladas*[6] *and Valladas* et al.[7] *The U-238, Th-232, and K-40 contents of the dated specimens (columns 2 to 4) were measured by neutron activation analysis.*[8] *For each radioisotopic measurement, a combined statistical and systematic error of ± 10% is essentially caused by uncertainties in the reference standard. The alpha sensitivity (column 5) was determined by comparing the TL-α and TL-β signals induced by α and β particles from Pu-238 and Y/Sr-90 sources, respectively.*[9] *The internal dose-rate (column 8) of each flint was computed from its contents U, Th and K and from the specific dose-rates given by Adiamec and Aitken.*[10] *The external dose-rates (column 12) were deduced from measurements taken in the field by SO_4Ca : Dy dosimeters. The gamma dose-rate (column 10) has been corrected to take into account the absorption of the gamma rays passing through the sample. The cosmic dose-rate (column 11) was estimated after taking into account the thickness of the sediment fill.*[11] *The ED values (column 16) were measured with an automatic apparatus by the additive method; they include the supralinearity correction. This latter was measured from the TL growth curve regenerated after the natural powder was reheated for 1.5 hrs. at 350 °C.*[12] *Following Aitken's*[13] *recommendations, the statistical and systematic errors were calculated separately for each flint. Each of the tabulated overall errors represents the mean square average of the two (column 19).*

Sector	U (ppm)	Th (ppm)	K (%)	α–sens. μGγ/a/1000 α	Alpha	Beta	Total dose mGg/a				External	+ –	Total dose	+ –	Ed (Gy)	+ –	Age (ka)	+ –
							Internal	+ –	Gamma	Cosmic								
V	0,93	0,24	0,06	23,2	386	192	584	44	606	160	766	60	1350	75	88,9	4,7	65,9	5,5
V	0,73	0,22	0,07	19	251	167	423	31	635	160	795	60	1218	67	85,5	7	70,2	6,8
VI	0,81	0,27	0,07	18,7	274	180	462	29	295	160	455	51	918	58	68,6	3	74,8	6,6
VI	1,45	0,2	0,07	14,6	366	273	653	48	294	160	454	50	1107	70	87,4	2,4	79	7,1
VI	1,54	0,18	0,08	21,6	570	291	870	68	302	160	462	52	1331	86	84,1	4,8	63,2	6,3
VI	1,43	0,15	0,08	22,8	558	272	841	62	299	160	459	51	1300	81	75	15	57,7	8,4
VI	1,02	0,22	0,01	20,5	368	165	538	42	302	160	462	52	1000	66	80,6	5,9	80,6	8,1
VI	0,86	0,19	0,19	14,8	227	281	519	29	294	160	454	50	972	58	67	2,7	68,9	6
VI	1,32	0,17	0,02	22,5	511	211	732	57	297	160	457	51	1189	76	90	3,5	75,7	7

the underside, which has impregnated the underlying sediment, leaving a very clear and well-preserved imprint.[14] They are thus not randomly located on the lithic tools and may be interpreted as adhesives used to haft the flakes and points, as shown by two of the Levallois artefacts presented in this paper: a flake (n°1780, fig. 5) and a Levallois point (n°2023, fig. 6). The position of these residues in the mesial and proximal parts leaves the edges and the distal end free, *i.e.*, the point on a tool or the cutting parts of tools. These observations are corroborated by microscopic analysis (magnification 50X-200X) of more than 200 Levallois artefacts by examination of large flakes, elongated flakes and points from layer VI 3d'. However, the position of the black material depends upon the technological types (large or elongated flakes, points) and especially on their likely function(s) and means of use (cutting tools, scraping tools, etc.). For example, the black material could be located along the debitage axis of an artefact (figs. 5-6), oblique to the axis (fig. 4) or laterally along an edge of the artefact (fig. 4).

Pursuant to these observations, it appears that this material likely played a role in the grasping of such tools. While it is impossible to know the precise use of such material, we infer that it could have been used as adhesive to attach a handle composed of animal or vegetal material or as a sleeve to provide protection for the hand of the user. It is also probable that the treatment or preparation of this material, as well as the amount applied to the surface of tools, would vary depending on its ultimate purpose: as a adhesive or as a protective.[15]

6. VALLADAS, 1992.
7. VALLADAS *et al.*, 1994.
8. JORON, 1974.
9. VALLADAS and VALLADAS, 1982.
10. ADIAMEC and AITKEN, 1998.
11. PRESCOTT and HUTTON, 1994.
12. MERCIER *et al.*, 1992.
13. AITKEN, 1985.
14. Around thirty flakes from this archaeological level have a very fine black material systematically located on their dorsal and ventral faces, but which did not leave imprints on the underlying sediment where they were found. The limited amount of material available was insufficient for geochemical analyses.

15. Functional analyses carried out in the framework of a doctoral thesis on more than a hundred tools from layer VI3d' corroborate the use of this black material as an adhesive or hafting material which corresponded in every case to the prehensile parts of the tools. In addition, a recent study of a laminar flake embedded in an ostrich pelvis (layer VI3d') shows the role of this black material in tool maintenance (BONILAURI *et al.*, 2007).

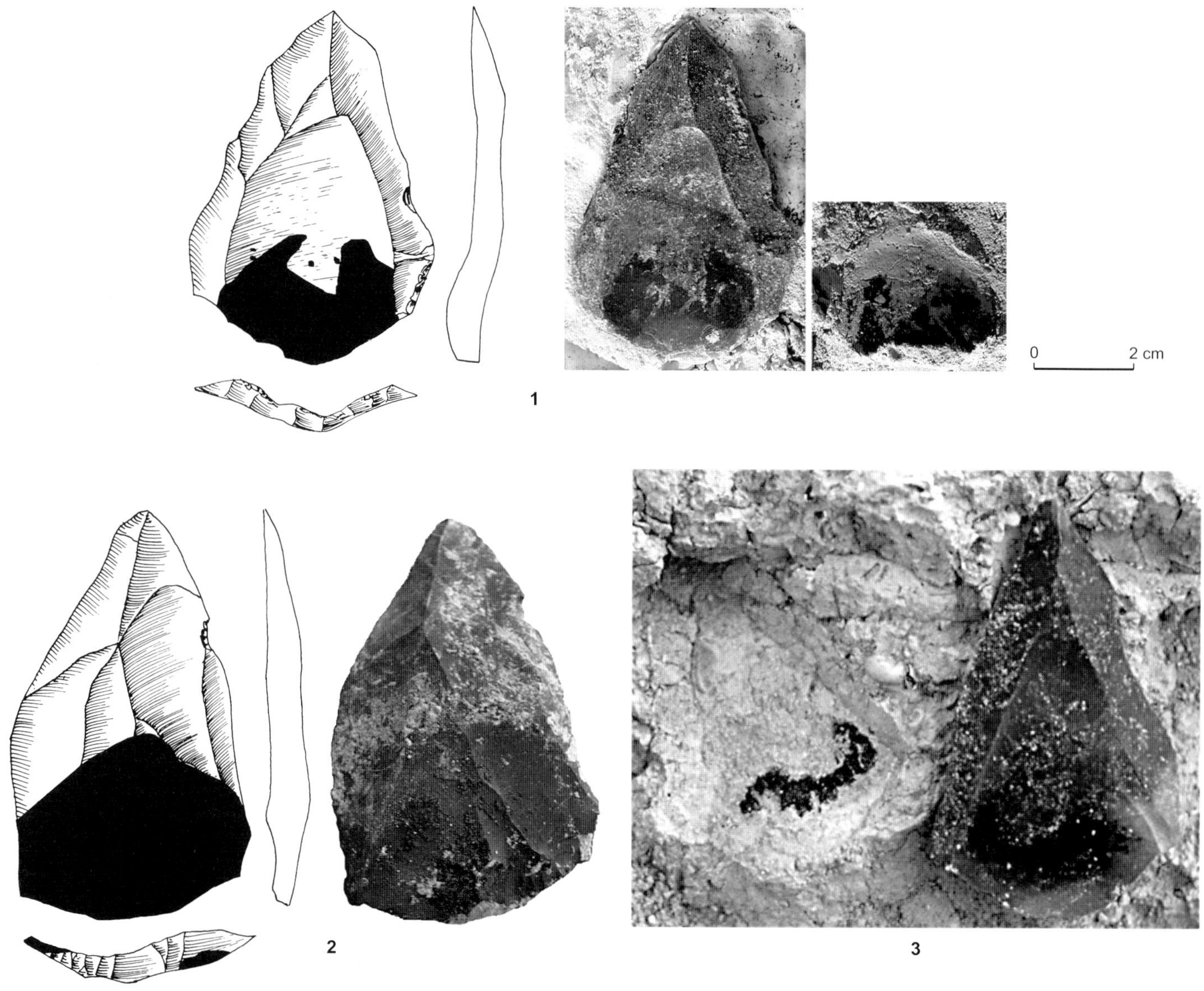

Fig. 2 – *Layers IV 3, Levallois artefacts with deposit on the surface and underlying these aheste artefacts, as an imprint.* ***1, 3,*** *Levallois point;* ***2,*** *Levallois flake triangular.*

GEOCHEMICAL ANALYSIS OF BITUMEN: IDENTIFICATION OF THE GEOGRAPHIC SOURCE

Geochemical analyses were carried out on samples of macro-residues of bitumen taken from the surface or from imprints of three Levallois artefacts (n°1780, n°1781, n°2023, table 2), an archaeological cobble fragment (n°1782, table 2), and bitumen samples collected from Bichri tar sands (n°170 and n°430, table 2).

A crust of black organic matter (150 mg for n°1780 and 4.7 mg for n°2023), located on both faces of the Levallois artefacts (figs. 5-6), was scraped from the surface. This isolated material was compared to a small ball of oil-stained sand recovered from the same layer. This free ball, which was not associated with any artefact, was suspected to be the pristine tar sand that was used to prepare the bituminous layer found on the

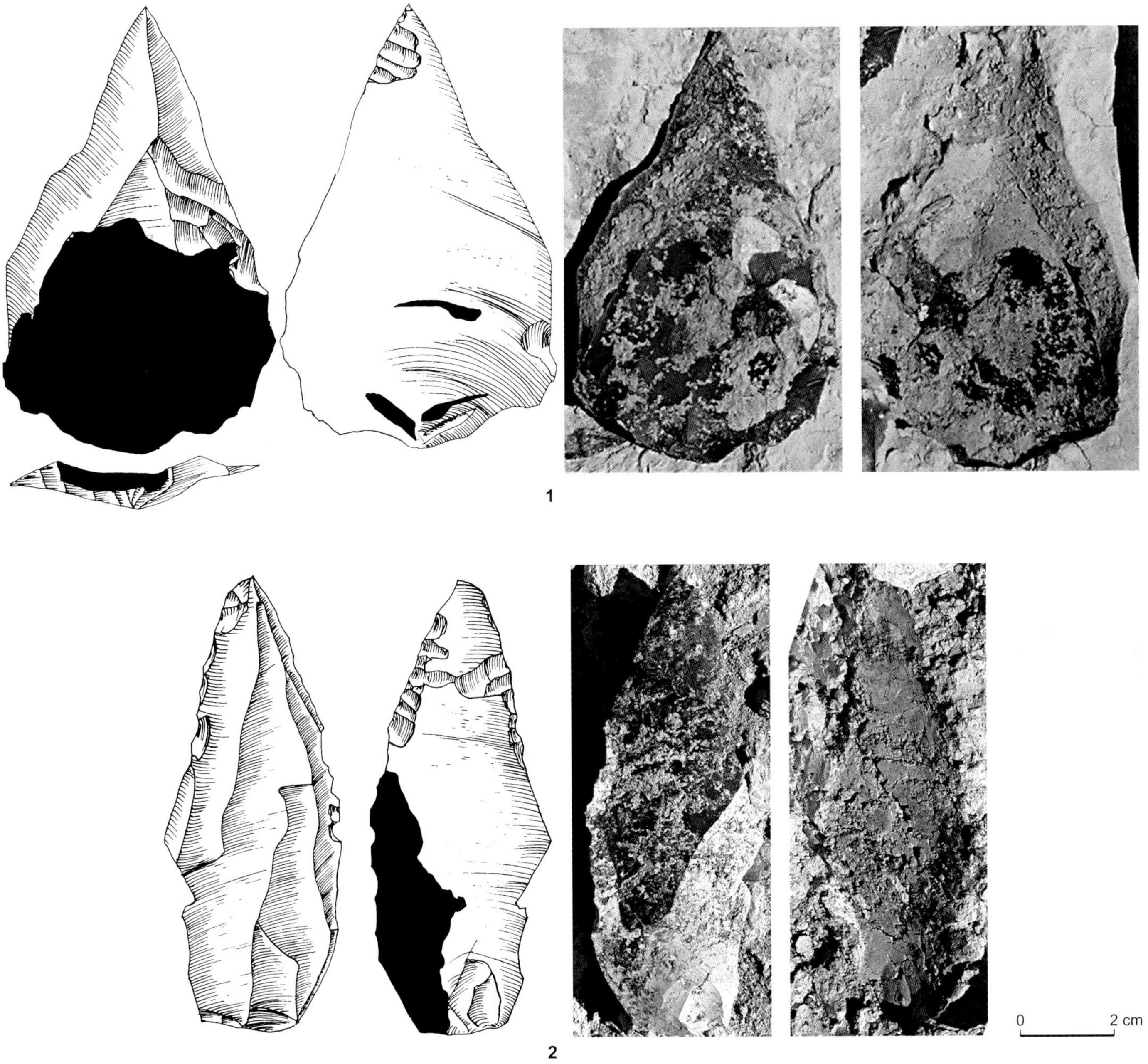

Fig. 3 – *Layers IV 3, Levallois artefacts with deposit on the surface and underlying these artefacts, as an imprint. The black surface represented on the line drawing* ***1*** *corresponds to the addition of visible bitumen traces on the artifact and the imprint. For the laminar artifact* ***2*** *the traces are present as dissociated grains which cannot be clearly seen in the photo. The line drawing shows the surface containing the bitumen grains, observable with a simple 2x magnifying lens.*

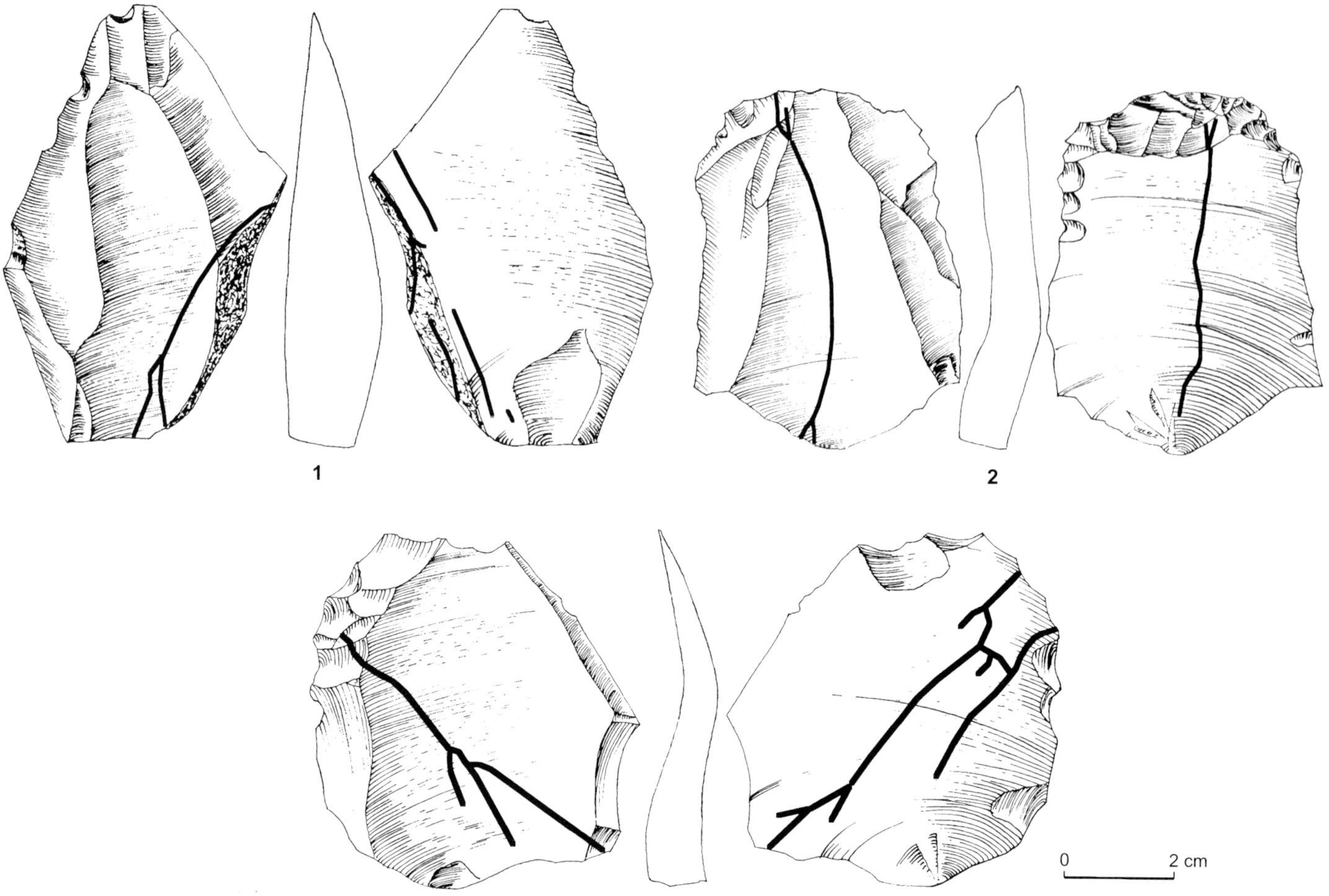

Fig. 4 – *Layers IV 3, Levallois artefacts with microscopic traces, marked with a thick black line.* ***1****, triangular Levallois backed flake, the lines of bitumen from one side of the piece to the other are in relation to the cortical back;* ***2****, Nahr Ibrahim, lines from one side to the other are parallel to the two edges of the flake;* ***3****, sub-quadrangular Levallois flake, bitumen lines are strictly parallel to single regular cutting edge of the flake.*

artefacts. Gross parameters (Hydrogen Index = HI expressed in mg of hydrocarbons per g of Total Organic Carbon, Oxygen Index = OI expressed in mg CO_2 per g of Total Organic Carbon and Tmax in °C of the S2 pyrolysis peak maximum) recorded by Rock-Eval pyrolysis[16] are fully consistent with previous data collected on archaeological bituminous mixtures excavated from various sites from the Near East and the Gulf.[17] Following Rock-Eval screening analysis, the samples were extracted with dichloromethane and this extractable organic matter (= true bitumen) was subsequently separated into four fractions (saturated and aromatic hydrocarbons, resins, and asphaltenes) and the weights of each fraction were recorded. Comparison of TOC values and total extract yields shows that the organic matter present is largely soluble in dichloromethane. This extract exhibits a gross composition (table 2) of archaeological bitumen, *i.e.*, crude oil which has been severely altered by weathering (photo-oxidation, biodegradation, oxidation, water washing, evaporation, etc.). δD values of asphaltenes are very sensitive to weathering processes (by oxidation as seen in the Bichri tar sands analyzed, *e.g.*, δD may shift from -90 to -50‰ / SMOW).[18] However, $\delta^{13}C$ values are reasonably stable

16. ESPITALIÉ and BORDENAVE, 1993.
17. CONNAN et DESCHESNE, 1996; CONNAN *et al.*, 1998.
18. CONNAN and NISHIAKI, 2003.

Table 2 – *References and geochemical data on the three Umm el Tlel samples and on the two related Bichri tar sands. Significance of abbreviations: TOC = Total Organic Carbon (in % by weight / sample), HI = Hydrogen Index, HC = Hydrocarbons formed by cracking between 300 and 600°C, OI = Oxygen Index, CO_2 = CO_2 formed and released below 400°C, Tmax = temperature of the maximum of the S2 pyrolysis peak, Total extract = Total extractable organic matter isolated by dichloromethane, saturates = C_{15+} saturated hydrocarbons, aromatics = C_{15+} aromatic hydrocarbons, δ^{13}Casp = δ^{13}C of asphaltenes (in ‰ / PDB), Ts = 18α-22,29,30-trisnorneohopane, Tm = 17α-22,29,30-trisnorhopane, GCR = Gammacerane, 30αβH = 17α,21β-hopane.*

Umm el Tlel campaign	Lab number	Field reference	Sample type	Sample description	TOC (weight % / sample)	HI (in mg HC / g TOC)	OI (in mg CO_2 / g TOC)	Tmax (°C)	Total extract (weight % / sample)	% saturates	% aromatics	% resins	% asphaltenes	δ^{13}Casp (in ‰ / PDB)	Ts/Tm	GCR/30 αβH
UM.01	1780	AL 157 n° 60 - VI 3d'	Black material on a Levallois point	Black grains with excavation earth and quartz grains (Bichri sand?)	21,3	570	8	430	24,5	3,2	7,8	17	64,2	-28,5	0,9	3,8
UM.01	1781	AJ 158 n° 5 - VI 3d'	Black material on a Levallois flake	Two black lumps					49	8,1	8,1	18,2	58,6	-28,8	0,9	3,4
UM.01	1782	AL159 n° 518 - VI 3d'1	Small fragments of a black pebble	Thin flat crust with numerous quartz grains (Bichri tar sand?)	10,8	552	15	429	9	3	7,9	15,2	68,2	-28,5	0,4	5,1
UM.04	2023	AJ 154-VI 3c' - seat	Black material on a Levallois point	Some black grains in a beige matrix						4,3	8,5	36,2	51,1	-28,4	0,31	2,6
Bichri outcrops	170		Samples from surface	Tar sand										-28,4	0,58	2,6
Bichri quarry	430	BS07	Surface of the accumulation	Tar sand										-28,7	0,34	2,2

Fig. 5 – *Levallois flake 1780 (square AJ 158 n°5) with imprint of bitumen deposited on the flake's surface and on the soil.*

and only vary within a narrow range, estimated at 1.5‰ / PDB.[19] δ^{13}C values of asphaltenes are therefore good genetic parameters to correlate archaeological bitumen to parent oil seeps where they were collected as documented in several published recent studies.[20] In the present study, δ^{13}C values of the three Umm el Tlel artefact samples (n°1780, n°1781, n°1782, table 2) agree with the values recorded in 21 samples of tar sands from

19. *Ibid.*

20. *e.g.*, CONNAN and CARTER, 2007; STERN *et al.*, 2007; CONNAN *et al.*, 2006.

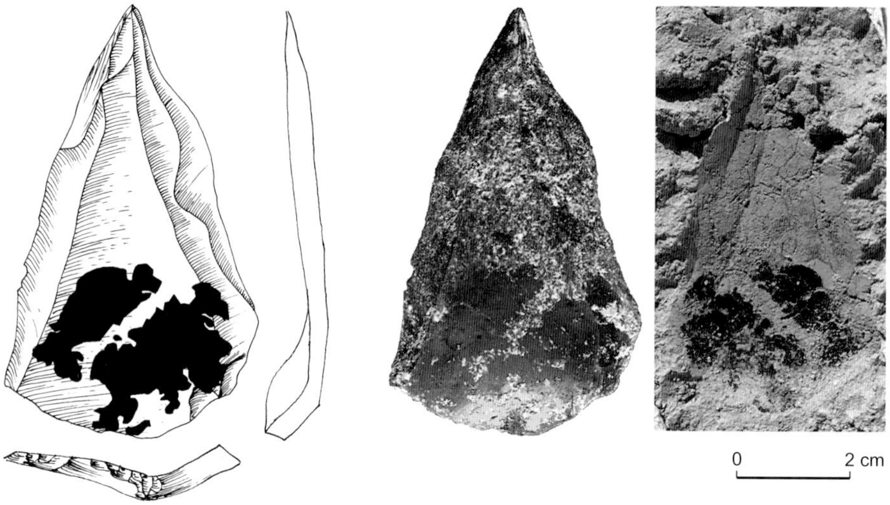

Fig. 6 – *Levallois point 2023 (square AJ 154 banquette) with imprint of bitumen deposited on the point's surface and on the soil.*

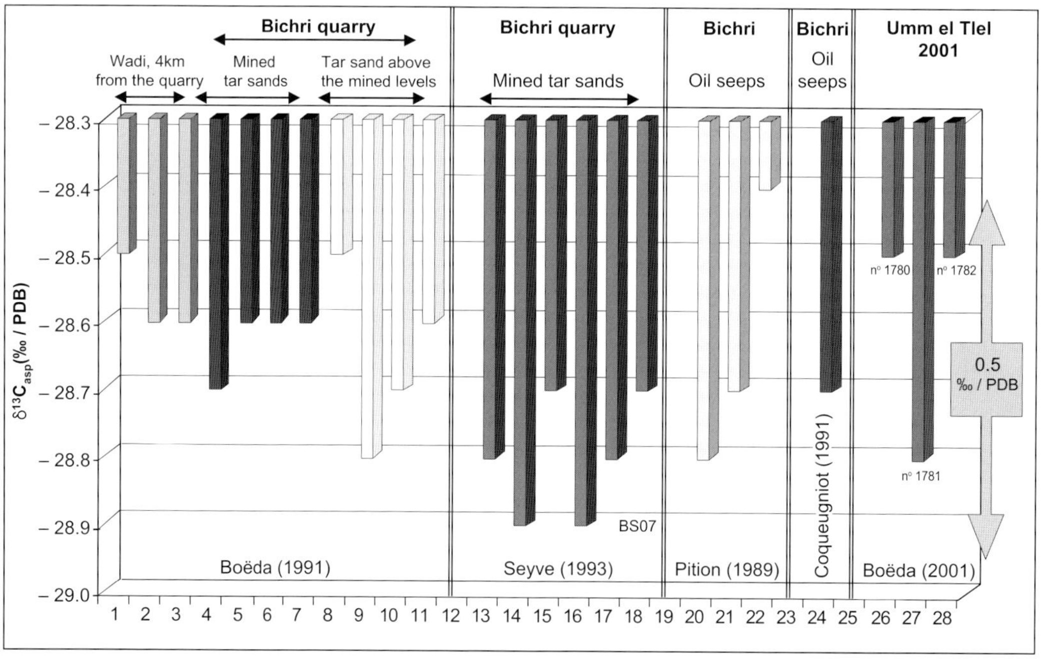

Fig. 7 – *$\delta^{13}C$ of asphaltenes (in ‰ / PDB) from various tar sands of the Djebel Bichri and from samples of the Umm el Tlel 2001 campaign. Number 1 to 11, 13 to 18, 20 to 22, 24 and 26 to 28 corresponds to sample numbers. Boëda (1991), Seyve (1993), Pition (1989) and Coqueugniot (1991) are referring to several sampling campaigns of tar sands in the Djebel Bichri. Boëda (2001) is the Umm el Tlel excavation campaign.*

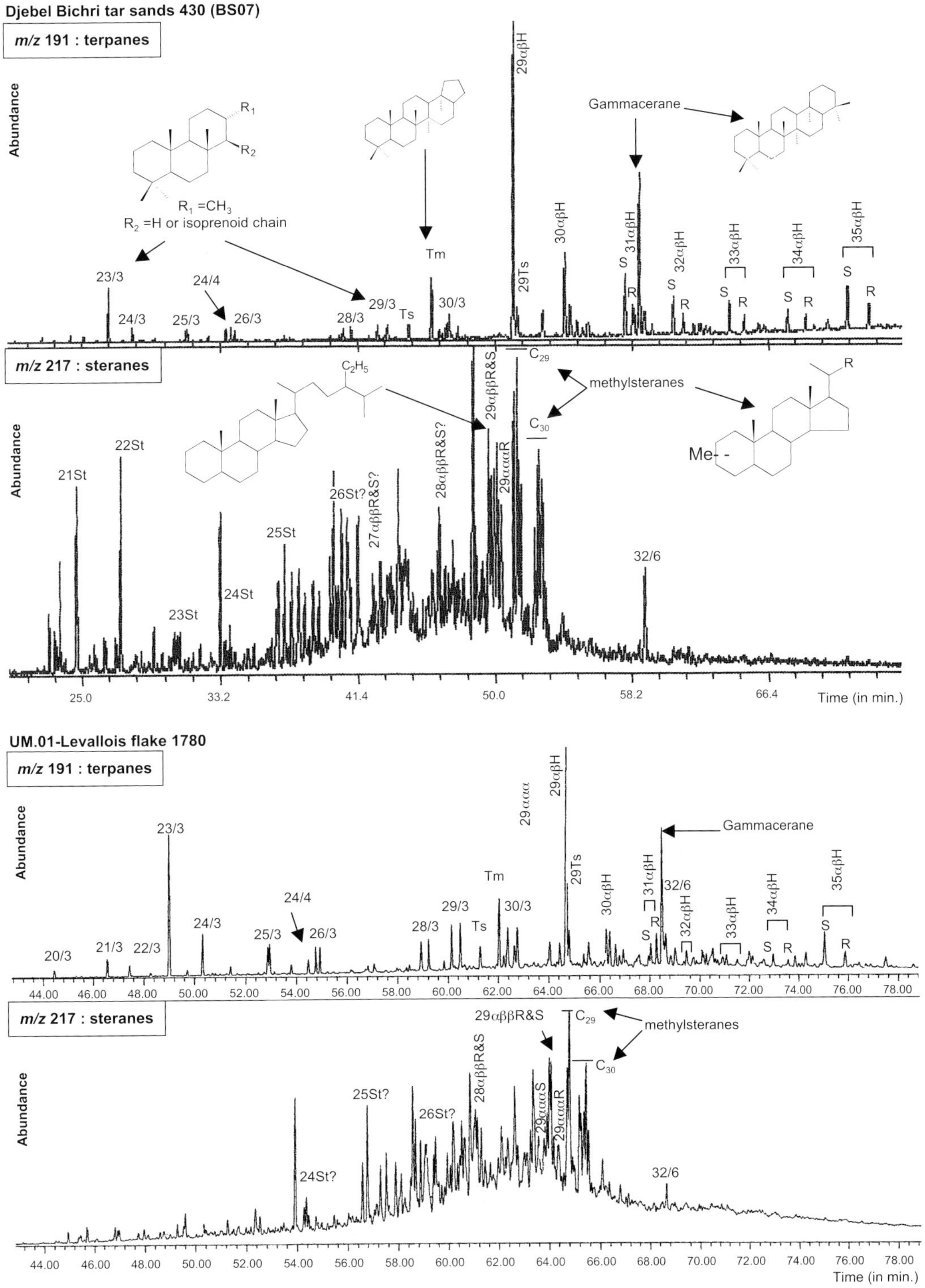

Fig. 8 – *Distribution patterns of steranes (m/z 217) and terpanes (m/z 191): comparison of a surface tar sand (n°430) from the Djebel Bichri to a bitumen (n°1780) isolated from a Levallois flake, dated 70,000 BP. Significance of abbreviations: C23/3 = C_{23} tricyclopolyprenane, 29αβH = 17α,21β-30-norhopane, 31αβHS = 17α,21β-22S-30-homohopane, 35αβHS = 17α,21β-22S-29-pentakishomohopane, 21St = 5α,14β,17β-pregnane, 23St = C_{23} sterane, C29αββR = 5α,14β,17β-20R-24-ethylcholestane, 32/6 = C_{32} hexahydrobenzohopane. Fingerprints are not directly comparable for the two acquisitions were not carried out at the same time. Umm el Tlel pattern was compared to older data of our data bank.*

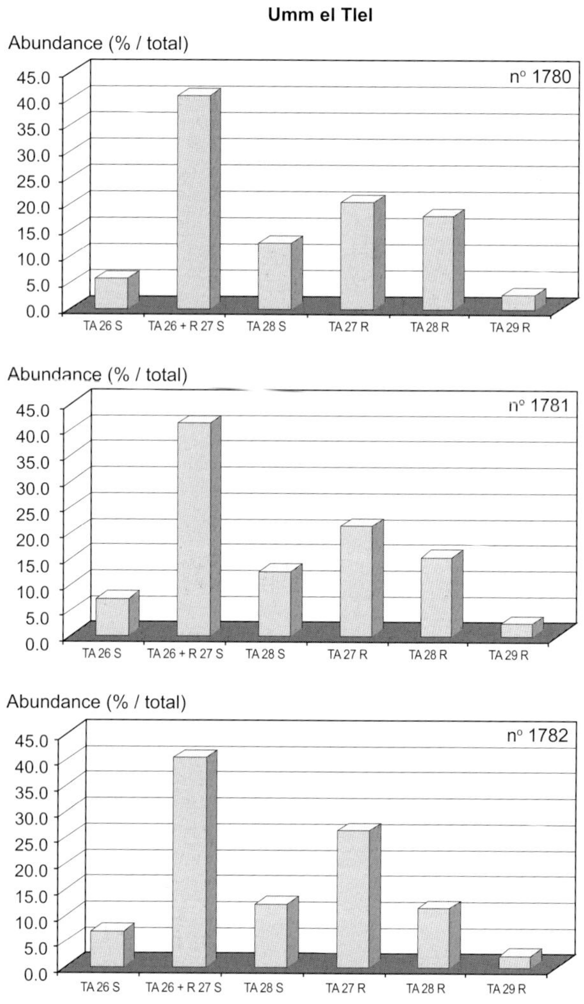

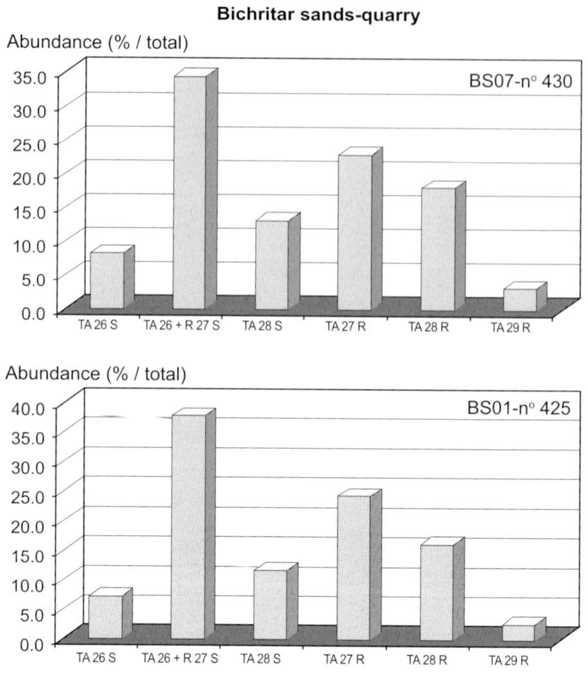

Fig. 9 – *Distribution patterns of triaromatic steroids calculated from the m/z 231 mass fragmentograms and expressed in % individual compound / total triaromatic steroids (identification after* RIOLO *et al.).*[21] *Significance of abbreviation: TA26S =* C_{26} *triaromatic steroid 20S.*

various locations in the Djebel Bichri (fig. 7). These values are significantly different from those recorded in samples of the other famous source of bitumen in the area namely Hit. $\delta^{13}C$ values of Hit bitumen vary between -28.2‰ and -28.3‰ as shown in a previous paper.[22] Within the asphaltenes of both geological tar sands and archaeological bitumen, the variations in $\delta^{13}C$ values do not exceed 0.5‰ / PDB. Therefore, the associated bitumen from Umm el Tlel samples likely came from the Djebel Bichri tar sand deposits.

To further substantiate this correlation, the molecular spectra of the C_{15+} alkanes and C_{15+} aromatics were acquired from GC-MS analysis. Sterane (mass fragmentogram *m/z* 217) and terpane (mass fragmentogram *m/z* 191) distribution patterns confirm the fossil character of the black organic matter and its origin from the Djebel Bichri tar sands. In fact among the 21 samples quoted in figure 7, only two samples (n°430 and n°170, fig. 8) show almost identical patterns as those observed in Umm el Tlel samples (fig. 8). Sample n°1780 was chosen as characteristic of the bitumen, but sample n°2023 exhibits the same diagnostic molecular fingerprints. Fragmentograms, reproduced in figure 8, have not been acquired at the same time for the tar sand data (n°430). However, the data were available from our data archives and permitted identification of comparable diagnostic peaks in each sample allowing to correlate the archaeological bitumen to tar sand. These patterns are characterized by the following diagnostic molecular features: occurrence of the complete series of tricyclopolyprenanes,[23]

21. RIOLO *et al.*, 1986.
22. CONNAN and NISHIAKI, 2003.
23. AQUITO NETO *et al.*, 1983.

very high concentration of gammacerane, reduced 17α,21β-hopane ($C_{30}\alpha\beta H$), and altered distribution of C_{27}-C_{29}-steranes by biodegradation that enhanced the concentration of C_{29}-C_{30}-methylsteranes. Comparison of these patterns with unaltered tar sand samples found in freshly mined samples in the Bichri quarry allowed us to ascribe the observed molecular changes to biodegradation from weathering processes. Geological samples that exhibit similar alteration patterns have crumbly consistencies that result from weathering and oxidation. This stage of degradation permits the flow of water through oil-stained sand resulting in biodegradation of the tar. Among the most severely altered samples, the Umm el Tlel samples are more biodegraded than their geological counterparts (n°430 and n°170) as shown by their Ts to Tm and GCR to $C_{30}\alpha\beta H$ ratios from GC-MS analysis (table 2). These ratios increase due to the preferential removal of the $\alpha\beta$-hopane family by biodegradation. As a consequence this observation suggests that Middle Palaeolithic people preferred to use tar sands outcropping at the surface and did not dig into the tar accumulation to obtain better preserved bitumen. An abundance of surface tar sand in the vicinity of their settlement provided sufficient material for their needs.

GC-MS analysis of C_{15+} aromatics confirms the source of the bitumen from Umm el Tlel. C_{15+} aromatics, although highly biodegraded, still contain several classes of diagnostic biomarkers, namely benzohopanes,[24] and mono- and triaromatic steroids.[25]

Well preserved triaromatic steroids also show similar GC-MS results (fig. 9) on Umm el Tel samples and Bichri tar sands from the mined quarry.

EXPERIMENTATION AND COMPARISON WITH ARCHAEOLOGICAL MATERIAL (LAYER VI 3D')

Assessment of the nature of micro-residues required a different analytical approach as there was insufficient sample material for geochemical analysis by GC-MS. Micro-residues were compared to bituminous micro-traces present on the three artefacts analyzed geochemically (n°1780, n°1781, n°2023, table 2), and to micro-traces obtained by experimentation. The bitumen utilized for these experiments is from the same source that was used during the Mousterian, *i.e.*, the Djebel Bichri tar deposits, about 40 km east of the site.

24. HUSSLER *et al.*, 1984.
25. RIOLO *et al.*, 1986; LEMOINE, 1996.

The experimental protocol applied was designed:

- to test hypotheses concerning bitumen use. Based on microscopic observations, it is highly probable that, during the Mousterian, bituminous mixtures were used either as an adhesive to attach handles or as a protective sleeve,
- to assess the use of these tools for butchering animals (skin and meat removal, disarticulation), cutting reeds (a common observation at the site) and scraping wood. These tools were used until their edges became dull, at which time they were disposed of or the cutting edge was, in some cases, renewed,
- to observe and analyze bitumen on flake surfaces, in particular in areas adjacent to handles or sleeves, using an optical reflecting microscope, with magnifications ranging from 50X to 200X.

Experiments were conducted as follows: after having heated bitumen cobbles mixed with quartz and gypsum to render it malleable, the material was applied in large quantities onto five pieces to obtain sleeves, and in lesser amount on five other flakes to serve as adhesive between the flint and wooden handles. Analytical results of these experiments have demonstrated that micro-residues of bitumen have different forms, namely:

- more or less thick and wide traces (fig. 10c) or elongated shapes (fig. 11c),
- very fine residues that are bonded together (fig. 12c),
- balls or cobbles (fig. 13c),
- thin plates varying in extension and fill (fig. 14c).

These micro-traces, varying in thickness and extent, are fairly reflective to light and can be either isolated or partially or completely associated.

These experimental data were then compared to microscopic observations on:

- the three Levallois artifacts for which the material was geochemically identified as bitumen (fig. 10a, fig. 11a, fig. 12a, fig. 13a, fig. 14a),
- Levallois artefacts from layer VI 3d' (within complex VI 3) where black micro-residues could not be analyzed by geochemical techniques (fig. 10b, fig. 11b, fig. 12b, fig. 13b, fig. 14b).

Results of these comparisons show that all archaeological residues are similar to experimental micro-residues (figs. 10-14) and are therefore very likely bitumen traces.[26]

Finally, the presence of the same observable microscopic traits on the 200 artefacts analyzed clearly attests to the generalized use of bitumen, which is abundant as raw material in the vicinity of the site.

26. In this article, only the results from bitumen analysis are presented. Research on the use and comparative observation of vegetal bitumen with the black micro-residues present on European artifacts is currently in progress.

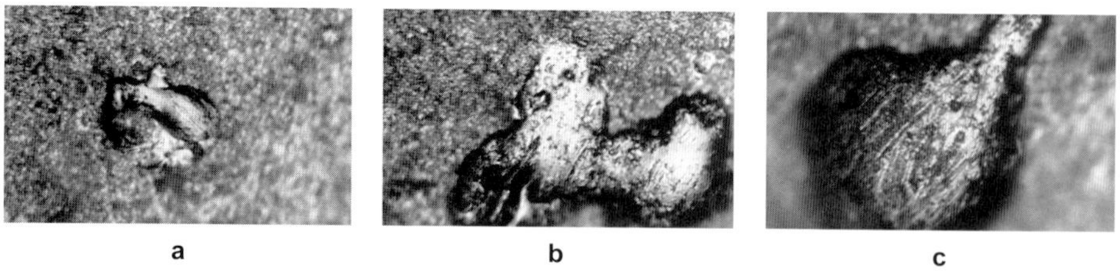

Fig. 10 – *Microphotographs (magnification of 200X) of residues of black material and bitumen. Relatively thick and wide micro-residues: **a**, archaeological bitumen; **b**, black archaeological material; **c**, experimental bitumen.*

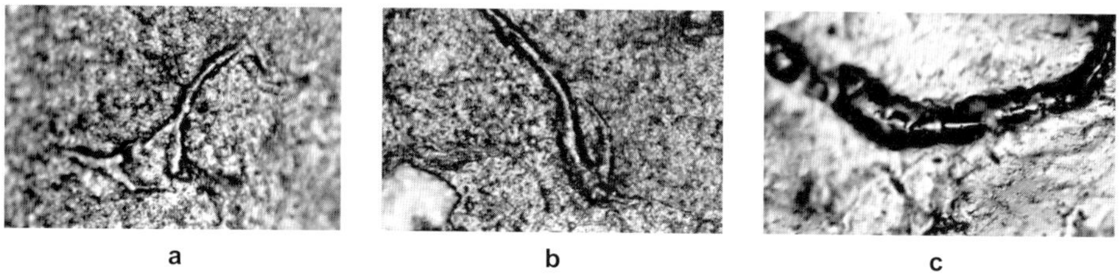

Fig. 11 – *Microphotographs (magnification of 200X) of residues of black material and bitumen. Elongated micro-residues: **a**, archaeological bitumen; **b**, black archaeological material; **c**, experimental bitumen.*

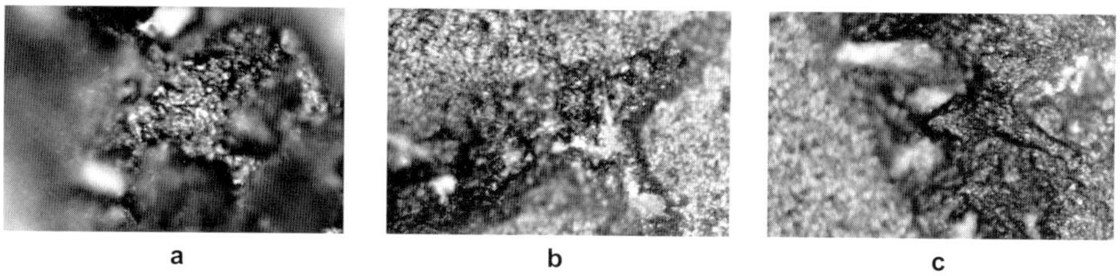

Fig. 12 – *Microphotographs (magnification of 200X) of residues of black material and bitumen. Fine micro-residues bonded together: **a**, archaeological bitumen; **b**, black archaeological material; **c**, experimental bitumen.*

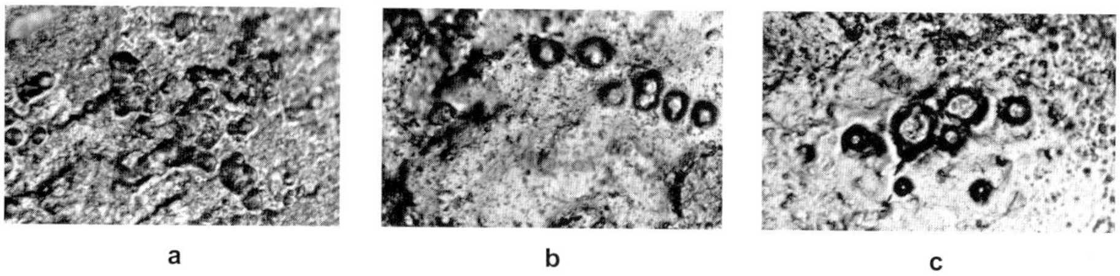

Fig. 13 – *Microphotographs (magnification of 200X) of residues of black material and bitumen. Balls: **a**, archaeological bitumen; **b**, black archaeological material; **c**, experimental bitumen.*

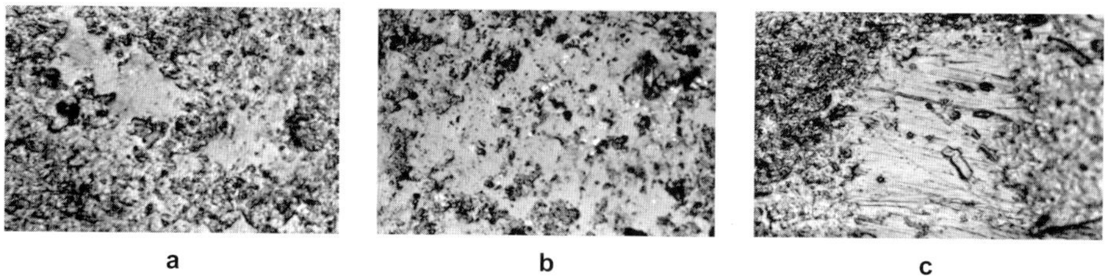

Fig. 14 – *Microphotographs (magnification of 200X) of residues of black material and bitumen. Fines deposited plates: **a**, archaeological bitumen; **b**, black archaeological material; **c**, experimental bitumen.*

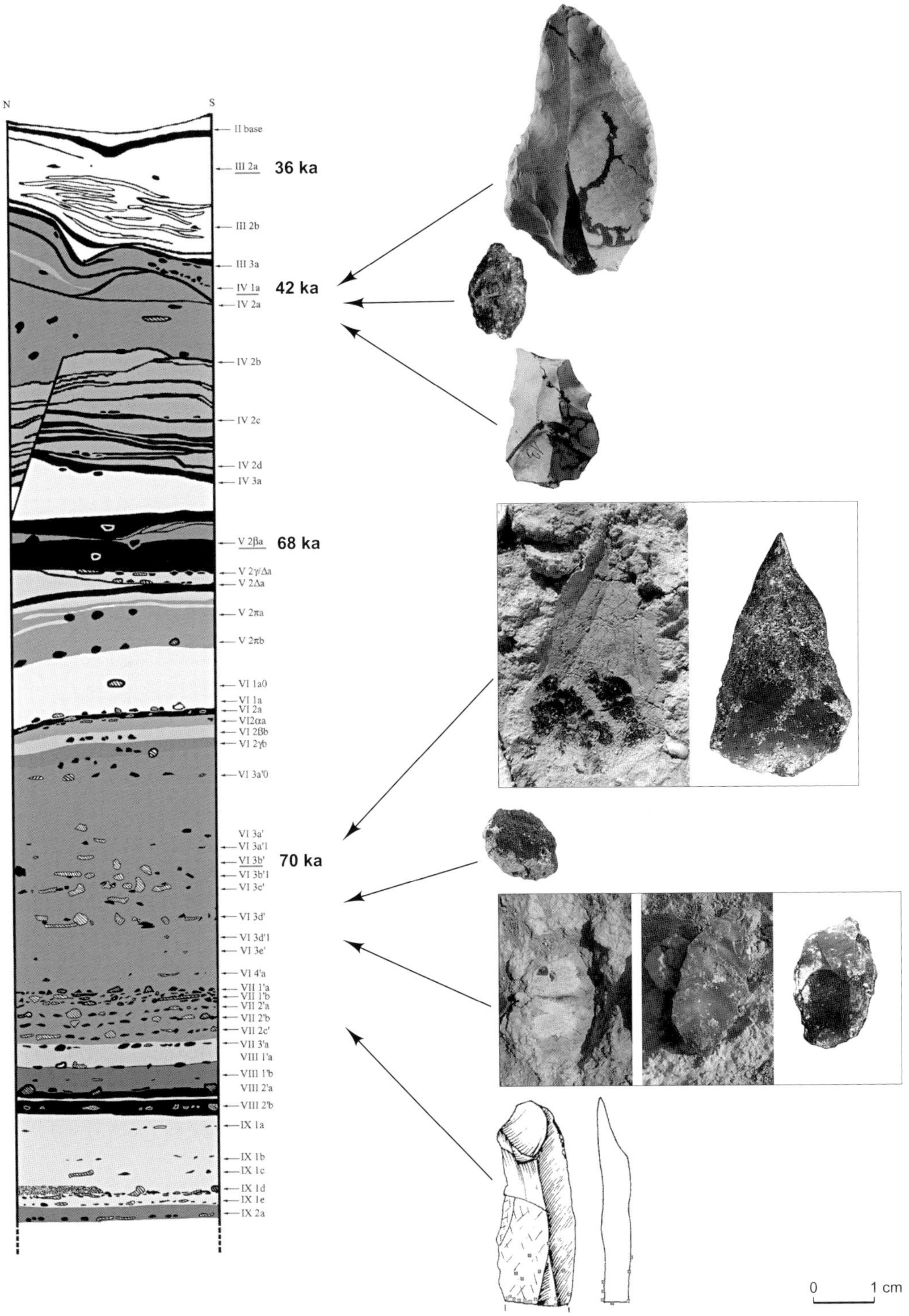

Fig. 15 – *Stratigraphy of Mousterian layers excavated through 2008. Layers with bituminous artifacts belong to an initial complex comprising layers VI1a, VI2a and IV2b*[27] *and to a second complex including all layers between VI 3a0 and VII 2c'.*[28]

27. BOËDA *et al.*, 1996 and 1998.
28. CONNAN, 1999.

CONCLUSIONS

These new analyses confirm definitively the use of bitumen during the Middle Palaeolithic at Umm el Tlel at the beginning of MIS 4, as well as its importance in the lithic technological system. Far from being a rare occurrence, it appears rather to be an omnipresent element in the use of predetermined Levallois artefacts. This raw material likely comes from sources more than 40 km east of the Bichri Mountains.

Nevertheless, this use cannot be generally applied to all 60 Mousterian layers currently known at Umm el Tlel, as it is absent in 15 layers dated between 71,000 and 42,000 BP. This absence does not have an obvious explanation. These layers are not distinguished by unusual techno-typological characteristics or by specific activities. The only pertinent element is that these 15 archaeological layers are all, based on typological, technological and use-wear analyses, culturally different; the four layers showing bitumen use at 40,000 BP are culturally identical, as are the eight layers of complex VI 3 dated to 71,000 BP (fig. 15). These observations, resituated in their sedimentary context, yield several interesting correlations, of which the most significant is that at each sedimentological change, we observe a change in the technocultural facies.[29] The corollary is that a single technocultural facies can be repeated x times in a single ensemble. This is the case for complexes IV 2 and VI 3 where the bitumen has been found. In other words, the use of bitumen could have been related to populations established in this region over a long period of time and who managed resources including bitumen throughout a vast peripheral territory although these unique technocultural facies are only evidence of nomadic populations circulating from one halt to another.

Very recent discoveries from underlying Mousterian complex VII seem to attest to an even older use of bitumen. All of these data confirm that bitumen was an integral part of the Near Eastern Mousterian technological world from its earliest phase. More generally, taking into account the data available for this period, we see that hafting and use of other tars, such as birch-bark pitch in Europe[30] were commonly known in different Neanderthal populations and perhaps Anatomically Modern Human ones. This use simply reflects a level of technological development in certain Mousterian populations that is too often underestimated.

29. These changes took place regardless of paleoenvironmental context. In other words, climatic change seems to have been a motor for displacement, but the type of climate does not appear to have been a determining element, likewise for site function.

30. GRÜNBERG, 2002; MAZZA *et al.*, 2006; MODUGNO *et al.*, 2006.

ANNEX: GEOCHEMICAL METHODS FOR BITUMEN ANALYSIS

SAMPLE PREPARATION

Samples for TOC and Rock-Eval analysis were ground, and material passing through a 60 mesh (250 micron) sieve was used for analysis. Samples were not high-graded prior to grinding.

TOTAL ORGANIC CARBON

Approximately 0.10 g of ground rock is accurately weighed and then treated with concentrated Hydrochloric Acid to remove carbonates. Sample weight may be varied depending on sample richness. The samples are left in acid for a minimum of two hours. The acid is removed from the sample with a filtration apparatus fitted with a glass microfiber filter. The filter is placed in a LECO crucible and dried at 110°C for a minimum of one hour. After drying the sample is analyzed with a LECO 600 Carbon Analyzer.

ROCK-EVAL

Total organic carbon (TOC) and Rock-Eval pyrolysis were performed to assess source rock quality and thermal maturity. In Rock-Eval pyrolysis, ground samples are heated in an inert environment to measure the yield three groups of compounds (S1, S2, and S3), measured as three peaks on a pyrogram. Sample heating at 300°C for 3 min. produces the S1 peak by vaporizing the free (unbound) hydrocarbons. High S1 values indicate either large amounts of kerogen-derived bitumen (as in an active source rock) or the presence of migrated hydrocarbons. The oven then increases in temperature by 25°C/minute to 600°C, and the S2 and S3 peaks are measured from the pyrolytic degradation of the kerogen in the sample. The S2 peak is proportional to the amount of hydrogen-rich kerogen in the rock, and the S3 peak measures the carbon dioxide released (to 390°C) providing an assessment of the oxygen content of the rock. The temperature at which the S2 peak reaches a maximum, "Tmax", is a measure of the source rock maturity. Accuracy of Tmax is 1-3°C, depending on the instrument, program rate and sample size. Tmax also varies by organic matter type. Tmax values for samples with S2 peaks less than 0.2 mg HC/g rock are often inaccurate and should be rejected unless a definitive kerogen peak is noted from the pyrogram. A summary of operating conditions and analytical results from Rock-Eval pyrolysis follows.

Operating Conditions
S1: 300°C for 3 min.,
S2: 300°C to 600°C at 25°C/min; held at 600°C for 1 min.,

S3: trapped between 300 to 390°C,
S4: oxidation at 580°C for 12 min.

RESULTS

S1: free oil content (mg hydrocarbons per gram of rock),
S2: remaining hydrocarbon potential (mg hydrocarbons per gram of rock),
S3: organic carbon dioxide (mg CO_2 per gram of rock),
TOC: total organic carbon content (wt. %),
Tmax: temperature at maximum evolution of S2 hydrocarbons,
Cal. %Ro: calculated vitrinite reflectance based on Tmax,
Ratios: hydrogen index (HI), oxygen index (OI), production index (PI), S2/S3, and S1/TOC.

EXTRACTION AND LIQUID CHROMATOGRAPHY

Samples were extracted using dichloromethane as a solvent. Aliquots of the extracts were separated by liquid chromatography as a preparatory step for biomarker analysis. The asphaltene fraction of the sample was recovered by precipitation in pentane. The pentane soluble portion of the sample (the maltenes) was further separated by liquid chromatography into C_{15+} saturates, C_{15+} aromatics and resin fractions with fractions quantified by gravimetric analysis.

BIOMARKERS

Samples are prepared by performing liquid chromatography (LC) on oil or extracts samples to isolate C_{15+} saturates and C_{15+} aromatics. C_{15+} saturates are diluted with 5 µl and C_{15+} aromatics/whole extracts/whole oils with 10 µl of carbon disulfide per mg of fraction. β-cholane and o-terphenyl-d14 are used as saturate and aromatic internal standards, respectively. Samples are analyzed on an HP6890GC / 5973MSD system equipped with a high-resolution column (60 m, DB-1 phase, 0.2 mm *i.d.*, 0.2 µm film). The GC is temperature programmed from 80°C (2 min. hold) to 320°C at 3.5°C/min. with a final hold time of 20 min. There is a 7.5 min. solvent delay on acquisition and selected ion monitoring (SIM) is the standard method for analysis. Reports of peaks and standard geochemical parameters are included for both the saturate and aromatic fractions.

ACKNOWLEDGMENTS

We thank the French Minister of Foreign Affairs, the General Direction of Antiquities and the Museums of Syria for their financial support, and Total E&P Syria for their logistical support.

Éric BOËDA – Stéphanie BONILAURI
CNRS UMR 7041 – ArScAn, équipe AnTET
Maison de l'Archéologie et de L'Ethnologie
21, allée de l'Université
92023 Nanterre Cedex – FRANCE
eric.boeda@wanadoo.fr
stephanie.bonilauri@wanadoo.fr

Jacques CONNAN
Laboratoire de Biogéochimie Moléculaire
CNRS UMR 7177
Université Louis Pasteur
25, rue Becquerel
67200 Strasbourg – FRANCE
connan.jacques@orange.fr

Dan JARVIE
Worldwide Geochemistry, LLC
P.O. Box 789
Humble, Texas 77347 – USA
danjarvie@wwgeochem.com

Norbert MERCIER
Institut de Recherche sur les Archéomatériaux
CNRS UMR 5060
Centre de Recherche en Physique Appliquée à l'Archéologie (CRP2A)
Maison de l'Archéologie
Université de Bordeaux
33607 Pessac Cedex – FRANCE
Norbert.Mercier@u-bordeaux3.fr

Mark TOBEY
Encana Oil & Gas, USA Inc., 370 17th St., Suite 1700
Denver, CO 80303 – USA
Mark.Tobey@encana.com

Hélène VALLADAS
Laboratoire des Sciences du Climat et de l'Environnement
LSCE/IPS, UMR CEA-CNRS-UVSQ
Centre des faibles Radioactivités
Avenue de la Terrasse
91198 Gif-sur-Yvette Cedex – FRANCE
helene.valladas@lsce.ipsl.fr

Heba al SAKHEL
Musée national de Damas
Ministère de la Culture
Direction générale des Antiquités et des Musées
Shoukry al-Qouwatly St.
Damas – SYRIE
h.alsakhel@ifporient.org

BIBLIOGRAPHY

ADIAMEC G. and AITKEN M.J.

1998 Dose-rate conversion factors: update. *Ancient TL* 16: 35-70.

AITKEN M.J.

1985 *Thermoluminescence Dating*. London: Academic Press.

AQUITO NETO F.R., TRENDEL J.-M., RESTLE A., CONNAN J. and ALBRECHT P.

1983 Occurrence and formation of tricyclic terpanes in sediments and petroleums. *In*: BJORØY M. (ed.), *Advances in Organic Geochemistry 1981*: 659-667. Chichester: John Wiley and Sons Ltd.

BOËDA É. et MUHESEN S.

1993 Umm El Tlel (El Kowm, Syrie): étude préliminaire des industries lithiques du Paléolithique Moyen et Supérieur, campagne 1991-1992. *In: Cahiers de l'Euphrate* 7: 47-91. Paris: ERC.

BOËDA É., CONNAN J. and MUHESEN S.

1998 Bitumen as hafting material on Middle Paleolithic artefacts from the El Kowm Basin, Syria. *In*: AKAZAWA T., AOKI K. and BAR-YOSEF O. (eds.), *Neandertals and Modern Humans in Western Asia*: 181-204. New York: Plenum.

BOËDA É., GRIGGO C. et NOËL-SORIANO S.A.

2002 Différents modes d'occupation du site d'Umm el Tlel au cours du Paléolithique moyen (El Kowm, Syrie centrale). *Paléorient* 7,2: 13-28.

BOËDA É., CONNAN J., DESSORT D., MUHESEN S., MERCIER N., VALLADAS H. and TISNERAT N.

1996 Bitumen as a hafting material on Middle Palaeolithic artefacts. *Nature* 380: 336-338.

BOËDA É., BONILAURI S., CONNAN J., JARVIE D., MERCIER N., TOBEY M., VALLADAS H., AL SAKHEL H. and MUHESEN S.

2008 Middle Palaeolithic bitumen use at Umm el Tlel around 70 000 BP. *Antiquity* 82: 853-861.

BONILAURI S., BOËDA É., GRIGGO C., AL-SAKHEL H. et MUHESEN S.

2007 Un éclat de silex moustérien coincé dans un bassin d'autruche à Umm el Tlel (Syrie centrale). *Paléorient* 33,2: 39-46.

CONNAN J.

1999 Use and trade of bitumen in antiquity and prehistory: molecular archaeology reveals secrets of past civilizations. *Phil. Trans. R. Soc. Lond. B* 353: 33-50.

CONNAN J. and CARTER R.

2007 A geochemical study of bituminous mixtures from Failaka and Umm an-Namel (Kuwait), from the Early Dilmun to the Early Islamic period. *Arab. arch. epig.* 18: 139-181.

CONNAN J. et DESCHESNE O.

1996 *Le bitume à Suse – Collections du Musée de Louvre*. Pau / Paris: Coédition Elf Aquitaine / RMN.

CONNAN J. and NISHIAKI Y.

2003 The bituminous mixture of Tell Kosak Shamali on the Euphrates (Syria) from the Early Ubaid to the Post Ubaid: composition of mixtures and origin of bitumen. *In*: NISHIAKI Y. and MATSUTANI T. (eds.), *Tell Kosak Shamali-The Archaeological Investigations on the Upper Euphrates, Syria, Chalcolithic Technology and Subsistence, vol. II*: 283-306. Tokyo: The University Museum and the University of Tokyo.

CONNAN, J., NISSENBAUM A., IMBUS K., ZUMBERGE J. and MACKO S.

2006 Asphalt in Iron Age excavations of the Philistine Tel Miqne-Ekron city (Israel): origin and trade routes. *Organic Geochemistry* 37,17: 1768-1786.

CONNAN J., LOMBARD P., KILLICK R., HØJLUND F., SALLES J.-F. and KHALAF A.

1998 The archaeological bitumens of Bahrain from the Early Dilmun period (c. 2200 BC) to the sixteenth century AD: a problem of sources and trade. *Arab. arch. epig.* 9: 141-181.

ESPITALIE J. and BORDENAVE M.L.

1993 Rock-Eval pyrolysis. *In*: BORDENAVE M.L. (ed.), *Applied Petroleum Geochemistry*, Chapter II-2: 237-261. Paris: Éditions Technip.

GRÜNBERG J.-M.

2002 Middle Palaeolithic Birch-Bark Pitch. *Antiquity* 76: 15-16.

HUSSLER G., CONNAN J. and ALBRECHT P.

1984 Novel families of tetra-and hexacyclic aromatic hopanoids predominant in carbonate rocks and crude oils. *Org. Geochem.* 6: 39-49.

JORON J.L.

1974 *Contribution à l'analyse des éléments en traces dans les roches et les minéraux par activation neutronique: Application à la caractérisation d'objets archéologiques*. Paris: Université Paris-Sud 11. Thèse non publiée.

LEMOINE S.

1996 *Oxydation abiotique de bitumes pétroliers en milieu naturel: aspects moléculaires et macromoléculaires*. Strasbourg: Université Louis Pasteur 1. Thèse non publiée.

MARTISON P.-M., PISIAS N.-G., HAYS J.-D., IMBRIE J., MOORE T.C. and SHACKLETON N.-G.

1987 Age dating and the orbital theory of the ice ages, development of a high resolution 0 to 300,000-year old chronostratigraphy. *Quaternary Research* 27: 1-29.

MAZZA P.P.A., MARTINI F., SALA B., MAGI M., COLOMBINI M.P., GIACHI G., LANDUCCI F., LEMORINI C., MODUGNO F. and RIBECHINI E.

2006 A new Palaeolithic discovery: tar-hafted stone tools in a European Mid-Pleistocene bone-bearing bed. *Journal of Archaeological Science* 33: 1310-1318.

MERCIER N., VALLADAS H. and VALLADAS G.

1992 Observations on paleodose determination with burnt flints. *Ancient TL* 10: 28-32.

MODUGNO, F., RIBECHINI, E. and COLOMBINI M.P.

2006 Chemical study of triterpenoid resinous materials in archaeological findings by means of direct exposure electron ionisation mass spectrometry and gas chromatography/mass spectrometry. *Rapid Commun. Mass Spectom.* 20: 1787-1800.

PRESCOTT J.-R. and HUTTON J.-T.

1994 Cosmic ray contributions to dose rates for luminescence and ESR dating: large depths and long-term time variations. *Radiat. Meas.* 23: 497-500.

RIOLO J., HUSSLER G., ALBRECHT P. and CONNAN J.
1986 Distribution of aromatic steroids in geological samples: their evaluation as geochemical parameters. *Org. Geochem.* 10: 981-990.

STERN B., CONNAN J., BLAKELOCK E., JACKMAN R., CONINGHAM R.A.E. and HERON C.
2007 From Susa to Anuradhapura: reconstructing aspect of trade and exchange in bitumen-coated ceramic vessels between Iran and Sri Lanka from the third to the ninth centuries AD. *Archaeometry* 50,3: 409-428.

VALLADAS G. et VALLADAS H.
1982 Effet de l'irradiation alpha sur des grains de quartz. *In*: *Second specialist seminar on thermoluminescence dating*: 171-178. Strasbourg : Conseil de l'Europe (*PACT* 6).

VALLADAS G., MERCIER N. and LETUVE R.
1994 A simple semi-automatic TL apparatus of new design. *Ancient TL* 12: 39-40.

VALLADAS H.
1992 Thermoluminescence dating of flint. *Quat. Sci. Reviews* 11: 1-5.

From Neolithic to Chalcolithic in the Southern Caucasus: Economy and Macrolithic Implements from Shulaveri-Shomu Sites of Kwemo-Kartli (Georgia)

C. Hamon

Abstract: *In Transcaucasia, the emergence of the first neolithic cultures can be dated to the middle of the 6th millennium BC. The Shulaveri-Shomu culture, which occupies the middle course of the Kura valley, is among the best known Neolithic culture of Transcaucasia. As for the subsistence economy, the presence of a wide range of grinding and abrading tools is mentioned, but no real study of their technological evolution and functional implications has yet been carried out. The present work aims at better defining the techniques of cereal processing and the exploitation of mineral and animal resources through the stone assemblages of four sites of the Kwemo-Kartli region in Georgia. The implication of macrolithic tools in the technical system of the Shulaveri-Shomu culture is also discussed in terms of relative chronology of the sites and comparison with other neolithic and chalcolithic cultures from Transcaucasia. Using the macrolithic implements, the question of the relationship between the Shulaveri-Shomu tepe culture and the southern regions of the Near and Middle East is discussed.*

Résumé : *En Transcaucasie, l'émergence des premières cultures néolithiques date du VI*e *millénaire BC. Celle de Shulaveri-Shomu, qui occupe le cours moyen de la Kura, compte parmi les cultures néolithiques les mieux connues de la région. Dans le domaine de l'économie de subsistance, un large éventail d'outils de broyage et de polissage est mentionné mais jusqu'à présent aucune étude de leur évolution technique et de leurs implications fonctionnelles n'avait été conduite. Par une analyse des assemblages en pierre de quatre sites de Kwemo-Kartli en Géorgie, le présent article vise à mieux définir les techniques de préparation des céréales et d'exploitation des ressources animales et minérales. L'implication de l'outillage macrolithique dans le système technique de la culture de Shulaveri-Shomu est également discuté en termes de chronologie relative et de comparaison avec d'autres cultures néolithiques et chalcolithiques de Transcaucasie. La question des relations entre celle de Shulaveri-Shomu et les cultures des régions plus méridionales du Proche et Moyen-Orient est discutée.*

Keywords: *Caucasus, Neolithic, Shulaveri-Shomu, Macrolithic implements, Subsistence economy.*
Mots-clés : *Caucase, Néolithique, Shulaveri-Shomu, Macro outillage lithique, Économie de subsistance.*

INTRODUCTION

In the southern Caucasus or Transcaucasia, the emergence of the first Neolithic cultures is dated to the end of the 7th millennium cal. BC or to the very beginning of the 6th millennium cal. BC.[1] Two main processes of neolithisation have been recognized, respectively in the western and eastern parts

1. Badalyan *et al.*, 2007; Lyonnet, 2007; C. Chataigner, personal communication.

Manuscrit reçu le 14 avril 2008, accepté le 25 novembre 2008

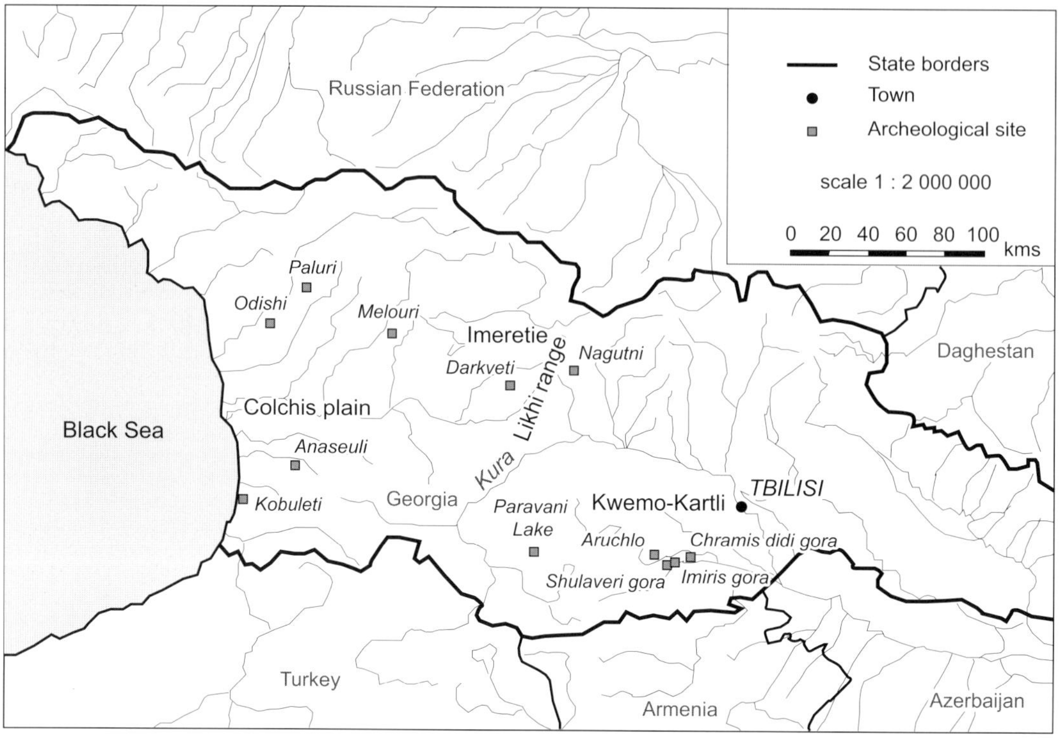

Fig. 1 – *Map of localization of the Shulaveri-shomu sites studied in Georgia.*

of present-day Georgia. Their territorial expansion broadly follows the major physical and climatic regions, showing a close relationship between these Neolithic populations and their environment. The lack of connections between these two entities can be partially explained by the landscape itself. Eastern and western Georgia are physically separated by the Likhi range which forms a boundary between the eastern open steppe plain, with its arid continental climate, and subtropical climate of the Black Sea shore and the rainy mountainous regions (fig. 1).

DEFINITION OF THE NEOLITHIC IN TRANSCAUCASIA

The chronological framework of the Georgian Neolithic and Chalcolithic (also called "Eneolithic") is still discussed today. In South Caucasus, the Neolithic has been diversely and successively determined by:

- a certain type of pottery (which by definition excluded pre-ceramic populations),
- a specific flint and obsidian industry (different from Mesolithic industries),
- the adoption of a sedentary way of life (which excluded pastoral and semi-nomadic populations),
- a subsistence economy based on agriculture.

This explains why some sites have been attributed successively to the Neolithic or the Chalcolithic.[2] These approximate definitions, together with a lack of stratigraphy or reliable calibrated radiocarbon dates, have generated two important problems in research on the southern Caucasus:

- differentiation between Mesolithic and pre-ceramic Neolithic populations, characterized by a specific lithic industry including axes and grinding tools, especially in western Georgia,
- differentiation between the end of the Neolithic and the beginning of the Chalcolithic, reinforced by continuity of occupation on the "gora" or "tepe" (tell settlements) and by uniformity in material culture, especially on the Shulaveri-Shomu sites (metal objects excluded).

Our study adopts the following definition of the Neolithic: a transformation of lifestyle, with a production economy based

2. A synthesis of the discussions on this topic can be found in CHATAIGNER, 1995 and KIGURADZE and MENABDE, 2004

on plant and animal domestication.[3] In some cases, this was accompanied by the introduction of new techniques (ceramics), the widespread use of some others (stone and bone polishing) and by change in settlement patterns. Throughout this paper, evidence of metal production is considered diagnostic of the Chalcolithic.

THE NEOLITHIC OF EASTERN AND WESTERN GEORGIA

Located in eastern Georgia, the Shulaveri-Shomu culture is one of the best known Neolithic cultures of Transcaucasia. Its central geographical distribution along the middle course of the Kura valley, between eastern Georgia and south Azerbaijan, makes it of prime importance for the study of the Neolithic of the whole Transcaucasian area. Preceded by the Nagutni preceramic culture, the Shulaveri-Shomu culture is characterized by circular mud-brick houses, domestic animals and cereals. The material culture includes handmade pottery with incised and relief decoration, an obsidian industry based on the production of regular blades with a high proportion of burins and scrapers, and large numbers of bone and antler implements.[4] The architecture and material culture are evidence for exchange and technical transfers between the Shulaveri-Shomu culture and Anatolian or Near Eastern Neolithic cultures.

Compared to the Shulaveri-Shomu culture, the adoption of a Neolithic economy in western Transcaucasia follows a completely different pattern since agriculture, pottery manufacture and new social structures do not appear simultaneously. Different cultural groups, broadly corresponding to the major landscape zones (Black Sea shores, Colchida plain and Imeretie) share similar economic features. Their definition, relationships and dating are still unclear and their specificities are difficult to interpret in terms of chronology or regional factors. Although most sites are open-air settlements (Odishi, Anaseuli I and II, Kobuleti, Melouri), much of the fauna and plant remains come from rock shelters (Darkveti). Pre-ceramic occupations are characterized by two types of lithic production: flake production (blades, chisels and scrapers) is found in the highlands, whereas blade production characterizes the Colchis plain and the Black Sea shores (Anaseuli, Melouri).[5] Pottery appears in the upper levels of the Imeretian rock-shelters and in the so-called Late Neolithic of the Black Sea shores (Odishi). Blades and large numbers of polished tools are found together with querns and grinders. The lack of organic remains explains our partial knowledge of the economy of these populations. Taking into consideration the material culture (the association of microlithic elements, polished adzes and ceramics), these groups probably participated in the process of neolithisation of the eastern shores of the Black sea.[6] Nevertheless, they show very few relationships with the south-eastern Shulaveri-Shomu culture.

RESEARCH OBJECTIVES

The aim of this article is to discuss the role of macrolithic implements in the subsistence economy and techniques of the Shulaveri-Shomu culture. Our study is based on a typo-technological and use-wear analysis[7] of macrolithic implements found on four sites of this culture.

This work follows developments in research at an international level over the last fifteen years on the functional analysis of macrolithic tools.[8] In the Near East, the pioneer work of M.C. Nierlé and K. Wright[9] laid the grounds for a typological and technological classification of macrolithic implements, regardless of their function. Despite the recent work of L. Dubreuil on the function of Natufian macrolithic tools,[10] this functional approach has still to be applied to Neolithic contexts.[11]

In this study, we have adopted an international terminology proposed in a recent article for the study of so-called "macrolithic tools" and their typological, technological and functional analysis.[12] The term "macrolithic tool" or "macrolithic implement" includes grinding, pounding, crushing and polishing tools, also called "ground stone tools"[13] or "non-chipped stone tools", depending on the authors and the geographical area concerned.[14]

Cereal and plant processing constitute a primary focus of research. This question is all the more crucial as the cereals cultivated in Transcaucasia are much more diverse than

3. Following A. LEROI-GOURHAN, 1988.
4. KIGURADZE, 1986; KIGURADZE and MENABDE, 2004.
5. KIGURADZE and MENABDE, 2004.
6. KUSHNAREVA, 1997: 24-29; LYONNET, 2004: 91; NIEBIERIDZE, 1972, 1978 and 1986.
7. HAMON, 2006 and 2008.
8. ADAMS, 2002; DE BEAUNE, 2000; DUBREUIL, 2004; FULLAGAR, 1998; PROCOPIOU, 1998; RISCH, 2002.
9. NIERLÉ, 1982; WRIGHT, 1992.
10. DUBREUIL, 2004.
11. WRIGHT, 2008.
12. ADAMS *et al.*, in press.
13. WRIGHT, 1992; ADAMS, 2002.
14. See for more details in HAMON, 2008.

Table 1 – *Table of chronological correspondance between the different levels of Kwemo-kartlian sites, Transcaucasian and Anatolian sites and cultures after Kiguradze 1986 and 2001.*

Stages	Shulaveri	Imiris	Khramis Didi	Aruchlo I	Transcaucasia	Western Anatolia	South Anatolia	North Mesopotamia
I	Shulaveri IX-IV					Hacılar VI-II	Çatal Hüyük X-IX	Samarra
II	Shulaveri III-I	Imiris VII-VI						
III		Imiris V				Hacılar I		Hassuna Ia-c
IV		Imiris IV-I	Khramis Didi IX-VI		Shomu tepe, Toire tepe			Hassuna- Halaf culture
V			Khramis Didi V-I	Aruchlo I	Kul tepe I, Tekhut			

in Anatolia and Mesopotamia. Cereal processing may have required specific grinding slabs or mortars, adapted to tasks such as dehusking, pounding or grinding. Consequently, study of grinding and pounding equipment can provide information about food preparation and change in dietary habits through the sequence.

Macrolithic tools can also tell us about the activities taking place on Shulaveri-Shomu sites. For instance, can we identify temper or colour processing on grinding and pounding tools? What is the role of abrasion in the shaping of bone and stone tools? We will examine how the identification of such activities can be interpreted in terms of economy, site status and chronological trends.

Finally, we will consider the similarities between the macrolithic implements from the Kwemo-Kartli region and those from other Neolithic cultures of Transcaucasia, as well as their relationship with Chalcolithic cultures of south-east Georgia. Is there any continuity in the practices and techniques used between the two periods? Can we distinguish regional specificities in the macrolithic tools? Their role in the Shulaveri-Shomu technical system and economy is discussed here through comparison with Neolithic cultures from western Transcaucasia and the Near East. Comparisons between the Shulaveri-Shomu complex and the contemporaneous Halaf and Hassuna cultures in north Mesopotamia, Jarmo in the Zagros and Hacılar in Anatolia have been made frequently (table 1). According to Kiguradze, the finds from the earliest layers of Shulaveri gora bear some close resemblances to material from the site of Hacılar.[15] Some ceramic decoration patterns at Imiris gora are similar to Umm Dabaghiyah productions.[16] The anthropomorphic figurines of the upper levels of Khramis Didi gora also evoke examples of the Hassuna, Halaf and Samarra cultures.[17] Despite some similarities in the productions of all these cultures, the nature of their relationship is still not clearly established. What about macrolithic tools? Is there, for example, a complete similarity between the Shulaveri-Shomu cereal processing and that of Anatolia or Mesopotamia or do some distinctions reflect regional variations linked to the local economy and tradition?

15. KIGURADZE, 1986.
16. KAVTARADZE, 1981.
17. GLONTI *et al.*, 1975.

THE SHULAVERI-SHOMU CULTURE AND THE NEOLITHIC OF TRANSCAUCASIA

ORIGIN AND DEVELOPMENT OF THE NEOLITHIC IN TRANSCAUCASIA

The first Neolithic of central and south-east Georgia

The first Neolithic of central Georgia is characterized by pre-ceramic settlements with an elaborate and non-Mesolithic flint or obsidian industry. Because of a lack of reliable radiocarbon dates, the Nagutni sites (Shida-Kartli) were first considered Late Mesolithic but are now attributed to the Early Neolithic. Recent evidence suggests they date to before the Shulaveri-Shomu culture,[18] even if a parallel development of the two cultures cannot be excluded for the end of the 6th millennium BC cal. The economic and functional status of these sites is still under discussion, and closely resembles the western sites of the Paluri group.

The sites of Paravani I and II may correspond to seasonal obsidian exploitation on the shores of Lake Paravani.[19] The origin of these groups is still undetermined. Do they correspond to the emergence of a local Neolithic culture that would have been swept away by the Neolithic communities from the Near East, at the origin of the development of the Shulaveri-Shomu culture? Or did they coexist with the first Shulaveri-Shomu populations in the region? Further investigations in Georgia

18. After KIGURADZE and MENABDE, 2004; C. Chataigner, personnal communication.
19. KIGURADZE and MENABDE, 2004.

and Armenia should provide new elements for a discussion of these issues.

The Shulaveri-Shomu culture

The Shulaveri-Shomu culture occupied the middle course of the Kura (Mtkvari) from the 6th to the beginning of the 5th millennium cal. BC. Today it corresponds to south-east Georgia and west Azerbaijan but it also extends southward into the Armenian regions of the Ararat valley.[20] This territory broadly prefigures the geographical distribution of the later Sioni (Chalcolithic) and Kuro-Araxe (Early Bronze Age) complexes. Georgian sites of the Shulaveri-Shomu complexe were first considered Chalcolithic, but this culture is now included in the Neolithic, except for the latest layers where metal artefacts have been found, as for example in the upper levels of Khramis Didi gora and Aruchlo I. In fact, the Shulaveri-Shomu culture has a similar economic basis to contemporary Neolithic cultures of eastern Anatolia.

THE SHULAVERI-SHOMU CULTURE SITES OF THE KWEMO-KARTLI REGION

Sites and excavations

In Georgia, the Shulaveri-Shomu culture is mostly known through excavations conducted in the Kwemo-Kartli region (fig. 1). The sites are generally located on the most fertile land alongside rivers, not far from the foothills. The sites of Shulaveri, Imiris and Khramis Didi gora were excavated by a team from the National Museum of Georgia, directed first by O. Djaparidze and A. Djavaxishvili and then by T. Kiguradze and M. Menabde.

In Aruchlo I, excavations and sondages were first conducted under the direction of T. Chubinishvili between 1966 and 1976 and D. Gogelia between 1978 and 1985. Since 2005, new excavations are in progress.[21]

The chronology of the Kwemo-Kartlian sites is based on ten levels dated to the Neolithic, and on two upper levels dated to the beginning of the Chalcolithic.[22] These levels have been recognized on four main sites: Shulaveri gora and Imiris gora (Kwemo-Shulaveri group), Khramis Didi gora (Kachagani group) and Aruchlo I (Nachiduri group). Some smaller or little-known sites such as Dangreuli gora (only known through two small trial trenches) or Gadačrili gora (Shulaveri group) complete our knowledge of this regional entity. Most of these sites have been partly disturbed by later occupations dating from the Bronze Age, the Roman period or the Middle Ages.

Stratigraphy and dating

The sites generally cover less than 1 ha, except for Khramis Didi gora whose surface area is 4 to 5 ha.[23] Nine (Shulaveri and Khramis Didi gora) to seven (Imiris gora and Aruchlo I) layers of occupation or destruction have been recognised on the mounds. The height of the stratigraphy is quite stable, as the nine Neolithic layers represent 4 to 5 m of a total stratigraphy of 7 m at Shulaveri gora, and there are about 6,5 m at Khramis Didi gora. In Aruchlo I, the six building horizons correspond to about 5 m.[24]

According to Kiguradze's ceramic study, five stages can be identified. The lowest levels of Shulaveri gora (levels IX-IV) are the earliest of this culture, and the upper levels (levels III-I) are considered to be contemporaneous with the lowest level of Imiris gora (levels VII-VI). The upper levels of this site (levels IV-I) are themselves contemporary with the lowest level of Khramis Didi gora (levels IX-VI).[25] While the ceramic seriation has not been discussed since then, there has been some debate about the position of Aruchlo I in the sequence. On the basis of early but uncertain radiocarbon dates[26] and some specific features of the lithic material, the site of Aruchlo I was considered very early. However, following analysis of all the finds and a reassessment of the radiocarbon dates, it is now suggested that Aruchlo I occupies a rather late position in the sequence.[27]

The ^{14}C dates from the four main sites of Kwemo-Kartli and also from Shomu Tepe, Toïre Tepe and Gargalar Tepesi in Azerbaijan are confusing, due to variation in sampling, calculation and laboratories (table 1).[28] According to these dates, the Shulaveri-Shomu culture began in the second half of the 6th millennium BC and ended in the middle of the 5th millennium. Recent dates and excavations at Arastashen[29] (Armenia)

20. KIGURADZE, 1986; KIGURADZE and MENABDE, 2004.
21. HANSEN *et al.*, 2006.
22. KIGURADZE, 1986; KIGURADZE and MENABDE, 2004.
23. KIGURADZE, 1986.
24. KUSHNAREVA and CHUBINISHVILI, 1970; quoted in KIGURADZE, 1986.
25. KIGURADZE, 1986; KIGURADZE and MENABDE, 2004.
26. After CHATAIGNER, 1995: 35-44 and KIGURADZE, 1986: Aruchlo I level VII: 6250-5645 cal. BC, Shulaveri gora level IX: 5715-4895 cal. BC, Imiris gora level VI-I: 5635-5305 cal. BC, Khramis Didi gora level VI: 5545-5265 cal. BC.
27. CHATAIGNER, 1995.
28. *Ibid.*
29. BADALYAN *et al.*, 2007.

and Göy Tepe (Azerbaijan)[30] reveal that the Shulaveri-Shomu culture appeared at the beginning of the 7th millennium BC rather than in the second half of the 6th millennium BC.

Settlements and architecture

The architecture is characterized by use of plano-convex mud-bricks for the building of circular, oval or semi-oval buildings with diameters ranging from 1,5 to 4,5 m.[31] Some of the forty buildings excavated at Shulaveri gora have conical corbelled clay roofs. Adobe stucco was found on outer and inner sides of the walls. The larger buildings (2 to 5 m in diameter) are interpreted as living areas whereas the smaller ones (1-2 m diameter) are considered as "artisan" areas where storage and production activities may have taken place.[32] The same type of architecture has been found in Imiris gora, but with an important "innovation": the buildings are sometimes organised around a central courtyard (Imiris gora, Aruchlo I).[33] Some circular or oval hearths have been found in these courtyards. A system of three protective ditches of varying depth and width has been discovered around the site of Aruchlo I, located at the intersection of two rivers, as well as around Imiris gora.

Funerary customs

Little information is available concerning funerary practices. Tombs were discovered at Shulaveri gora and Aruchlo I. Their dating is not always clear. At Shulaveri gora, four graves (among which one double inhumation) were discovered; as they cut into the mud-brick buildings, they could be later than the Neolithic.[34]

At Aruchlo I, two tombs have been found right beneath the floor of some houses or in an enclosure between levels III and V: a teenager was found in tomb 10 with no associated finds; whereas two bone objects (a point and a knife) were discovered in tomb 11. Three other tombs were located next to half-sunken huts around the tell and cannot be precisely dated to the Shulaveri-Shomu culture. Covered by small heaps of stones, they all show obvious traces of ochre. A red-coloured pot was found in tomb 7, and another in tomb 8 in association with a cornelian bead. Two bone points and two carnelian beads were found next to a 50-55 years old woman in tomb 9.[35]

30. *Ibid.*
31. KIGURADZE, 1986; CHATAIGNER, 1995; DJAVAKHISHVILI, 1973; KUSHNAREVA, 1997.
32. KIGURADZE and MENABDE, 2004.
33. DJAVAKHISHVILI, 1973; KIGURADZE, 1986.
34. JANELIDZE, 1975: 11.
35. CHATAIGNER, 1995: 82; CHUBINISHVILI and GOGELIA, 1981: 39; KIGURADZE, 1986: 66-67.

SHULAVERI-SHOMU SUBSISTENCE ECONOMY

During the 6th millennium cal. BC, Kwemo-Kartli was characterised by a dry continental to a semi-arid climate.[36] According to palynological data,[37] there was "an extensive savanna on the plains, dense gallery forests along river courses and oak-hornbeam forests on the foothills" favourable for irrigation. These climatic conditions favoured the development of an agricultural system involving the cultivation of a broad range of domestic plants and the breeding of animals.

From the earliest occupation layers, bones of domesticated goat, sheep, cow, pig and even dog are present. They represent about 80% of the fauna at Shulaveri gora and 90% at Khramis Didi gora. There is also a wide range of wild mammals including deer, wolf, fox, aurochs, horse and gazelle, as well as turtles, birds and fishes.[38]

Nine species of wheat (*Triticum diccoccum, T. monococcum, T. aestivum, T. turgidum T. compactum, T. Durum, T. spelta, T. macha, T. paleo-colchicum*) have been identified together with barley (*Hordeum vulgare and Hordeum distichum*), oat (*Avena sativa*), rye (*Secale cereale*) and millet (*Panicum sativum*).[39] According to plant remains from Aruchlo I and Khramis Didi gora, *T. aestivo-compactum* was dominant, followed by *T. diccoccum* or *T. spelta,* whilst the other species of wheat were much rarer.[40] Barley (*Hordeum vulgare*) and legumes such as lentils (*Lens esculenta*), peas (*Pisum sativum*) and beans (*Vicia*) were also present. The earliest traces of grapevine (*Vinis vitifera*) were found at Shulaveri gora.[41]

One of the most crucial questions concerns the appearance of domesticated cereals in Transcaucasia. Although some endemic species were apparently domesticated locally (*T. macha, T. paleo-colchidum*), the question remains open for the main types of cereals (*T. diccoccum, T. monococcum, T. aestivum*). The great variety of endemic wheat species in Transcaucasia could have favoured local domestication of cereals, even if these cereals were already cultivated in the Near East. According to different authors,[42] following various hypotheses on the appearance of a Neolithic economy in Transcaucasia (an imported Neolithic economy with or without movement of population, local domestication and spontaneous

36. JANELIDZE, 1984.
37. CONNOR and SAGONA, 2007: 28.
38. KIGURADZE, 1986; KIGURADZE and MENABDE, 2004.
39. LISITSYNA and PRISCHEPENKO, 1977.
40. RUSSISHVILI, 1990.
41. LORDKIPANIDZE, 1991.
42. AMIRKHANOV, 1987; KUSHNAREVA, 1997: 155-158; LYONNET, 2004; WECHLER, 2001.

adoption of a Neolithic economy), these species are considered either to have been imported from the Near East or to have been domesticated locally. As for the Shulaveri-Shomu culture, the two hypotheses are still debated.[43]

The exceptional diversity of the cereal species cultivated on Shulaveri-Shomu sites is of great interest in terms of cereal processing and dietary habits. The possible correlation between the use of specific cereals and the grinding and pounding equipment deserves to be examined in more detail, as regards dehusking or grinding and more generally the preparation of cereals.

SHULAVERI-SHOMU MATERIAL CULTURE AND TECHNICAL SYSTEM

Ceramic production

The handmade pottery of the Shulaveri-Shomu culture is made from local clay. Sand, "chamotte" (or grog), basalt and plant materials are used as temper in the lowest layers, whereas plant materials are dominant in the later pottery:

- in phase I (layers XI-IV), there are coarse, ovoid pots with a protruding base,
- in phase II, incised patterns appear on fine pottery,
- in phase III, there are coarse brown to red ceramics with flat bases and relief decoration,
- in phase IV, some fragments of painted pots are found in disturbed layers,
- in phase V (upper levels of Khramis Didi gora), fine red polished pottery becomes predominant.[44]

Regional variations can also be seen in ceramic production, such as the Shulaveri gora and Khramis Didi gora groups.

Chipped stone industry

The obsidian industry comes from different sources including the Trialeti mountains.[45] The proportion of obsidian in the assemblages increases over time, as at Shulaveri gora where obsidian represents 82% of the chipped stone industry in the lower level IX and 98% in the upper level I.

In phases I to III, the blade industry is relatively standardized, as shown by the high quantity of conical cores and long, wide parallel macroblades found on the sites. Blades and segments of blades measure up to 15 cm. In phase IV, flake production increases to become predominant in phase V.[46]

Lithic tools change very little throughout the sequence. Burins, scrapers, splintered pieces, drills and denticulated blades make up the standard tool-kit. At Shulaveri gora, half of the lithic artefacts are tools and these are mainly chisels, scrapers and splintered pieces. Agricultural activities are reflected by sickle blades and adzes.[47]

Bone industry

The bone industry is rather standardized throughout the sequence and composed of a majority of spatulas, awls and spoons together with needles and arrows. Some antler digging sticks (or hoes) are also present from phase III onwards. At Imiris gora, spatula and awls are shaped by abrasion and polishing. A genuine diversity in bone and antler production can be underlined at Khramis Didi gora, with a large series of perforated axes, maceheads and splintered tools in antler. At Aruchlo, biconical needles and perforated antler objects constitute the main items. The typology and technology of Shulaveri-Shomu bone and antler tools appear broadly related to the standard production found on Near Eastern Neolithic sites.[48]

Anthropomorphic figurines

An abundant assemblage of anthropomorphic figurines has been found, especially in phase IV and V (over 60 examples at Khramis Didi gora). These figurines are generally seated women represented in a realistic way. They are interpreted as votive items dedicated to a cult of fertility and are reminiscent of similar examples produced in the Near East.

Metal

The earliest metal objects are in arsenical copper and come from Khramis Didi gora, levels VI and VII, and from Aruchlo I. Their appearance coincides with a general deterioration of the lithic industry and stone toolkit.[49] Thus the latest levels of these two sites can be considered Chalcolithic or at least transitional between the Neolithic and the Chalcolithic.

43. LOMBARD et CHATAIGNER, 2004.
44. KIGURADZE, 1976.
45. BADALYAN *et al.*, 2004.
46. LOMBARD et CHATAIGNER, 2004.
47. CHATAIGNER, 1995.
48. *Ibid.*; LOMBARD et CHATAIGNER, 2004; KUSHNAREVA, 1997.
49. KAVTARADZE, 1999.

TECHNOLOGICAL AND FUNCTIONAL CHARACTERIZATION OF MACROLITHIC IMPLEMENTS

In fact, grinding (cereal and plant processing, temper, ore or colouring grinding) or abrading stages (adzes, bone tools) are involved in most of the "chaînes opératoires" of production and transformation of the Shulaveri-Shomu implements. Kiguradze had already discussed some characteristics of grinding stones, massive pounder-mortar tools, "palettes", "hammer stones", "rubbing stones", "grooved stones" and stone vessels.[50]

This study includes 568 macrolithic tools coming from four main sites of the Shulaveri-Shomu culture: 70 from Shulaveri gora, 170 from Imiris Gora, 285 from Khramis Didi gora and 43 from Aruchlo (table 2).

Unfortunately, the stratigraphic context of the objects is not always mentioned in the inventories, and thus it is impossible to relate stone objects to building levels on all the sites. In addition to this problem, the finds of the latest excavation at Khramis Didi gora have not been completely analysed, as finds were still being processed in August 2006.

The classification of the macrolithic implements has been made first by activity (grinding, pounding, hammering, polishing and rubbing), then by the position of the tools (working

Table 2 – *Number of tools by activities, types and sites. (In grey, the most important characteristics of the assemblages.)*

Category	Type	Shulaveri	Imiris	Aruchlo I	Khramis Didi	Total
Grinding	quadrangular querns	2	5	2	1	**10**
	flat basaltic querns	2	2			**4**
	ovoid querns	2	4	3		**9**
	narrow querns		2			**2**
	flat grinders		8		5	
	short grinders (sandstone / granite)	9	8	2	2	**21**
	short grinders (basalt)	5	7	3	1	**16**
	mortars		3	2		**5**
	grinding slabs	1	7	2	1	**11**
	pestles	5	21	12	30	**68**
	crushing cobbles		13	3	9	**25**
	hand stones	7	11	1	14	**33**
Hammering	anvils	2	4			**6**
	hammer stones (metamorphic rocks)	17	29	2	75	**123**
	hammer stones - rubbers	1			1	**2**
	hammer stones diverse	9	8		6	**23**
	splintered tools		3	1	4	**8**
Polishing	grooved abrader (basalt)		2		2	**4**
	grooved abrader (discoid)		2		1	**3**
	hand abraders	1	3	1	2	**7**
	prismatic polishers		1		2	**3**
	sharpeners		5	1	4	**10**
Adzes and axes	adzes	1	4	3	52	**60**
	adzes rough-out	1			26	**27**
	small axes				3	**3**
	chisels				6	**6**
Limestone items	stone vessels				4	**4**
	limestone mortars				8	**8**
	limestone rings				4	**4**
	limestone hammers (1 ex. basalt)				6	**6**
	limestone picks or smoothers		2		1	**3**
	limestone splintered tools		1		2	**3**
Perforation	perforated tools	1	7	3	7	**18**
Indet	palettes		4	1		**5**
	undetermined	4	3	1	6	**14**
	Total	**70**	**169**	**43**	**285**	**554**
Grinding	ochre traces	**3**	**15**	**3**	**21**	**42**
		4,3 %	**8,8 %**	**7,0 %**	**7,3 %**	**7,4 %**

Table 3 – *Average dimensions of the main categories of tool.*

Tools	Length (cm)	Width (cm)	Thickness (cm)
Quadrangular querns	55	28	12
Ovoid querns	30	26-29	10
Narrow querns	29	14	5
Grinders	17-26	10-14	2,5-8
Quern-grinders	32	14	3,5
Mortars	20-30	22-26	20
Pestles Ia	6-10	5-6	4-5
Pestles Ib	11-15	-	4-7
Grinding slabs	14-27	15-21	4-7
Hammer stones	4-13	4-10	8,5 max.
Crushing cobbles	9-17	7-11	4-11
Handheld polishers	7-13	5-9	
Prismatic polishers	5,5-9,5	4-5	5
Grooved abraders I	14	10	8
Sharpeners (max.)	15	5	3
Hand stones	5-15	4-10	4,5-6
Adzes	5-20	3-7	1,5-5
Chisels	7	3	1,5
Palettes	14	12	3
Limestone mortars	7-14	6-15	4-8
Stone rings	4-7	2-3	1-3,5
Limestone hammers	7,8-9,8	5,2	4,4-5,8
Perforates cobbles	6-13	5-12	3-5,8
Splintered tools	9-15	3-5	2-3

50. KIGURADZE, 1986.

Table 4 – *Querns main characteristics (raw material, shape, dimensions).*

Site	N°	Type	Raw material	Conservation	Length	Width	Thickness	Form	Working surface	Technology	Figure
Aruchlo	260	quern	basalt	complete	44	34	16	ovoid	concave	cobble	
	241	quern	basalt	complete	25	20	10	ovoid	flat	cobble	21c
	259	quern	basalt	fragment	30	32	10	quadrangular	concave	back and shaped sides	
	240	quern	basalt	half	21	17	7	ovoid	concave	cobble, shaped sides	21b
	239	quern	basalt	half	25	28	12	quadrangular	concave	cobble, shaped sides	21a
Imiris	64	quern	basalt	complete	27	27	9	quadrangular	concave (6,4 cm)	shaped sides, pecked back	8f
	230	quern	basalt	complete	57	28	16	quadrangular	concave (11 cm)	cobble, retouched	8a
	228	quern	basalt	complete	42	15	5	ovoid		cobble, pecked sides	
	129	quern	basalt	complete	55	28	10	ovoid	concave (5,2 cm)	cobble, pecked sides	8c
	130	quern	basalt	complete	29	32	10	quadrangular	concave	cobble, shaped sides	8e
	232	quern	basalt	complete	26	32	10	quadrangular	concave (9,6 cm)	cobble, shaped sides	
	217	quern	basalt	fragment	16	27	8	half-circular	plano-concave	shaped sides	
	163	quern	basalt	fragment	31	25	6	ovoid	concave	cobble	8b
	95	quern	basalt	fragment	9	10	3,6	quadrangular	concave	cobble, pecked sides	
	164	quern	sandstone	fragment	31	21	11	quadrangular	concave	back and shaped sides	
	165	quern	metamorphic rock	fragment	32	22	12	quadrangular	concave	pecked sides and back	8d
	218	quern	basalt	fragment	20	15	5	quadrangular	concave	cobble, shaped sides	
	143	quern	basalt	fragment	25	31	10,8	triangular	concave		
	229	quern	basalt	half	-	-	-	ovoid	concave	cobble, retouched	
Khramis Didi	481	quern	basalt	fragment	27	23	7,5	ovoid	concave	cobble, retouched	14b
	480	quern-mortar	basalt	complete	37	34	12	quadrangular	concave	reuse of a quern as a mortar	14a
Shulaveri	10	quern	basalt	complete	20	18	8	triangular	concave	cobble, shaped sides	
	62	quern	basalt	fragment	25	28	6	ovoid		central pecking and traces of ochre	4a
	4	quern	basalt	fragment	11	10,6	3	quadrangular	concave		4b
	50	quern	sandstone	fragment	16	19	5,4	quadrangular	concave	cobble	
	53	quern	basalt	fragment	30	31	12	quadrangular	concave	cobble, shaped sides	4c
	51	quern	basalt	fragment	24	19	6	triangular	concave	cobble, shaped sides	4d

or passive) and finally by the various techno-morphological criteria used by most authors (shapes in plan and section, the working surface, etc.).[51] Whenever possible, we mention the correspondence between our description and Wright's terminology.[52] The average dimensions of the tools are indicated in table 3.

QUERNS AND GRINDERS

Querns constitute the lower part of the grinding equipment, in association with a "grinder" manipulated with two hands in a back-and-forth movement. The majority of the 75 grinding tools are made of vesicular basalts (66%), although sandstones were also used (24%). 85% of the querns and 60% of their upper parts were made of basalts.

51. ADAMS, 2002; HAMON, 2006; NIERLÉ, 1982; WRIGHT, 1992.
52. WRIGHT, 1992.

Querns

The 25 querns have been divided into three main types, on the basis of their shape (table 4, figs. 4, 8, 14 and 21).

• Quadrangular grinding slabs (ten examples) are generally broken: fragments generally range from one to two thirds of the tools. These large quadrangular querns often show an important concavity (up to 11 cm). Two variants are attested:
  – "saddle-shaped querns"[53] (seven examples). These tools are generally made on basalt cobbles, and their back is naturally convex. Flat slabs are included in this variant. They are particularly thin and regular, and scarcely concave on the working surface, but a distinct ledge of 2 to 3 cm wide is visible on the edge of the object and can show traces of flaking (fig. 4a),
  – "saddle-shaped slabs" (three examples). These tools are generally preserved for two thirds of their length and show regular concave surfaces (fig. 8b). The

53. *Ibid.*

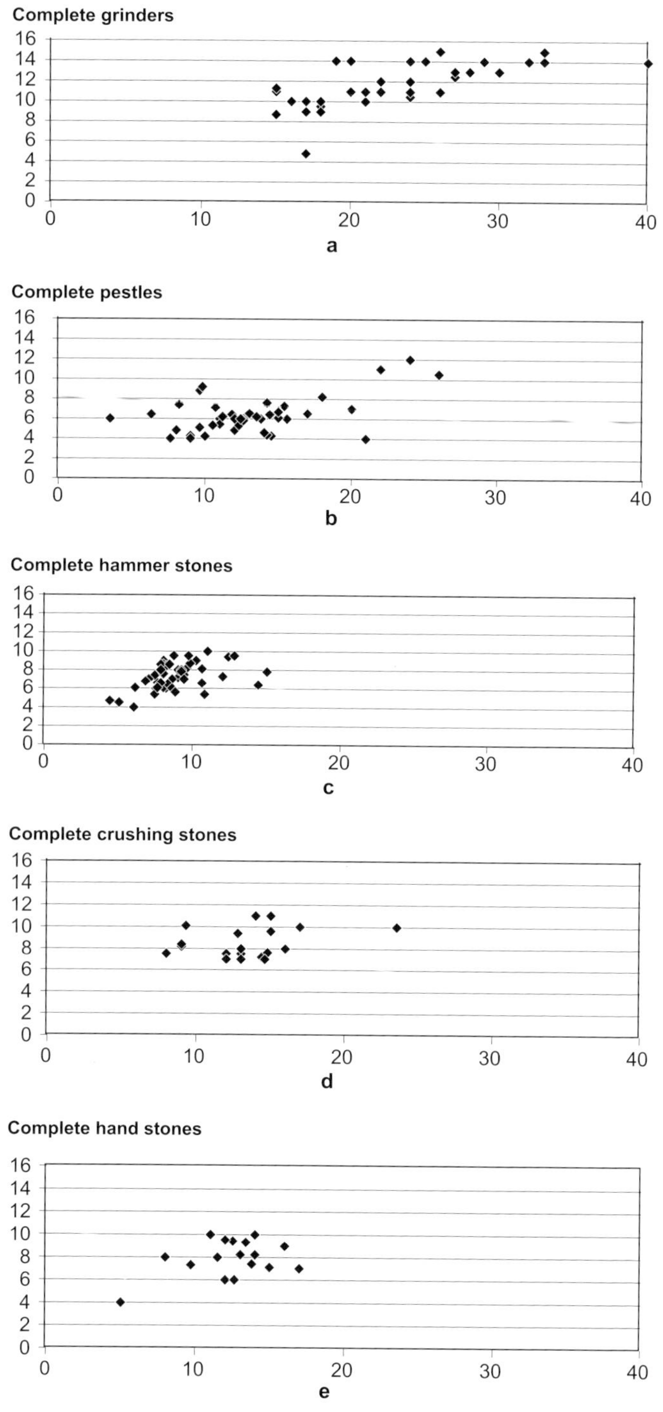

Fig. 2 – *Dimensions (length / width in centimetres) of entire tools from five important categories:* ***a****, grinders;* ***b****, pestles;* ***c****, hammer stones;* ***d****, crushing tools;* ***e****, hand stones.*

basalts chosen are generally more compact than the vesicular ones. Different stages of pecking can be distinguished on the working surface. These tools have often been used or reused for the grinding of pigments.

• Ovoid (nine examples) or triangular (three examples) querns, with two distal edges and a strong concavity (figs. 3a, 5a, c). The back of the tools show polished zones. One is illustrated for Shulaveri gora.[54]

• Narrow querns of quadrangular shape (two examples). They are broken in two and have a semi-circular section. The working surface is slightly concave and the back of the tool is hammered.

Quern variability can be partially attributed to morphological change in a tool during its use, such as the deformation of the working surface and the presence or not of an edge on its ends and periphery. All types of querns have been used or re-used for colour processing. According to functional data, these cases must be interpreted as a re-use of grinding stones that were first used for cereals. Another example of re-use at Imiris gora is represented by a quern with a groove on its back.

Grinders

Most of the 50 grinders are of "short type": their length is less than the width of the corresponding lower tool (table 5, figs. 3b, 5, 9, 14 and 22). Coarse sandstones, calcareous sandstones and compact vesicular basalts are used for their manufacture. The tools are semi-circular to ovoid in shape ("bread-shaped") and semi-circular to flat quadrangular in sections[55] (fig. 2a).

The sides are shaped by flaking and the ends often show two or three steps of flaking. Pecking was used in order to smooth the back and side edges and allow a better grasp. The flat to plano-convex working surfaces were often pecked transversely. A polishing zone of 2 to 3 cm wide occupies the ends and sides if not the whole periphery of the working surface of these grinders. Some of the tools have two working surfaces on opposing sides.

One isolated example of a grinder from Imiris (n° 148) is quadrangular in shape and in section. Two symmetrical grooves, parallel to the sides, were made by a longitudinal pecking and create a very specific section. The function of this arrangement is uncertain. Another example of ovoid grinder from Imiris (n° 198) (fig. 9f) has been re-used: its opposite face shows traces of polishing on the central part.

54. Kiguradze, 1986: 22, fig. II: 37.
55. Hand stones 43-44 of Wright's terminology (Wright, 1992).

Table 5 – *Grinders main characteristics (raw material, shape, dimensions).*

Site	N°	Type	Raw material	Conservation	Length	Width	Thickness	Form	Working surface	Technology	Figure
Aruchlo	243	grinder	basalt	complete	28	13	6	ovoid	flat		22a
	244	grinder	basalt	complete	26	15	4,5	ovoid	flat	shaped, pecked sides, back	22b
	272	grinder	granite	complete	22	11	4	quadrangular	convex	pecked sides	
	257	grinding tool	basalt	fragment	21	16	6	ovoid		cobble	23j
	245	grinder	limestone	fragment	22	16	7	ovoid	flat	shaped, pecked sides, back	22c
	242	grinder	sandstone	half	16	18	6	quadrangular	flat	shaped sides	
	283	short grinder	basalt	fragment	20	14	6	ovoid	flat	cobble, shaped sides	20b
Imiris	85	grinder	basalt	complete	24	12	6	ovoid	convex	cobble, shaped sides	
	227	grinder	basalt	complete	33	14	3,6	ovoid		pecked sides	
	189	grinder	basalt	complete	32	14	3,5	ovoid		pecked sides	9g
	188	grinder	basalt	complete	33	15	3,1	quadrangular	flat	pecked back and sides	9h
	187	grinder	basalt	complete	24	11	4	quadrangular	plano-concave	shaped, pecked sides, back	9b
	199	grinder	basalt	complete	29	14	5	quadrangular		pecked sides	9e
	142	grinder	basalt	complete	40	14	6,5	quadrangular		pecked sides	
	131	grinder	basalt	complete	22	12	6	ovoid	convex	pecked sides	
	132	grinder	basalt	complete	18	10	4	ovoid	convex	cobble, pecked sides	
	141	grinder	basalt	complete	27	13	4	ovoid	plano-convex	pecked back, sides	
	144	grinder	undetermined	complete	24	11	5	ovoid	plano-convex	pecked sides	
	147	grinder	sandstone	complete	18	10	7	ovoid		shaped, pecked back	
	152	grinder	basalt	complete	18	9	6	ovoid	plano-convex	shaped back, pecked sides	
	157	grinder	basalt	complete	27	13	3,8	ovoid	flat	shaped, pecked sides, back	9a
	159	grinder	sandstone	complete	15	9	3	ovoid		pecked back	9d
	166	grinder	basalt	complete	30	13	3,6	ovoid	plano-concave	pecked sides	9i
	146	grinder	sandstone	complete	20	14	2	quadrangular	convex	shaped sides, pecked back	
	83	grinder	basalt	fragment	17	12	3,5	ovoid	flat		9k
	84	grinder	sandstone	fragment	17	11	3,7	ovoid		pecked sides	9j
	112	grinder	basalt	fragment	11	7	3,9	quadrangular		shaped sides	
	72	grinder	basalt	fragment	13	11	5,2	quadrangular	convex	pecked back	
	148	grinder	metamorphic rock	fragment	12	14	5,6	quadrangular	concave		
	198	grinder-grooved abrader	basalt	complete	8	7	6,4	ovoid	plano-convex	one grinding surface and one polishing surface	9f
	226	grinder	sandstone	complete	24	10	6,5	ovoid	-		9c
	151	grinder	sandstone	complete	12	5	5	ovoid	concave	possible reuse as a grinding slab on the opposite side	9l
Khramis Didi	482	grinder	basalt	complete	24	10,5	4	ovoid	convex		
	485	grinder	basalt	complete	25	14	4	ovoid	plano-convex	shaped sides	
	501	grinder	sandstone	complete	17	10	6,5	ovoid	plano-convex	pecked sides	
	484	short grinder	basalt	complete	26	11	2,3	ovoid	flat	pecked sides	14c
	325	grinder	basalt	complete	15	11	4,6	quadrangular	plano-convex	cobble, pecked sides	14e
	31	grinder	sandstone	complete	22	11	4	quadrangular	plano-convex	cobble, pecked sides	
	486	short grinder	basalt	complete	24	14	4	quadrangular	plano-convex	pecked sides	14d
	361	grinder	basalt	half	17	14	3,5	ovoid	plano-convex	shaped, pecked sides	
Shulaveri	3	grinder	sandstone	complete	17	9	4	ovoid	flat		5d
	43	grinder	coarse sandstone	complete	17	10	3,5	ovoid	plano-convex	cobble, shaped sides	5b
	47	grinder	sandstone	complete	21	10	6,2	ovoid	plano-convex	shaped back, pecked sides	5h
	49	grinder	basalt	complete	20	11	8	ovoid	plano-convex	back and shaped sides	
	61	grinder	sandstone	complete	20	11	4,5	ovoid	plano-convex		
	58	grinder	sandstone	complete	18	9	4	ovoid	plano-convex	pecked back	5c
	59	grinder	sandstone	complete	17	4,8	4,8	ovoid	plano-convex	pecked back	5a
	60	grinder	sandstone	complete	16	10	4,5	ovoid	plano-convex	pecked back	
	48	grinder	sandstone	complete	21	11	6	ovoid	convex	pecked sides	5f
	5	grinder	basalt	complete	15	11,3	6,7	ovoid	convex	cobble	
	11	grinder	basalt	complete	19	14	6,6	ovoid	convex	cobble	5e
	41	crushing cobble	basalt	fragment	10	8	4	ovoid	plano-convex		7k
	44	grinder	fine sandstone	fragment	11,4	8	4,1	ovoid	plano-convex	pecked sides	5i
	46	grinder	granite	fragment	19	12	5,2	ovoid	plano-convex	shaped back, pecked sides	5g

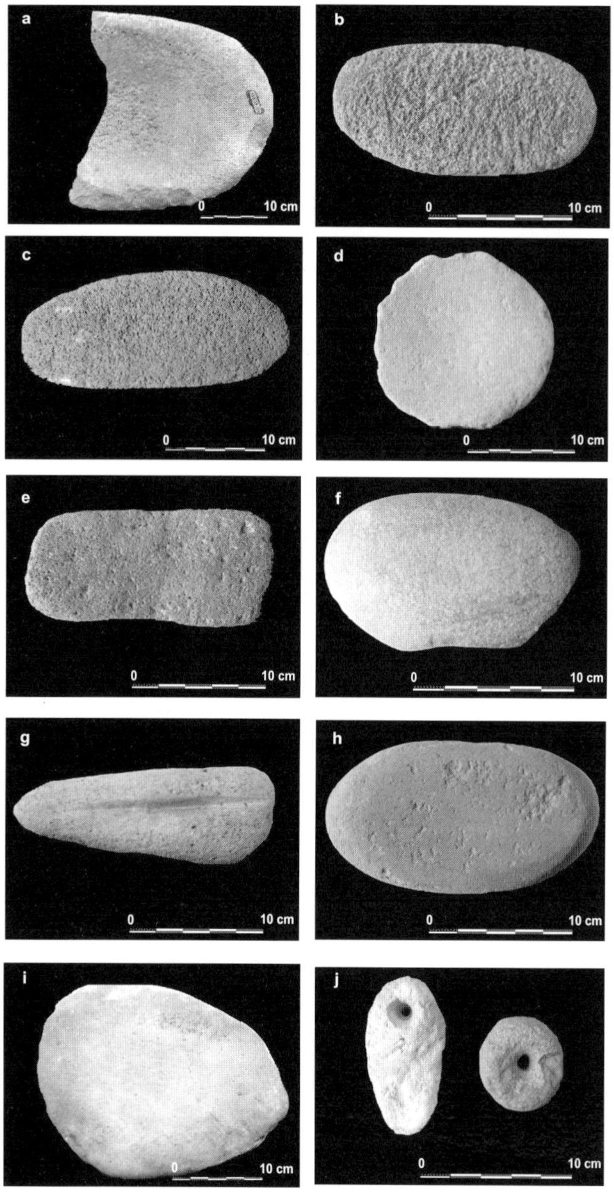

Fig. 3 – *Stonetools samples from Shulaveri-shomu Kwemo-kartlian sites:* ***a****, Shulaveri gora, ovoid quern fragment in sandstone—notice the pecking in the centre and the ochre traces on the sides of the working surface;* ***b****, Shulaveri gora, short grinder in coarse sandstone with transversal pecking striations;* ***c****, Khramis Didi gora, flat ovoid quern in basalt—both sides are used;* ***d****, Imiris gora, circular grinding slab—both sides are used and show polish and impacts and their working surface;* ***e****, Shulaveri gora, short pestle in basalt—both ends are crushed;* ***f****, Imiris gora, crushing cobbles—notice the covering fine impacts of percussion on the sides and the coarser impacts in the centre;* ***g****, Imiris gora, grooved abrader in sandstone;* ***h****, Imiris gora, hand stone in sandstone—notice the impacts of percussion, the covering facett of polishing and the transversal striations;* ***i****, Imiris gora, "palette"—notice the traces of ochre and the long striations of cutting and scraping;* ***j****, Imiris gora, perforated stones in limestone (left) and basalt (right)—notice the impacts of preparation and the mechanical and circular striations of perforation.*

Other grinding tools

A third type of grinding tool can be isolated from among the undetermined objects (two examples). They could not be classified with certainty either as a grinder or a quern. These are flat, ovoid tools with a quadrangular, regular section. They are long and narrow, fairly standardized, and present two working surfaces on the opposite faces (figs. 3c and 9i-j). The working surface is plano-concave and is particularly abrasive since it is made of vesicular basalt. The sides of these tools are regularly pecked, the edge retouched and the working surface smoothed. This morphology is difficult to interpret in terms of function. As a quern, its function would have been quite different from the large quadrangular or ovoid ones, but as a grinder no association can be found with any of the lower slabs present.

POUNDING TOOLS: MORTARS AND POUNDERS

J. Adams defines the mortar as "a basin that confines an intermediate substance that is worked with a pestle in some combination of crushing, stirring, or pounding strokes."[56] As well as mortars and pestles, an original type of grinding tool has been found on the Shulaveri-Shomu sites: we will use by default the term "grinding slab" to describe it. It belongs to Adams' general category of netherstones.[57]

Mortars

Granite and basalt were used for five mortars (table 6, figs. 6 and 13a). Massive examples were found in Shulaveri gora and Imiris gora. The two examples have been broken at half of their length, in the middle of the depression.[58] On a natural block, a flat surface has been chosen to create a circular depression about 10 cm deep and wide. A narrow strip 2 to 3 cm wide occupies the periphery of the depression; it is probably due to the shaping of the hole by hammering and pecking. The breaking of the mortar may be the consequence of a strong gesture of pounding in the depression, where the tool is the thinnest. Inside the depression, use-wear reveals impacts of percussion and circular grinding. One quern has been gradually transformed into a so-called "mortar" (n° 480) (fig. 14a). The initial working surface has been gradually transformed in order to obtain a deep working surface, the concavity of which starts at

56. ADAMS, 2002: 126.
57. *Ibid.*: 143.
58. KIGURADZE, 1986: fig. II: 34.

Table 6 – *Mortars main characteristics (raw material, shape, dimensions).*

Site	N°	Type	Raw material	Conservation	Length	Width	Thickness	Form	Working surface	Technology	Figure
Aruchlo	270	mortar	basalt	complete	20	19	10	triangular	concave	cobble	20a
Imiris	233	mortar	granite	half	28	26	20	triangular	cavity 10 cm deep	cobble	13a
Shulaveri	54	mortar	metamorphic rock	fragment	20	22	10,5	ovoid	concave	cobble, shaped sides	6a

the four corners of the tool. The working surface shows circular grinding and even vertical crushing in its centre.

Pestles

Two types of pestles in basalts (50%), sandstones (11%) and metamorphic rocks (34%) were found on the Shulaveri-Shomu sites (table 7, figs. 6, 11, 15 and 20). Two variants can be distinguished amongst these 68 tools:

- The first type (11 examples) presents ends with fine percussion and smoothing of the edges.
  - The tronconical type with a circular section is shaped on all its sides (two examples) (fig. 15b). Both ends are flattened because of the repetition of a crushing action, but one end is much wider and presents crushed edges.
  - The cylindrical type with a circular section shows no differences between its two ends (nine examples) (fig. 11b). The impacts of percussion have often overlapped the edges and spread onto the sides of the pestle (fig. 3e). The impacts are rather coarse and deep, regular and well delimited. In one case, one of the ends has been violently crushed and shows a large and deep impact of percussion in its centre. These pestles are also frequently used on their faces and can be used as crushing hand stones. In this case, one to four of their sides present small cavities created by percussion and associated with other traces of use such as striations and polishing. The centre of one pestle (n° 173 from Imiris) (fig. 11a), has been thinned by pecking in order to obtain an object with two prominent ends and a central part for handling.
- The second type includes used cobbles of more varied shapes (57 examples) and smaller dimensions, although the largest one measures 26 x 10 x 5 cm (figs. 2b and 23b). The two ends show coarse impacts of percussion. Some samples show actual shaping and narrowing of the central part of the object for a better grip. Sometimes, polishing can be observed on the sides.

Grinding slabs

Amongst eleven grinding slabs, eight special tools in sandstone or quartzitic sandstone were found at Imiris gora (table 8, figs. 3d and 10). These lower tools have been used in a circular gesture of grinding. We have not yet been able to identify the associated upper tools. These opposite faces of these circular tools are tronconical in section. They are complete except in one case. Their morphology and shaping are particularly standardized. One of the working surfaces is always larger in diameter than the other. The side is shaped by the removal of regular flakes from one face to another. The edge is also sometimes retouched. The slightly concave working surfaces are characterized by a smoothing of their periphery and some rare impacts of percussion located in their centre. One example from Khramis Didi gora (n° 488, fig. 17a) has been used both as a grinding table and as an anvil, as suggested by the small cavity created by percussion in the centre of one face.

PERCUSSION TOOLS

Anvils and hammer stones are characterized by a wide range of actions of percussion on the matter. Anvils can be used as lower tools for multiple activities, as shown by the numerous impacts of percussion, the polish and the incisions on the surface.[59] Hammer stones show different zones of percussion, on their ends, sides and faces. Crushing cobbles are characterized by impacts of percussion concentred on a small zone on one or both faces, used primarily for the crushing of large fragments into smaller ones.

Anvils

The six anvils found in Shulaveri and Imiris gora are massive cobbles of various raw materials: limestone, granite, sandstone or metamorphic rocks (table 9). They are also of various dimensions but never exceed 20 cm in length and width. The upper convex faces show rare impacts of percussion that damage the natural surface and are concentrated in the centre of one

59. See the "lithic anvils" of ADAMS, 2002: 157.

Table 7 – *Pestles main characteristics (raw material, shape, dimensions).*

Site	N°	Type	Raw material	Conservation	Length	Width	Thickness	Form	Technology	Figure
Aruchlo	275	pestle	metamorphic rock	complete	14,2	4,4	8,7	cylindrical		20d
	261	pestle	sandstone	complete	22	11	7,5	ovoid		20j
	252	pestle	basalt	complete	7,6	4	4,2	quadrangular		
	253	pestle	metamorphic rock	complete	13,5	6,2	4,5	quadrangular		
	255	pestle	metamorphic rock	complete	12	6	6	quadrangular		23b
	264	pestle	sandstone	complete	9,8	9,2	6,4	quadrangular		20c
	266	pestle	sandstone	complete	10,7	7,1	5	quadrangular		
	282	pestle	basalt	complete	6,3	6,4	5,1	tronconical		20h
	254	pestle	metamorphic rock	fragment	6	5,3	4,5	ovoid		23a
	262	pestle	metamorphic rock	fragment	6	3,4	2	ovoid		
	256	pestle	basalt	fragment	6,5	3	2,6	quadrangular		23c
	263	pestle	metamorphic rock	fragment	12,6	6,4	3,4	quadrangular		20e
Imiris	125	pestle	basalt	complete	13	7	6,6	cylindrical		11d
	223	pestle	metamorphic rock	complete	18	8	6,8	cylindrical		
	118	pestle	metamorphic rock	complete	12	6	4,2	ovoid		11b
	172	pestle	metamorphic rock	complete	10	9	6,1	ovoid		
	173	pestle	basalt	complete	16	6	4,7	ovoid		11a
	215	pestle	basalt	complete	11	5	5,2	ovoid		
	219	pestle	basalt	complete	24	12	8	ovoid		
	96	pestle	basalt	complete	9	4	4,6	quadrangular		11e
	190	pestle	basalt	complete	14	6	5,6	quadrangular		
	214	pestle	basalt	complete	11	5	5	quadrangular		
	211	pestle	basalt	complete	21	4	5	quadrangular		11c
	213	pestle	basalt	complete	8	7	6,4			
	109	pestle	basalt	fragment	8	6	4	ovoid		
	195	pestle	metamorphic rock	complete	18	8	5,5	cylindrical		
	80	pestle	sandstone	complete	17	9	5	cylindrical		12m
	116	pestle / crushing tool	basalt	complete	11	8	5,8	ovoid		12c
	138	pestle	metamorphic rock	complete	17	7	6	ovoid		
	124	pestle	basalt	complete	7	4	4,5	cylindrical		11f
	153	pestle	basalt	complete	-	-	-	ovoid		
Khramis didi	313	pestle	metamorphic rock	complete	14,5	4,3	4,5	cylindrical		
	403	pestle	metamorphic rock	complete	12	4,8	4,1	cylindrical		
	404	pestle	metamorphic rock	complete	12,6	6	5,5	cylindrical		
	499	pestle	metamorphic rock	complete	17	6,5	6,2	cylindrical		15a
	312	pestle	basalt	complete	11	6	4,6	cylindrical		
	310	pestle	metamorphic rock	complete	12,3	5,3	3,4	ovoid		
	346	pestle	basalt	complete	15	6,1	6	ovoid		
	373	pestle	basalt	complete	10	4,2	2,9	ovoid		
	442	pestle	granite	complete	11,8	6,4	4,8	ovoid		
	466	pestle	metamorphic rock	complete	8	4,8	3,5	ovoid		
	491	pestle	basalt	complete	3,5	6	4,6	ovoid		
	494	pestle	basalt	complete	15	6,7	3,4	ovoid		
	498	pestle	metamorphic rock	complete	20	7	6	ovoid		
	330	pestle	basalt	complete	9	4	4,2	quadrangular		
	395	pestle	basalt	complete	12,6	5,8	5	quadrangular		
	399	pestle	basalt	complete	14	4,6	4,4	quadrangular		
	322	pestle	metamorphic rock	complete	12,4	6	4,4	triangular		
	478	pestle	basalt	complete	9,6	5,1	4,2	tronconical		
	372	pestle	basalt	fragment	7,8	4,8	4,5	quadrangular		
	331	pestle	basalt	half	-	-	-	ovoid		
	366	pestle	basalt	complete	-	-	-	quadrangular		
	329	pestle-crushing stone	metamorphic rock	complete	18	9	8,2	ovoid		15g
	347	pestle	basalt	complete	4,4	5,4	4,5	ovoid		
	400	pestle-hammmerstone	sandstone	complete	4	3,2	6	ovoid		15j
	457	pestle	basalt	complete	15,3	6,4	3,4	ovoid		
	500	pestle-handstone		complete	21	9	5	cylindrical		15f
Shulaveri	33	pestle	metamorphic rock	complete	14,2	7,6	5,2	cylindrical	cobble	6c
	57	pestle	basalt	complete	15,4	7,3	4,3	quadrangular	both ends crushed	6b
	36	pestle	sandstone	complete	26	10,5	5,5	ovoid		
	63	pestle	metamorphic rock	complete	14,4	6,4	7,4	quadrangular		
	25	pestle	sandstone	complete	11,2	6,2	3,6	triangular	both ends crushed	6d

Table 8 – *Grinding slabs main characteristics (raw material, shape, dimensions).*

Site	N°	Type	Raw material	Conservation	Length	Width	Thickness	Form	Working surface	Technology	Figure
Aruchlo	247	palette	sandstone	complete	18	14	4	quadrangular	plano-concave	-	23g
	285	palette	metamorphic rock	complete	16	12	2,2	ovoid	plano-concave		20k
Imiris	79	palette	metamorphic rock	complete	13,6	10	3	quadrangular	concave	cobble	11g
	160	palette	sandstone	complete	26	19	4	triangular	plano-concave	cobble	10a
Aruchlo	258	grinding slab	sandstone	complete	27	15	6	quadrangular	concave	-	23i
Imiris	135	grinding slab	sandstone	complete	8	7	6	circular	-	cobble	10e
	231	grinding slab	sandstone	complete	19	21	4,5	tronconical	convex	shaped sides	
	133	grinding slab	sandstone	complete	17	15	6	circular	concave	shaped sides	
	65	grinding slab	sandstone	complete	20	18	7	tronconical	convex	shaped sides	10c
	75	grinding slab	sandstone	complete	14	15	4,7	tronconical	plano-concave	shaped sides	10d
	123	grinding slab	sandstone	complete	8	8	6	tronconical	plano-convex	shaped sides	
	74	grinding slab	sandstone	fragment	16	7	6	tronconical	flat	shaped sides	
	208	grinding slab	sandstone	complete	20	18	4	circular	shaped sides	VII,b	
Khramis Didi	488	grinding slab	sandstone	complete	23	18	6	circular	plano-convex	shaped sides	17a

Table 9 – *Anvils main characteristics (raw material, shape, dimensions).*

Site	N°	Type	Raw material	Conservation	Length	Width	Thickness	Form	Technology	Figure
Imiris	111	anvil	granite	complete	14	9	8	ovoid		
	134	anvil	granite	complete	13	12	7,6	ovoid	cobble	
	221	anvil	limestone	fragment	15	11	8,2	quadrangular		
Shulaveri	56	anvil	metamorphic rock	complete	-	-	-	ovoid	cobble	
	21	anvil	metamorphic rock	complete	8,5	7,5	2,4	quadrangular		7e
	31	anvil	sandstone	complete	16	13	8,5	quadrangular		

face. The centre of one thick fragment of limestone presents a circular cavity of percussion. Sometimes, one of the faces or one of the corners bears significant impacts of percussion.

Hammer stones

Six types are represented among the 148 hammer stones (table 10, figs. 7, 12 and 16). They are made of andesite, basalt, limestone, dolerite, sandstone, quartzite and one unidentified metamorphic rock.

• One hundred and twenty three hammer stones made on cobbles of metamorphic rocks (and more rarely limestone), are quadrangular or circular in shape. Dimensions are quite uniform, although there are two main groups (fig. 2c).

On some of them, one or more sides show important flaking created by percussion; these hammer stones were not used for a long time since the flaked zones are not completely crushed. Another category associates impacts of percussion with a coarse flaking from the use of one or two sides. One side and one or two corners show traces of crushing and smoothing. Crushing generally concerns the levelled edges:

• One spherical cobble made of a dense black metamorphic rock was broken in its larger diameter (5,6 x 6 cm). The edges and head of this half cobble wear some percussion impacts.

• The only example of a "hammer" was found broken in half; its dimensions are therefore rather small (5,8 x 4,3 x 2,1 cm) (fig. 19o). Despite a general quadrangular shape, the end of the hammer is thicker and it must have been used for percussion. Some lateral notches were prepared for hafting. A vertical perforation was made perpendicular to the edge by percussion.

• Few hammer stones were used as grinders. Made of metamorphic rock, they are circular in shape but quadrangular in section. Their ends are natural, their faces bear striations or polished zones and their sides are finely and completely pecked.

• One quadrangular example from Khramis Didi gora is made of vesicular basalt (9,8 x 8 x 7 cm) (fig. 16k). The face and sides are pecked and polished, whereas the ends are flattened by percussion.

• One cobble has been flaked symmetrically on both of its sides in order to obtain a pointed end (9,4 x 6,4 x 4 cm). The upper parts are macroflaked while the sides are microflaked (fig. 16m). The natural surface has been maintained on its base, certainly for handling purposes. The tool's morphology indicates that it was used to work material deeply.

Table 10 – *Hammer stones main characteristics (raw material, shape, dimensions).*

Site	Raw material	Form	Nb
Aruchlo	flint	ovoid	1
	metamorphic rock	ovoid	1
Khramis Didi	andesite	triangular	1
	basalt	ovoid	3
	basalt	semi-circular	1
	basalt	spherical	1
	limestone	ovoid	1
	sandstone	ovoid	2
	quartzite	ovoid	1
	metamorphic rock	circular	12
	metamorphic rock	cylindrical	1
	metamorphic rock	ovoid	20
	metamorphic rock	quadrangular	18
	metamorphic rock	semi-circular	2
	metamorphic rock	triangular	9
	tuf	quadrangular	1
Imiris	andesite	circular	1
	andesite	quadrangular	1
	basalt	cylindrical	1
	basalt	ovoid	2
	basalt	spherical	2
	limestone	quadrangular	1
	quartzite	ovoid	1
	quartzite	quadrangular	1
	metamorphic rock	quadrangular	22
	metamorphic rock	triangular	2
	metamorphic rock	spherical	1
	calcarious tuf	circular	1
Shulaveri	basalt	circular	1
	basalt	ovoid	4
	limestone	quadrangular	2
	limestone	spherical	1
	dolerite	circular	1
	dolerite	ovoid	1
	metamorphic rock	ovoid	3
	metamorphic rock	triangular	6
	metamorphic rock	spherical	1
	metamorphic rock	circular	2
	metamorphic rock	quadrangular	1

Crushing cobbles

Twenty-five crushing cobbles are made of sandstone, granite or metamorphic rocks (table 11, figs. 2d, 7k, 12 and 15).

They show at least two small cavities of percussion on two of their faces, and sometimes also on their sides (fig. 3f). These small cavities directly damage the natural surface, or occupy a pecked surface with polished zones. Such notches have thinned down the central part as if intended for hafting. The morphology and size of the cavities differ from one face to the other. The ends can also be flattened by repetitive percussion, or be damaged by macroflaking.

This type of tool is generally used for two purposes: as a pestle (impacts covering the ends) or as a hand stone (the faces associated with the small cavities have extensive polish and transverse striations). This suggests a double gesture of crushing and grinding. Most of these tools were used for ochre processing, as suggested by remains found in the interstices.

POLISHING AND ABRADING TOOLS

Different kind of polishers can be defined after their morphology and the location of the different use-wear traces.[60] They may be implied in a wide range of activities for the shaping and sharpening of different tools and objects. No netherstones, defined as "bottom tools against which something was worked",[61] were found on Shulaveri-Shomu sites. We refer here to the generic meaning of hand stones as a "large set of handheld tools."[62] They share the same general morphology as grinders, but their shape, dimensions and use-wear show a totally different kinetic and use.

Handheld polishers

Seven hand polishers are made of fine sandstone or tender limestone (table 12, fig. 17). Two quadrangular polishers with triangular sections present several concave faces. Sometimes the hand polishers are made of flat pieces of sandstone. Their upper face is generally concave and covered by a uniform use-wear: it presents a lateral edge. A transverse gesture of rubbing, with a slight movement of the hand, may have created this particular morphology.

Prismatic polishers

Three prismatic polishers have been found (table 12, fig. 17). They are made of homogeneous metamorphic rocks. All the faces of these very particular polishers present a uniform use-wear and a pronounced concavity, with a longitudinal distortion. Some corners are also levelled by polishing. Sometimes, longitudinal striations can be observed. Use-wear distribution on the surfaces suggests a transverse polishing movement.

60. ADAMS, 2002: 77.
61. *Ibid.*: 143.
62. *Ibid.*: 142.

Table 11 – *Crushing cobbles main characteristics (raw material, shape, dimensions).*

Site	N°	Type	Raw material	Conservation	Length	Width	Thickness	Form	Technology (cavities)	Figure
Aruchlo	273	crushing cobble	metamorphic rock	complete	15	9,6	5,7	ovoid	on the 4 sides	
	276	crushing cobble	basalt	complete	8,4	7,5	5,6	quadrangular	shaped sides	20q
Khramis Didi	371	crushing cobble	metamorphic rock	complete	13	7	4,3	ovoid	on the 4 sides, 2 ends crushed	15d
	374	crushing cobble	metamorphic rock	complete	14,8	7,6	5,4	ovoid	on the 4 sides, 2 ends crushed	
	443	crushing cobble	sandstone	complete	14,6	7	4,2	ovoid	on 2 opposite faces	
	489	crushing cobble	sandstone	complete	9	8,4	4,6	quadrangular	on 2 opposite faces	
	415	crushing cobble	sandstone	complete	8	7,5	7	spherical	on 1 face	
	490	crushing cobble	metamorphic rock	complete	12	7	5,4	ovoid	on 2 sides, 2 ends crushed	
	350	crushing cobble	basalt	complete	12,8	9,4	6,2	ovoid	on 2 sides, 2 ends crushed	15i
Imiris	161	crushing cobble	metamorphic rock	complete	9	10	5,5	circul	-	12j
	136	crushing cobble	metamorphic rock	complete	9	8	6,3	ovoid	on 2 opposite faces	12h
	137	crushing cobble	metamorphic rock	complete	13	7	5,5	ovoid	on 2 sides, 2 ends crushed	
	158	crushing cobble	metamorphic rock	complete	12	8	4,2	ovoid	on 2 sides, 2 ends crushed	
	193	crushing cobble	metamorphic rock	complete	14	7	5,3	ovoid	on 2 sides, 2 ends crushed	
	209	crushing cobble	basalt	complete	13	8	7,2	ovoid	on 2 opposite faces	
	210	crushing cobble	sandstone	complete	13	8	5,8	ovoid	on 2 sides, 2 ends crushed	
	225	crushing cobble	basalt	complete	15	11	5	ovoid	on 2 opposite faces	
	145	crushing cobble	basalt	complete	17	10	6	quadrangular	on 2 opposite faces	
	150	crushing cobble	metamorphic rock	complete	14	11	6	quadrangular	on 2 opposite faces	12e
	192	crushing cobble	metamorphic rock	complete	16	8	6	quadrangular	on 2 sides, 2 ends crushed	

Table 12 – *Polishers main characteristics (raw material, shape, dimensions).*

Site	N°	Type	Raw material	Conservation	Length	Width	Thickness	Form	Working surfaces	Figure
Khramis Didi	493	handpolisher	sandstone	complete	-	-	-	quadrangular	2	17g
	386	handpolisher	fine sandstone	complete	10	11	7,4	triangular	4	17b
Imiris	212	handpolisher	sandstone	complete	13	9	4,1	quadrangular	1	11h
	115	handpolisher	sandstone	complete	10	9	4,4	triangular	4	11i
	100	handpolisher	sandstone	fragment	7	5	2	quadrangular	1	11l
Shulaveri	52	handpolisher	sandstone	fragment	11	8	2,5	quadrangular	1	7m
Khramis Didi	376	prismatic polisher	sandstone	complete	5,6	4,8	5	prismatic	3	17c
	456	prismatic polisher	sandstone	complete	9,5	4,5	5	prismatic	2	17d
Imiris	93	grooved abrader	metamorphic rock	complete	7	6	3,2	quadrangular	1	11k
	127	grooved abrader	sandstone	complete	19	8	4	triangular	1	11j
Khramis Didi	367	grooved abrader	sandstone	complete	14	10,6	8,6	triangular	1	17e
	397	grooved abrader	quartzite	complete	9,6	9,2	3,2	circular	1	17f
Imiris	94	grooved abrader	undet.	complete	8	6	4,1	ovoid	1	
	103	grooved abrader	sandstone	complete	9	7	5,1	ovoid	1	

Grooved abraders

Three types have been recognized amongst the seven grooved abraders (table 12, fig. 17).

• The first type is made of basaltic cobbles with no particular shape (three examples in Imiris gora n° 93 and 94, fig. 11k and Khramis Didi gora n° 367, fig. 17e). A longitudinal groove with a "U" section and more or less parallel sides occupies the centre of this tool. Because of the poor quality of the raw material, they often look like natural grooves.

• Two discoid cobbles of quartzite (Khramis Didi gora n° 397: 9,2 x 9 x 3,2 cm) (fig. 17f) and sandstone (Imiris gora n° 103: 9 x 7 x 5,1 cm) have a slightly convex base. The "V" sectioned groove has parallel sides. The upper face and the groove are completely polished. Some longitudinal striations occur in the bottom of the grooves.

• In Imiris, one grooved abrader is of particular type (n° 127: 19 x 8 x 4 cm) (figs. 3g and 11j); a long, thin and "V" sectioned groove is located on the longitudinal axe of a cobble of fine sandstone. A regular, longitudinal incision is visible inside the groove.

Type 2 could correspond to a form of grooved abrader (with transverse grooves) as known in Anatolia and also in Aratashen[63] at the same period.

Sharpeners

Ten tools can be classified as sharpeners, even if their actual function may be diverse (figs. 13e, 16c and 19p).

Six long cobbles of metamorphic rock present some residues of metal oxides on their natural surfaces. They are associated with striations and distortions of the surfaces caused by polishing or re-sharpening gestures.

Three other small, flat fragments present groups of striations associated with levelled zones of polishing. The groups of striations indicate the orientation of the abrasion gesture. This kind of use has also thinned and narrowed the end of one tool.

One limestone cobble was also used as a sharpener (14 x 5 x 4,5 cm) (fig. 13e). One face of this sub-cylindrical object shows regular polishing and transverse striations. The natural curvature of the surface was distorted by use.

Hand stones

Thirty-eight hand stones of ovoid shape and of small size were made on thick cobbles of various rocks (five basalt, seven sandstone, six granite, eight metamorphic rock and one limestone) (figs. 7l, 12g, 15b, h and 16l). No shaping has been identified (fig. 2e). One or two opposite faces present polished areas with short transverse striations (fig. 3h). Traces of ochre are often visible on the surface of these tools. The edges are often completely smoothed, and indicate the same use as the working surface. Hand stone 80 (fig. 12m) from Imiris gora is a cobble of large dimensions with traces of use as an anvil and pestle. Five of the hand stones were also used as hammer stones, and six as pestles, like many crushing cobbles.

Other polishing tools

Two more examples belong to the category of polishing tools even if their function is not yet determined.

Tool 513 of Khramis Didi gora was first attributed to the category of adzes, due to its morphological and technical characteristics. It has been shaped from a green soft metamorphic rock, of a particular kind which has no equivalent in the series (11 x 4,2 x 1,6 cm). The pattern of use-wear demonstrates its re-use as a shaft straightener or smoother. The object shows several levelled zones characterized by an extensive polish associated with groups of striations. Each zone corresponds to one moment of the object's use. The sides have long linear striations, one end shows traces of microflaking and the centre of one face has been used for fine percussion.

Tool 114 from Imiris presents the same morphological and use-wear characteristics (16 x 5 x 1,8 cm). Made of soft limestone, this losange-shaped object presents an edge-like end and levelled zones, with an extensive polish and groups of fine striations indicating the direction of movement. Both faces show obvious impacts of superficial percussion.

ADZES AND CHISELS

Several types of tools can be grouped together in the family of axes and adzes. They are all made of dark green heterogeneous metamorphic cobbles. None of these objects have been found in Shulaveri gora, 5 come from Imiris gora, 75 from Khramis Didi gora and 17 from Aruchlo I.

Adzes

Ninety adzes or similar tools have been found: 4 in Imiris, 81 in Khramis Didi gora (fig. 16) and 4 in Aruchlo (fig. 20). They consist of several types of cobbles and blanks made of green or black metamorphic rocks, and more rarely quartzite. Several degrees of know-how can be distinguished in the shaping of these tools. Adzes have been shaped either on retouched natural cobbles or flakes. The tools are approximately triangular, and present a semi-circular to lenticular section.[64] Two unsymmetrical rough-outs had been abandoned. The edges are generally convex, rarely flat. The flaking is generally irregular and deep. The butt of the adzes is shaped by flaking. The surfaces are regularized by an extensive pecking, before fine polishing. The polishing is generally made on the first third of the adze. Facets of polishing show longitudinal striations on the faces and sides. They correspond to the different steps of shaping, use and re-sharpening of the adzes. The edge is polished transversely over approximately half a centimetre wide.

63. BADALYAN *et al.*, 2007.

64. N° 91 of Wright's terminology (WRIGHT, 1992).

The edge is often microflaked when not completely damaged by use. A fragment of adze has a vertical, cylindrical perforation at mid-length, for hafting.

Given the multiple stages of use, re-use and re-sharpening of the adzes, their "life" must have been quite long. It seems that these tools were used until they could not be re-sharpened at all.

An adze from Khramis Didi gora has been notched on its sides, in order to isolate two different parts: the blade itself and the base ready for hafting. This part was thinned by hammering. This adze is unique in the Shulaveri-Shomu sites but its technology resembles examples from the Neolithic site of Odishi in Colchida and also from the Chalcolithic sites of Imeretie.[65] Although the morphology is not identical, this tool could be the only clue of contact between the two Neolithic entities of eastern and western Georgia.

One category of "adze-pestle" has been distinguished in order to describe the last stage of use or re-use of axes or adzes (three examples in Aruchlo and one in Imiris, fig. 20f, g). The edge of these former adzes has been completely flattened by percussion: the ovoid surface presents all the characteristics of use as a pestle. The morphology is suitable for precise handling (approximately 9 x 5 x 4 cm). These tools show fine impacts of percussion and a smoothing of the periphery of the surface. What was initially a simple case of re-use finally creates a new category of tool in the Shulaveri-Shomu series.

Narrow edged adzes

On three flat quadrangular cobbles of metamorphic rock, one special type of tool has been shaped. The natural surface has been retained on the base of the cobble, perhaps for hafting or handling. The sides have been regularly flaked out in order to thin the tool gradually and to obtain an edge of less than 2 cm wide. This narrow edge is surrounded by a dense pecking and is polished on no more than 2 cm long, with very thin and dense striations orientated either transversely or longitudinally. Polished facets indicate the different stages of use and shaping of the edge. Only part of the tool has been shaped, mainly by regular flaking.

Chisels

The scheme of shaping and preparation of adzes has also been used for the making of six chisels made on small ovoid flat cobbles (fig. 16a-b). The chisels sometimes have bilateral microflaking on their sides. Some present fine traces of pecking and thin striations. A levelled edge is generally but not systematically made by flaking; it is either polished or microflaked, and may be associated with the removal of large secondary flakes. The other edge is either macroflaked or flattened by percussion. Chisels are only present in Khramis Didi gora, where stone vessels were found. Use-wear analysis has not been carried out on these tools, but they may have been used for shaping the stone vessel described below.

65. NIEBIERIDZE, 1978.

PALETTES

Four natural cobbles from Imiris (n° 79, 92, 160) (fig. 10a) and Aruchlo (n° 285) (fig. 20k) show coloured deposits and slightly distorted zones of use on one or two of their faces (figs. 3i and 10a). The raw material has not been identified. The coloured areas present long, deep regular striations orientated longitudinally. These tools can be interpreted as palettes used for the soft grinding and the mixing of black or red colours just before their use. The aspect of the deposit evokes the preparation of a kind of paste, which would explain the absence of actual traces of colouring powder. They constitute further evidence for the processing of ochre on the Shulaveri-Shomu sites, already attested by the presence of colouring matters on other crushing, pounding and grinding tools.

In Imiris, another tool (n° 160) (fig. 10a) can be identified as a palette: it is a massive ovoid cobble (26 x 19 x 4 cm) with a wide zone covered by longitudinal striations and peripheral traces of ochre.

LIMESTONE ITEMS

All the 28 items are made of soft limestone (table 13). They are only present in Khramis Didi gora (19 examples) (figs. 18 and 19).

Stone vessels

Two types of stone vessels have been found in Khramis Didi gora (n° 335, 406, 472, 479): one with a vertical rim[66] (three examples) and another in the shape of an open bowl (one example) (fig. 18). Their base is either flat or round, and they measure around 8 x 5 x 5 cm. The pots were probably first hollowed out by hammering and pecking, traces of which have been identified. The rims are generally shaped by carving with

66. The globular bowl 117 of Wrigth's terminology (WRIGHT, 1992).

Table 13 – *Limestone items main characteristics (raw material, shape, dimensions).*

Site	N°	Type	Raw material	Conservation	Length	Width	Thickness	Form	Figure
Khramis Didi	472	stone vessel	limestone	fragment	6,5	5	3,2	circular	
	335	stone vessel	limestone	half	8	8	5,8	circular	18j
	406	stone vessel	limestone	half	8,5	4	5,3	circular	18d
	408	stone vessel	limestone	complete	8	6,8	3,2	circular	18e
	385	mortar	limestone	complete	9,2	7	4,1	circular	18g
	479	stone vessel	limestone	fragment	11	5	4,2	circular	
	363	mortar	limestone	complete	12,2	10	11	ovoid	18b
	385	mortar	limestone	complete	9,2	7	4,2	ovoid	
	407	mortar	limestone	complete	14	15	6,5	ovoid	18a
	348	mortar	limestone	complete	7	6,4	4,2	quadrangular	18f
	360	mortar	limestone	complete	11,5	9,2	8	quadrangular	18h
	364	mortar	limestone	complete	11,7	10	7,1	quadrangular	18c
	419	mortar	limestone	complete	7,6	8	3,4	quadrangular	18i
	332	ring	limestone	half	3,8	1,7	2	semi-circular	19n
	473	ring	limestone	half	4	2,2	2,1	semi-circular	19m
	474	ring	limestone	half	7	3,4	3,4	semi-circular	19l
	475	ring	limestone	half	3,6	2,2	1,2	semi-circular	
	435	pick (or sharpener?)	limestone	complete	26,5	4,5	3,2	quadrangular	19a
	468	hammer	limestone	complete	8,8	5,2	5,8	quadrangular	19g
	349	hammer	limestone	complete	7,4	5,2	4,8	quadrangular	19e
	416	hammer	limestone	complete	9,8	5,2	4,4	quadrangular	

a fine chisel, as suggested by removal marks. Most of the rims were first shaped by subhorizontal retouch, and then levelled by polishing. Next, the inside of the pots was scraped, as shown by the oblique and horizontal grooves along the rims and sides. Fine pecking was used to flatten the sides and edges. Both the inside and the outside were smoothed by polishing.

"Mortars"

Small cavities have been observed on eight quadrangular blocks of limestone (n° 348, 360, 363, 364, 385, 407, 419) (fig. 18). A fine strip of pecking occurs on the periphery of these small-cavities, reflecting a preforming stage. The sides of the blocks are coarsely flaked to produce an approximate shape. These objects are interpreted as mortars, even if they have little in common with the so-called massive mortars. Their function certainly differs from the massive ones, but their interpretation is a mystery.

Stone rings

Four broken rough-outs of limestone rings and one nearly finished have been found (n° 332, 473, 474, 475) (fig. 19 l-m). They represent half or a quarter of the initial rings. Their section is quadrangular and there are two sizes. The two opposite faces are flat, the external side is convex, and a central edge has been shaped. A biconical perforation has been made by percussion, removal of flakes and scraping of the internal surface. Fine pecking and superficial scraping were also used for the surfaces. The function of these rings remains unknown, but they are clearly not ornaments.

One circular fragment (n° 470) is hammered on its external surface. Some linear traces of scraping are visible at the top of the perforation made by circular abrasion. This object is perhaps a preform of a macehead, a hammer or even a weight.

"Hammers"

Five so-called "hammers" in limestone have been identified (n° 349, 416, 468) (fig. 19d-g); one similar example is made of vesicular basalt. These objects present two "heads" of triangular shape and ovoid section, reminiscent of present-day geological hammers.

They are separated by a central groove for hafting almost 1 cm wide, and are shaped by pecking. Inside the groove, some thin linear traces suggest a system of strings to hold the object. The sides (one flat, one convex) have been shaped by regular flaking. Pecking and scraping complete the shaping. The pointed ends are slightly crushed and microflaked by use. On one object, an end has been regularly pecked and a perforation was attempted by circular abrasion: different systems of hafting may have existed for these types of limestone hammers.

On the basis of their morphology, size and use-wear, these tools can be interpreted as hammers for the shaping of stone vessels.

Picks

Among the three limestone picks or "smoothers", one double pointed tool (n° 435: 26 x 4,5 x 3,2 cm) (fig. 19a) has traces of polishing together with longitudinal striations and superficial grooves from rubbing. One of the pointed ends has been created by a longitudinal rather than circular movement of polishing.

DIVERSE OBJECTS

Massive block with small cavities

A natural massive block (52 x 59 x 28 cm) of hard limestone was found in Khramis Didi gora. The sides were coarsely flaked. A large perforation, 16 cm wide and 14 cm deep, has been made in the centre. Twelve cavities of 3 to 7 cm in diameter were set around this perforation. They were made by a repetitive crushing action.

Perforated objects

Limestone items are not included in this paragraph (see above).

Maceheads

The five maceheads are of two types and present approximately centred cylindrical perforation.

For the first type, large ovoid cobbles of siliceous rock or vesicular basalt have been selected. One of the ends has been flattened by crushing, and there is a large circular perforation in the centre of the tool. Because of the hypothetical position of the hafting and the location of hammering traces, this tool may have been used as a hammer or a mace. The second type is perfectly spherical (two examples: n° 156: 13 x 13 x 12,8 cm from Imiris gora and n° 470: 5,6 x 4,6 x 3,8 cm from Khramis Didi gora). Fragments of sandstone, basalt or limestone have been perfectly shaped by pecking, and regularized by polishing. Special techniques must have been used to give a shiny aspect to the surfaces. The cylindrical perforation was made by abrasion, according to the circular traces that can be observed at the bottom of the cavity. This object must have had a symbolic value as no traces of use have been detected (for example a sceptre).

Spindle-whorls

Two discoid cobbles of metamorphic rocks or basalt were perforated. The perforation is tronconical and under 5 cm in diameter. Although these objects are extremely rare at Khramis Didi gora, they can be interpreted as spindle whorls (fig. 19i).

Undetermined perforated cobbles (fig. 19j-n)

Nine basalt cobbles, mostly ovoid in shape, present a natural surface without any visible use-wear.[67] A tronconical perforation occupies the centre (fig. 3j). When the perforation has not been made by percussion, circular traces suggest the use of a mechanical system. They can be interpreted as weights.

A unique circular object was found at Khramis Didi gora. The first steps of a small circular perforation (less than 1 cm deep) are visible on its upper part while the whole external surface has been regularised.

A triangular object in limestone can be interpreted as an ornament.[68] Its faces are covered by polished and striated levelled surfaces, organized longitudinally (fig. 3j). A biconical perforation was made for suspension next to one end.

Splintered tools

Eight splintered tools have been found on all the sites except Shulaveri (table 14, fig. 13g-h). They are made of cylindrical or ovoid cobbles in metamorphic rock and present worked ends: macro and microflaking are visible on one end while crushing and percussion impacts are associated on the opposite end. Impacts of pecking are sometimes concentrated on the faces. These tools can be considered as splintered tools, used to remove superficial matter.

Sling-stones

Small cobbles of sandstone or limestone of regular dimensions (30 to 40g) do not show traces of use. They can be interpreted as sling-stones, for the hunting of small mammals or birds for example.

Ochre

One disc of ochre is present in Khramis Didi gora (n° 467: 6,2 x 5,8 x 2,4 cm) (fig. 19q). Several facets of abrasion are visible on its periphery, while striations show the orientation of the main direction of use. According to use-wear distribution,

67. Kiguradze, 1986: figs. 10-16.
68. *Ibid.*: fig. 35:1.

Table 14 – *Splintered tools main characteristics (raw material, shape, dimensions).*

Site	N°	Type	Raw material	Conservation	Length	Width	Thickness	Form	Figure
Aruchlo	251	splintered tool	limestone	complete	10,8	4,2	2,4	cylindrical	23d
Imiris	89	splintered tool	metamorphic rock	complete	15	5	3,4	cylindrical	13f
	90	splintered tool	limestone	complete	9	5	2,2	quadrangular	13g
	88	splintered tool	limestone	fragment	12	5	4	ovoid	
Khramis Didi	333	splintered tool	limestone	complete	10,6	2,5	2,6	cylindrical	19c
	496	splintered tool	limestone	complete	12,6	3,4	3,1	cylindrical	
	495	splintered tool	metamorphic rock	complete	15,4	4,5	3,2	ovoid	
	303	splintered tool	limestone	complete	12	3,2	2,2	quadrangular	16d
	304	splintered tool	andesite	complete	8	4,2	2,6	triangular	
	436	splintered tool	sandstone	fragment	7,7	2,8	1,6	ovoid	
Shulaveri	18	splintered tool	limestone	fragment	9,4	3,6	2,3	cylindrical	

this fragment of ochre was probably used both for the shaping by abrasion and the colouring of another object.

MACROLITHIC IMPLEMENTS AND THE SHULAVERI-SHOMU SITES OF KWEMO-KARTLI

MACROLITHIC IMPLEMENTS AND THE STATUS OF KWEMO-KARTLIAN SITES

Even if the four Kwemo-Kartli sites broadly share the same macrolithic toolkit, actual specificities can be identified through functional analysis. However, interpretation is difficult in terms of chronology because of the general lack of stratigraphic context for these objects.

The site of Shulaveri gora presents a high ratio of grinding tools, as if food preparation was still a major activity. In fact several categories of tool are rare if not completely absent in Shulaveri gora. The lack of pestles, polishing tools, splintered tools and crushing cobbles could reflect the minor importance of craft activities. The diversity of hammer stones fits with a multifunctional use and a lack of specialisation. Finally, the importance of cereal processing on the oldest site of Kwemo-Kartli would reflect the central role occupied by food processing for Shulaveri-Shomu populations that first settled.

While there are no polishers at Shulaveri gora, polishing and abrading activities have been identified at Imiris gora and Khramis Didi gora.

At Imiris gora, there is a specific grinding slab, as well as mortars and pestles. In fact lower tools are attested on the site. The particular status of grinding and pounding activities at Imiris gora is illustrated by a specific grinding slab (six examples), with only one or two examples on the other sites. These circular and double faced tools are of standard shape and were certainly used to produce a fine mineral powder (linked or not with grinding colours). Pestles also constitute an important part of the stone toolkit. Colour processing played an important role in activities at Imiris gora, as suggested by a wide range of objects used for this activity, such as pestles, grinding slabs and palettes.

The macrolithic tool assemblage of Khramis Didi gora is the most diversified from Kwemo-Kartli. Its two main particularities are the high number of adzes and the presence of an original limestone vessel. All the stages of shaping (by flaking, pecking and polishing) and use of the adzes are represented. While function of these tools as adzes is clear, their abundance is difficult to interpret. Does this over-representation correspond to additional clearance for the creation of new living or agricultural areas? Did these tools have a different function, such as wood working for other purposes (building, etc.)?

Limestone objects including rings, mortars, pots, spoons, spindle-whorls and hammers were found only in Khramis Didi gora. The presence of stone vessels could be linked to the important series of anthropomorphic clay figurines found on the site, as both productions reveal aesthetic expression and a high level of expertise. Given the complete absence of limestone vessels on the other sites of Kwemo-Kartli, this technique may have been imported from southern regions where stone vessels are well known.[69]

This diversity of macrolithic tool equipment in Khramis Didi gora is not an isolated feature, as shown by the rest of the material culture: this is especially noticeable for the bone and

69. See for example the halafian site of Tell Amarna, in Tunca and Molist, 2004.

antler tools (engraved and sculpted tools), clay figurines, and also for the chipped stone industry (quality of the retouch).[70]

The macrolithic tool assemblage of Aruchlo I is too small to isolate one particular category of tool, but it is representative of the Shulaveri-Shomu macrolithic equipment of Kwemo-Kartli.

MACROLITHIC TOOLS FUNCTION AND ACTIVITIES ON KWEMO-KARTLIAN SITES

A use-wear analysis was conducted at low magnification, in order to determine the exact function of each category of tool. Given the diversity of raw materials, our initial methodology had to be adapted.[71] The main problem concerned basalt tools for which we had no experimental reference examples. We refer to Dubreuil's work for a description of experimental use-wear on basalt tools.[72] Specific optical equipment was used for the functional analysis. As there were no suitable binocular microscopes in the various Georgian archaeological institutions, we decided to work with the "Proscope©" system. This enables observations up to x200 (with four lenses at x5, x20, x50, x100 and x200). Its handiness was particularly well suited to the rather mobile working conditions, despite the low quality of the photographs produced.

Eighty-nine objects out of 554 have been analysed: 20 objects from Shulaveri gora, 28 from Imiris gora, 33 from Khramis Didi gora and eight from Aruchlo I (table 15). Almost all categories of tools and objects have been chosen: grinders and querns, mortars, pestles, palettes, hammer stones, polishers, hand stones, perforated and grooved abraders (table 16).

Half of the querns and two-thirds of the grinders (79%) were used for cereal processing. 40% of the first and 15% of the second were used to grind a mineral material, including pigments. These proportions can be interpreted in terms of different re-use: it seems that querns were much more recycled than the upper tools, especially for the grinding of pigments or as anvils for their crushing. But their initial function was certainly for cereal grinding. Considering more precisely the morphology of grinding tools, all types of lower parts were used both for cereal and processing of pigments whereas a majority of the ovoid upper parts were used for cereal processing (78%) while quadrangular ones were used for mineral grinding (only two objects).

Table 15 – *Representativity of each functional category by sites in number of analyzed tools.*

Category	Site				
	Shulaveri	Imiris	Aruchlo I	Khramis Didi	total
Cereals	13	8	2	1	**24**
Vegetal grinding		3	1		**4**
Animal matter grinding	1				**1**
Hard mineral matter grinding				1	**1**
Mineral grinding	2	1	1	3	**7**
Abrasive powder grinding	1	3		3	**7**
Colour grinding	1	2	2	5	**10**
Colour grinding and rubbing		2		2	**4**
Palette with coloring		1			**1**
Colour items		2			**2**
Mineral matter polishing		1	1	2	**4**
Bone polishing		1			**1**
Wood or bone polishing		1			**1**
Wood sharpening				1	**1**
Undetermined sharpener				2	**2**
Pick		1			**1**
Link traces				1	**1**
Pick		1			**1**
Undetermined perforated tool				3	**3**
Mechanical perforation		1		1	**2**
Sharpener - splintered tool				1	**1**
Undetermined	2	1		7	**10**
Total	**20**	**28**	**8**	**33**	**89**

The function of mortars and pestles is more difficult to determine: mortars were used for mineral and plant processing, while pestles were mainly used for pigment transformation. A similar function can be mentioned for crushing cobbles and for grinding slabs which were used especially for the grinding of a very fine and abrasive mineral powder (which could correspond to a colouring powder). The importance of this colouring activity is confirmed by the fact that two thirds of the hand stones were used for it.

Three polishers were used for mineral abrasion (polishing of axes or adze-like tools), whereas one grooved abrader was used for bone abrasion.

According to our analysis, the activities are represented as follows: 24 tools were used for cereal grinding, 12 for mineral grinding and 19 for colouring grinding or preparation (including one palette, two fragments of colouring material and four tools used for the preparation of a colouring paste or for its application on a soft material). Bone grinding has been identified on two objects, while bone abrasion concerns only one grooved abrader. Four polishers were used for hard mineral polishing (such as axe or adze polishing). The motion and gesture were

70. KIGURADZE, 1986.
71. HAMON, 2006.
72. DUBREUIL, 2004.

Table 16 – *Representativity of each functional purpose by category of tool (number of objects).*

Category	Querns	Grinder	Mortar	Pestle	Grinding slab	Broyon	Crushing stone	Hammer stones	Hand abrader	Netherstone	Grooved abrader	Hand stone	Splintered tools	Ochre disc	Palette	Perforated item	Pick	Mortar-stone vessel	Undetermined	Total
Cereals	8	15				1														**24**
Vegetal grinding	1		1		1		1													**4**
Animal matter grinding								1			1									**2**
Hard mineral matter grinding				1																**1**
Mineral grinding	2	2	1				1					1								**7**
Abrasive powder grinding	1		1		2							1								**5**
Colour grinding	3	1		4			3	2				2			1					**16**
Colour items												2								**2**
Colour grinding and rubbing												1		1						**2**
Palette with colour															1					**1**
Mineral matter polishing									2	1		1								**4**
Wood sharpening											1									**1**
Wood or bone polishing								1												**1**
Undetermined sharpener								2												**2**
Pick				1																**1**
Link traces																			1	**1**
Mechanical perforation								1								1				**2**
Undetermined perforated tool																	1		1	**2**
Sharpener - splintered tool													1							**1**
Undetermined		1	3		1	1			1			1	1					1		**10**
Total	**15**	**19**	**6**	**6**	**4**	**2**	**5**	**7**	**3**	**1**	**2**	**9**	**2**	**1**	**2**	**1**	**1**	**1**	**2**	**89**

determined for eight tools (mainly perforated cobbles). Finally, the use of eight other tools is still undetermined.

Cereal processing, grinding of pigments and stone polishing are the three main activities represented. The respective proportions of these activities between the different sites underline several tendencies:

• In Shulaveri gora, cereal processing is dominant (65%), contrary to Khramis Didi gora where cereal processing is scarcely represented. This observation fits with the relatively diversified tool categories and morphologies on the latter site.

• The proportion of mineral matter grinding is equivalent on the four sites, even if this observation is based on a very small number of tools. This activity is not specific to one particular site.

• Pigment grinding is also rather stable on three sites but is quite rare in Shulaveri gora, which could imply that this activity had not yet appeared when Shulaveri gora was first occupied.

At Shulaveri gora, food preparation and cereal processing constitute the main activity carried out with macrolithic tools. At Imiris, a broad range of macrolithic tools was used for mineral and pigment grinding. At Khramis Didi gora, diversified macrolithic equipment appears with stone vessels and intensive use of adzes. Despite a common economic basis, real specificities can be isolated in the activities of the different occupations. The lack of stratigraphic contexts for the objects make these specificities difficult to interpret. Yet the earliest sites, Shulaveri gora and Imiris gora, show much evidence for cereal processing, with more diverse activities (pigment grinding) and macrolithic implements (grinding slabs and palettes in Imiris gora / adzes and limestone vessel at Khramis Didi gora) characterizing the later sites.

MACROLITHIC TOOLS EVOLUTION AND CHRONOLOGICAL IMPLICATIONS

Even if the exact stratigraphic position of the objects is not known, a seriation of the tool sub-categories on the four sites provides some elements for consideration of chronological trends in the macrolithic toolkit. Our discussion will rely on the seriation initially proposed by Kiguradze.[73]

73. KIGURADZE, 1986: 99.

Chronology of the grinding tools

We will first examine change in the shape of grinders on the four sites. Kiguradze has already observed that saddle querns and longer grinders appear in stage III of his classification of the Kwemo-Kartlian sites, and are still present in the next stages.

The distribution of flat and short grinders is not uniform between the sites, both in number and proportions. The chronology of querns is not as clear.

At Shulaveri gora, flat grinders are completely absent despite numerous grinding tools. All querns are concave and present a distal or peripheral edge, such as the saddle quern from level IV.[74] One massive mortar was also found in the same level.

At Imiris gora, both types of grinders (short and flat) coexist in equivalent proportions. A massive mortar was also found on the site. Querns of ovoid or quadrangular shapes are more or less concave and are delimited only in some cases by a distal edge. The morphology of querns is slightly different in Imiris gora, as ovoid flat tools are much more numerous in level V and quadrangular flat tools appear in level IV.[75]

At Khramis Didi gora, on the contrary, short grinders are rare compared to flat ones, whatever the raw material. Querns still show a certain concavity but no longer have a distal edge.

On all the sites, massive concave querns coexist with narrow flat ones.

In conclusion, a chronological development can be proposed from short grinders / ovoid concave querns with distal or peripheral edges at the beginning of the sequence to flat grinders / plano-concave querns without edges at the end of the sequence. Narrow grinding tools could be late, but their scarcity makes interpretation difficult. Given this general trend, it does seem that the relative frequency of each category of quern and grinder offers a reliable means of obtaining an approximate dating.

Macrolithic tools and the chronological position of Aruchlo

The seriation of macrolithic tools confirms the succession of Shulaveri gora and Imiris gora, and the late position of Khramis Didi gora in the chronological sequence.

The study of macrolithic tools from Aruchlo I provides new evidence for its chronological position in the sequence of the Shulaveri-Shomu sites from Kwemo-Kartli. Two querns from Aruchlo (one with a peripheral rim and one ovoid tool with a strong concavity) look very much like examples from Shulaveri gora. The relative frequency of tools used for pigment processing at Aruchlo I is similar to Imiris gora but lower than Khramis Didi gora. As at Imiris gora, pigment palettes have been found at Aruchlo. In addition, the function of the macrolithic implements from Arucho I is very close to the spectrum identified in Imiris gora. These elements indicate a degree of similarity between the macrolithic tools from Aruchlo I and the latest layers of Imiris gora. This could be interpreted as evidence for their relative contemporaneity.

In the light of these observations, a chronological trend can be proposed for macrolithic implements. At the beginning of the sequence, food procurement was still the main preoccupation of the Shulaveri-Shomu populations. The development of the settlements of Imiris gora and Aruchlo I induced a diversification of activities, among which pigment processing took an increasing part. This diversification of the macrolithic toolkit culminates in Khramis Didi gora. The development of this settlement required new territory, with forest clearance suggested by the presence of adzes. This development has enabled considerable craft and artistic production, like stone vessels and anthropomorphic figurines, which may have increased in parallel with reinforcement of exchange with southern cultures.

KWEMO-KARTLIAN MACROLITHIC TOOLKITS IN THEIR REGIONAL AND CULTURAL CONTEXT

An important and unsolved question is the place occupied by these Kwemo-Kartlian toolkits within the Neolithic and Chalcolithic of Transcaucasia. Can some comparisons help us understand the origin and specificities of these sites within a regional context?

Kwemo-Kartli among the Neolithic and Eneolithic cultures of eastern Transcaucasia

Some comparisons can be made with the help of illustrations of macrolithic artefacts from Transcaucasia.

For example, a quern with two small cavities from Shomu tepe is similar to one from Khramis Didi gora with a dozen small cavities on its face.[76]

But most comparisons are with Chalcolithic sites from eastern Transcaucasia. On the Early Chalcolithic site of Kültepe

74. KIGURADZE, 1986.
75. Chronological attribution after MASSON, 1982: 142.
76. After NARIMANOV, 1965.

(Naxcivan), the grinding tools are flat and ovoid with distal edges and one massive mortar has a central small percussion cavity.[77] Double perforated axes are also present and look very much like an example from Khramis Didi gora, even though such tools are also known from the Late Chalcolithic site of Tekhut[78] (Ararat plain).

These common features match the chronological sequence established by Kiguradze between the different horizons of the Shulaveri-Shomu culture.[79] The oldest layers of Khramis Didi gora (IX-VI) are broadly contemporary with Shomutepe, whereas the later layers of Khramis Didi gora (IV-I) correspond to the industry of KülTepe I and perhaps Tekhut as well.

All these observations seem to confirm that the upper layers of Khramis Didi gora are rather late in the sequence and date to the beginning of the Chalcolithic period.

East-West relationships during the Neolithic and early Chalcolithic

Very few grinding tools have been identified at Odishi (Colchida plain). One ovoid grinder is narrow and presents two distal edges,[80] characteristics that correspond to the equipment of stage 4 of the Shulaveri-Shomu culture, *i.e.* the Neolithic-Chalcolithic transition (Khramis V-I). On this sole basis, the site of Odishi should date to the end of the Shulaveri-Shomu culture, and thus corresponds to the end of the Neolithic in western Georgia.

West Georgian adzes are more massive and do not have the same shaping "chaîne opératoire" as in the Shulaveri-Shomu culture. The shaping of small natural cobbles characterizes the adzes of the Shulaveri-Shomu culture, whereas an actual sequence of shaping has been recognized on western sites. In addition to "shouldered" adzes of specific morphology, all the stages of adze shaping are found on western Neolithic sites: rough-out, flakes and even re-used adzes occur together in the assemblages. In fact, no correlation between western and eastern adzes can be established.

According to Lukin,[81] the site of Kistrik (Abkhazia) offers a more diversified toolkit. Pestles and splintered tools on cobbles, grooved abraders and small adzes share some features with Shulaveri-Shomu tools. Ovoid oblong querns, made on unshaped cobbles of basalt, are quite similar to the Neolithic ones from Kwemo-Kartli.

77. After ABIBULLAEV, 1963.
78. TOROSJAN, 1976.
79. KIGURADZE, 1986.
80. NIEBIERIDZE, 1986: fig. 12.
81. LUKIN, 1950: 257.

In conclusion, very few comparisons can be drawn between the Neolithic grinding tools from eastern and western Georgia. If some contacts between western and eastern Neolithic cultures of Georgia have occurred, they apparently concerned the subsistence economy rather than "craft" activities.

Kwemo-Kartli and the Neolithic cultures of the Near East

On the basis of published macrolithic implements, Shulaveri-Shomu tools from Kwemo-Kartli have no characteristics in common with the poor macrolithic tools of the steppe cultures of north Caucasus, where querns are small and there are no mortars.[82] Rather broad comparisons can be made between Shulaveri-Shomu and Near Eastern macrolithic tools, especially since macrolithic implements from the Kwemo-Kartlian sites cannot be attributed to precise layers. Nevertheless, a chronological trend in macrolithic implements can be established. The grinding and pounding tools, tronconic pestles, perforated cobbles and massive mortars from the first layers of the Shulaveri-Shomu culture also occur in aceramic and Neolithic horizons in central and eastern Anatolia.[83]

A closer look at possible similarities between Shulaveri-shomu and Halaf macrolithic tools enables us to define some common features, which can rarely be considered significant because of the ubiquity of macrolithic tool types throughout Anatolia and Mesopotamia.

Macrolithic implements from the upper layers of Shulaveri gora and from the layers of Imiris are apparently very similar to late Neolithic and early Halafian tools from Anatolia.

The basalt grinding and pounding equipment from the Late Neolithic and Early Halafian layers of Tell Sabi Abyad in Syria[84] is very similar to the material from Shulaveri and Imiris gora. On the Syrian site, conical or cylindrical pestles were used either with portable or massive mortars. One palette with red and black traces of pigment was also identified. Traces of ochre are mentioned for the majority of the macrolithic assemblage, as well as on the floors of several small rooms.[85]

Tronconical pestles, grinding slabs used for ochre processing, palettes, small polished chisels and one grooved cobble from Late Neolithic levels (II-VI) at Hacılar in south-western Anatolia[86] are similar to those found on the Shulaveri-Shomu

82. NEKHAEV, 1990.
83. See for example ESIN *et al.*, 1991; KORFMANN, 1982; PALMIERI, 1969; ROSENBERG *et al.*, 1995.
84. AKKERMANS, 1989: 285-286.
85. VERHOEVEN, 1999.
86. After MELLAART, 1970.

sites from Kwemo-Kartli. Mortars are also made in the same coarse rocks as at Shulaveri and Imiris gora. Palettes from Late Neolithic levels (VI-VII) of Hacılar[87] are very similar to Imiris gora ones, as are many of the palettes from the Halafian layers of Tell Kosak Shamali.[88]

Ochre processing seems quite an important activity in the intermediate layers of the Shulaveri-Shomu culture and early Halaf. Nevertheless, ochre processing on grinding stones and palettes has been widely recorded in the Near East, on a broad chronological and geographical scale. The use of ochre for painting, as in the decoration of Hacılar buildings at Hacılar,[89] is a possible explanation for the large quantities of ochre processed on Shulaveri-Shomu sites.

Common features are shared between the latest occupation layers of Khramis Didi gora and the middle to late Halaf culture. The presence of exclusively flat slabs in level VIII of Kurban Höyük confirms the chronological match between the end of the Shulaveri-Shomu culture and the middle Halaf culture.[90] A large slab with several small cavities from Khramis Didi Gora looks very much like a tool found in the Halaf phases I and II of Tilkitepe.[91] Limestone vessels, similar to the one found in Khramis Didi Gora, are also frequent in Halafian contexts.[92]

The connection between macrolithic tools of the Neolithic cultures of Kwemo-Kartli and Anatolia can only be established on a broad geographical and chronological scale. As for the internal seriation of Shulaveri-Shomu culture macrolithic tools, these comparisons are difficult to interpret, given the slow rate of change in these tools throughout the region, from the very beginning of the Neolithic to the beginning of the Chalcolithic.

Shulaveri-Shomu *versus* Sioni macrolithictools

The relationship between the Shulaveri-Shomu culture from Kwemo-Kartli and the following Chalcolithic Sioni culture in the eastern part of Georgia is not yet clear. The Sioni culture has been recently identified through pottery, but more work is required on its economy and origins. The earliest ceramics are mineral tempered, while later material is straw tempered. The pottery is flat-bottomed, with an ovoid to sinuous profile which becomes more angular at a later stage. Incisions on the rim and comb-made designs constitute the main decoration, but relief decoration is scarce even though comparable to Shulaveri. Painted pottery is found in small quantities on the sites, and some examples from Berikldeebi may have been wheel-made.[93] Depending on regional and probable chronological variations, the nature of the settlements varies (open-air settlements, rock-shelters, etc.), as does the architecture (buildings made of stone or mud-brick, semi-sunken structures). The sites are generally of short duration. Several features point to an important change in subsistence economy: a dominant flake industry, simple bone and macrolithic tools, and coarse pottery.[94]

Several authors have pointed out the possible relationship between the two cultural entities of Shulaveri-Shomu and Sioni, especially when considering the ceramic material.[95] The examination of Sioni macrolithic tools has raised the question of the subsistence economy of the Chalcolithic population, as well as its possible link to the Shulaveri-Shomu culture. Is there continuity between the two cultural entities in terms of stone-tool assemblages and techniques of plant food preparation?

In fact, Sioni grinding and pounding tools from the Aragvi valley share very few morphological and technical characteristics with the Shulaveri-Shomu material.[96] The Sioni grinding tools are made on short basalt cobbles, abundant along the riverbed of the Aragvi. They present very few traces of shaping and preparation, which could reflect a short use-life. They occur together with numerous small mortars (average diameter 12 cm) and crushing hand stones with small cavities. Unfortunately, large numbers of concretions in the cavities of many mortars made the application of use-wear analysis difficult. Several grinding stones were apparently used for cereal processing and a few mortars for processing other plants. A few pestles were used for pounding mineral matter. Pestles were thus used mainly for plant food preparation. These results are very similar to those obtained on several sites of the Chalcolithic culture of Maïkop (4th millennium),[97] and provide evidence for a decrease in cereal processing.

Several geographical factors could perhaps explain the differences between the two assemblages: raw material availability, a specific subsistence economy related to the hilly landscape, and so on... But the clear homogeneity of tool morphology and shaping suggests a different interpretation. In fact, macrolithic tools indicate a real shift in food preparation

87. MELLAART, 1978: fig. 174.
88. NISHIAKI, 2003: 121-183.
89. MELLAART, 1978: fig. 174.
90. ALGAZE *et al.*, 1984.
91. KORFMANN, 1982.
92. See for example Tell Amarna I in TUNCA and MOLIST, 2004.
93. LOMBARD et CHATAIGNER, 2004: 72.
94. *Ibid.*: 70-72.
95. MENABDE and KIGURADZE, 1981.
96. CHIKOVANI, 1999.
97. HAMON, 2007.

techniques and dietary habits between the Shulaveri-Shomu and the Sioni cultures. They must illustrate a deeper change in the subsistence economy and economic basis of the Sioni populations compared to the agricultural, sedentary way of life of the Shulaveri-Shomu populations of Kwemo-Kartli.

CONCLUSION

If cereal processing appears to be the main activity at Shulaveri gora, the importance of ochre processing at Imiris gora and the diversification of the stone toolkit (stone vessels and adzes) at Khramis Didi gora demonstrate the implication of macrolithic implements not only in the subsistence economy but also in the processes of craft production at the end of the Shulaveri-Shomu sequence. The functional specificities of each settlement can be interpreted in terms of both activities and chronological and cultural relationships. Change in the morphology of grinding tools through the sequence and the connections between the pounding equipment (mortars and pestles) from Kwemo-Kartli and material from eastern Anatolia and northern Mesopotamia all provide additional information on the economic and cultural background of these Georgian sites.

The results presented here only represent a first stage in the study of Neolithic and Chalcolithic macrolithic implements of Transcaucasia. Their significant role in domestic and craft activities raises new questions about the Shulaveri-Shomu culture, especially concerning the importance of pigment processing and its final use. But they also offer the opportunity to make broader comparisons, firstly with west Georgian Neolithic macrolithic implements, secondly with finds from Chalcolithic settlements of the Sioni culture, and lastly with material from various Neolithic and Chalcolithic cultural contexts in Anatolia and Mesopotamia.

ACKNOWLEDGMENTS

This study was funded by the French Ministry of Foreign Affairs, through a post-doctoral Lavoisier grant in 2005-2006. I would first like to thank B. Lyonnet (CNRS, UMR 7192) and Z. Makharadze (National Museum, Tbilisi) for their constant help and academic support. My thanks go also to colleagues from the Archaeology Department of the National Museum of Tbilisi: to M. Menabde who gave me access to the collections and to T. Kiguradze's documentation, and to N. Grigolava, M. Jalabadze, T. Meshveliani and E. Qoridze for their constant help in this work. I am also grateful to G. Mirtskhulava (Tbilisi Centre for Archaeological Science) and S. Hansen (Deutsche Archäeologisches Institut) for showing me the finds from Aruchlo I. The study of the Sioni macrolithic toolkits from the Aragvi valley would not have been possible without the collaboration of G. Chikovani, and the help of G. Kogodjouri from the Centre for Archaeological Science, Tbilisi. Lastly I would like to thank C. Chataigner and B. Lyonnet for all their help in the correction of this article, and M. Ilett for the great work on the correction of the English language.

Caroline HAMON
UMR 7041 ArscAn, Protohistoire européenne
Maison de l'archéologie et de l'ethnologie
21, allée de l'Université
92023 Nanterre cedex
FRANCE
caroline.hamon@mae.u-paris10.fr

Fig. 4 – *Shulaveri gora basalt querns:* ***a****, n° 62;* ***b****, n° 10;* ***c****, n° 50;* ***d****, n° 43.*

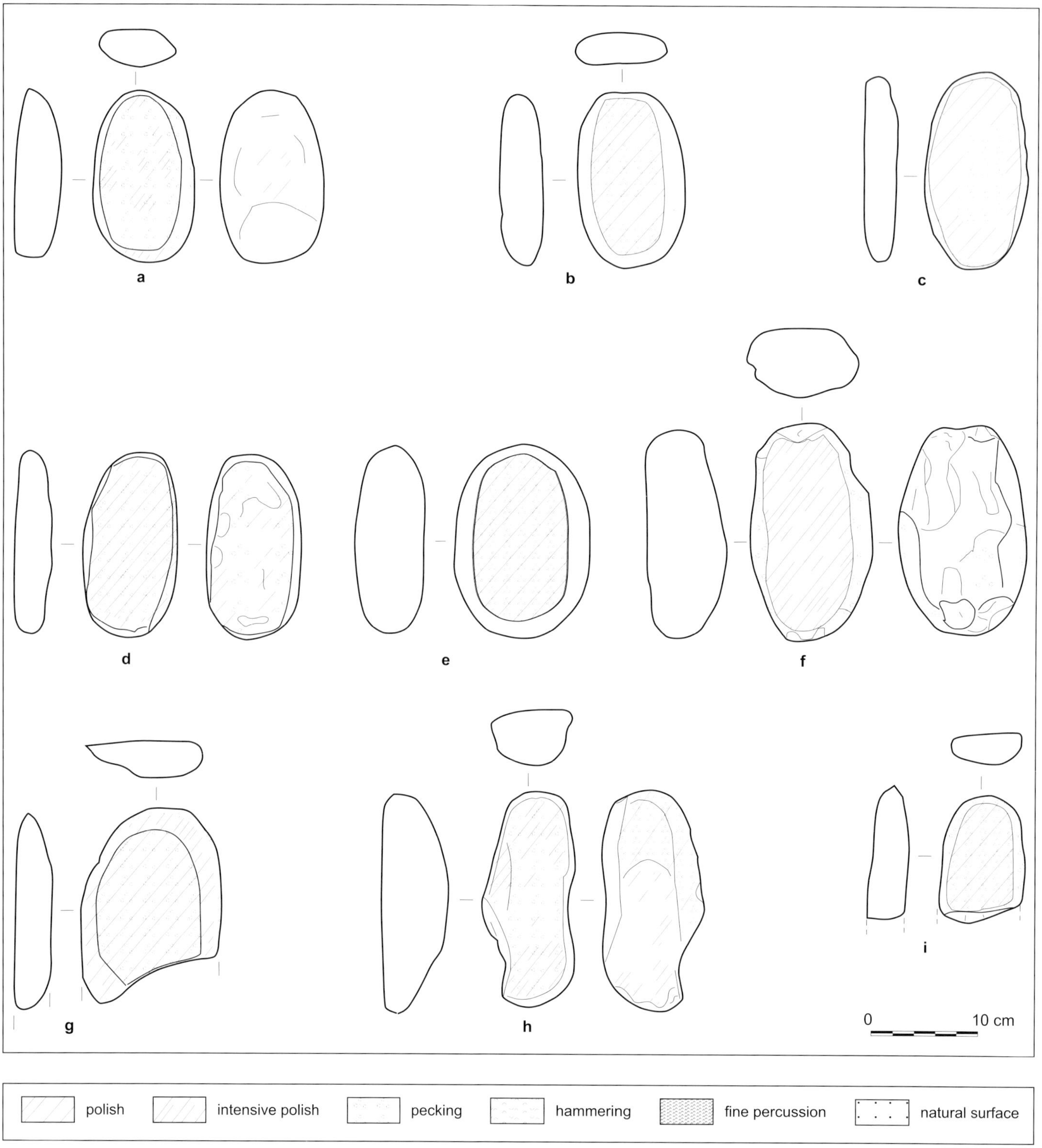

Fig. 5 – *Shulaveri gora short grinders:* ***a****, n°59;* ***b****, n°43;* ***c****, n°58;* ***d****, n°3;* ***e****, n°11;* ***f****, n°48;* ***g****, n°46;* ***h****, n°48;* ***i****, n°44.*

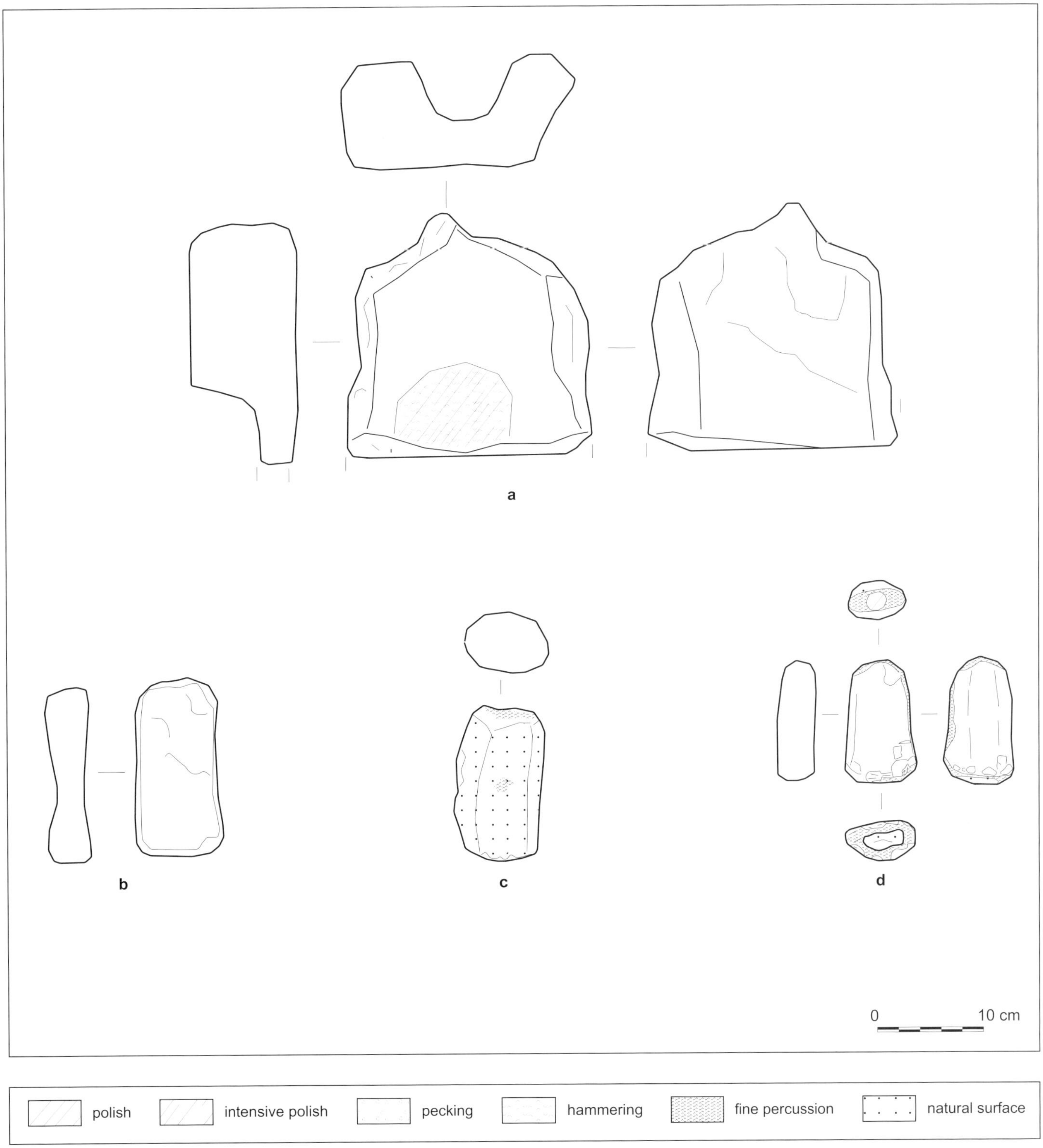

Fig. 6 – *Shulaveri gora, mortars and pestles:* ***a****, mortar 54;* ***b****, pestle 57;* ***c****, pestle 33;* ***d****, pestle 25.*

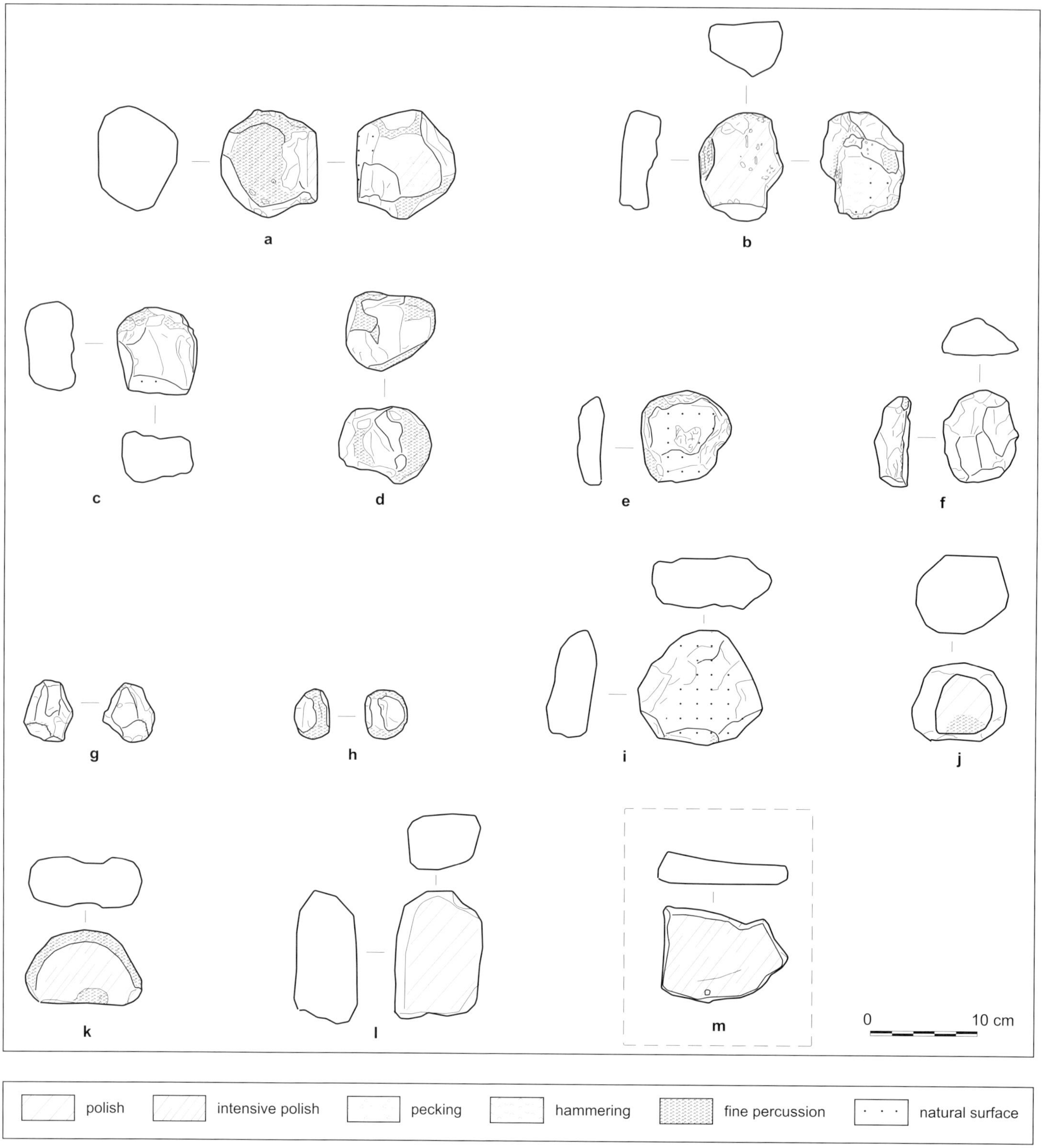

Fig. 7 – *Shulaveri gora hammer stones and polishers:* ***a****, hammer stone 6;* ***b****, hammer stone 12;* ***c****, hammer stone 15;* ***d****, hammer stone 29;* ***e****, hammer stone 21;* ***f****, hammer stone 13;* ***g****, hammer stone 16;* ***h****, hammer stone 2;* ***i****, hammer stone 19;* ***j****, "broyon" 34;* ***k****, crushing cobble 41;* ***l****, hand stone 35;* ***m****, hand polisher 52.*

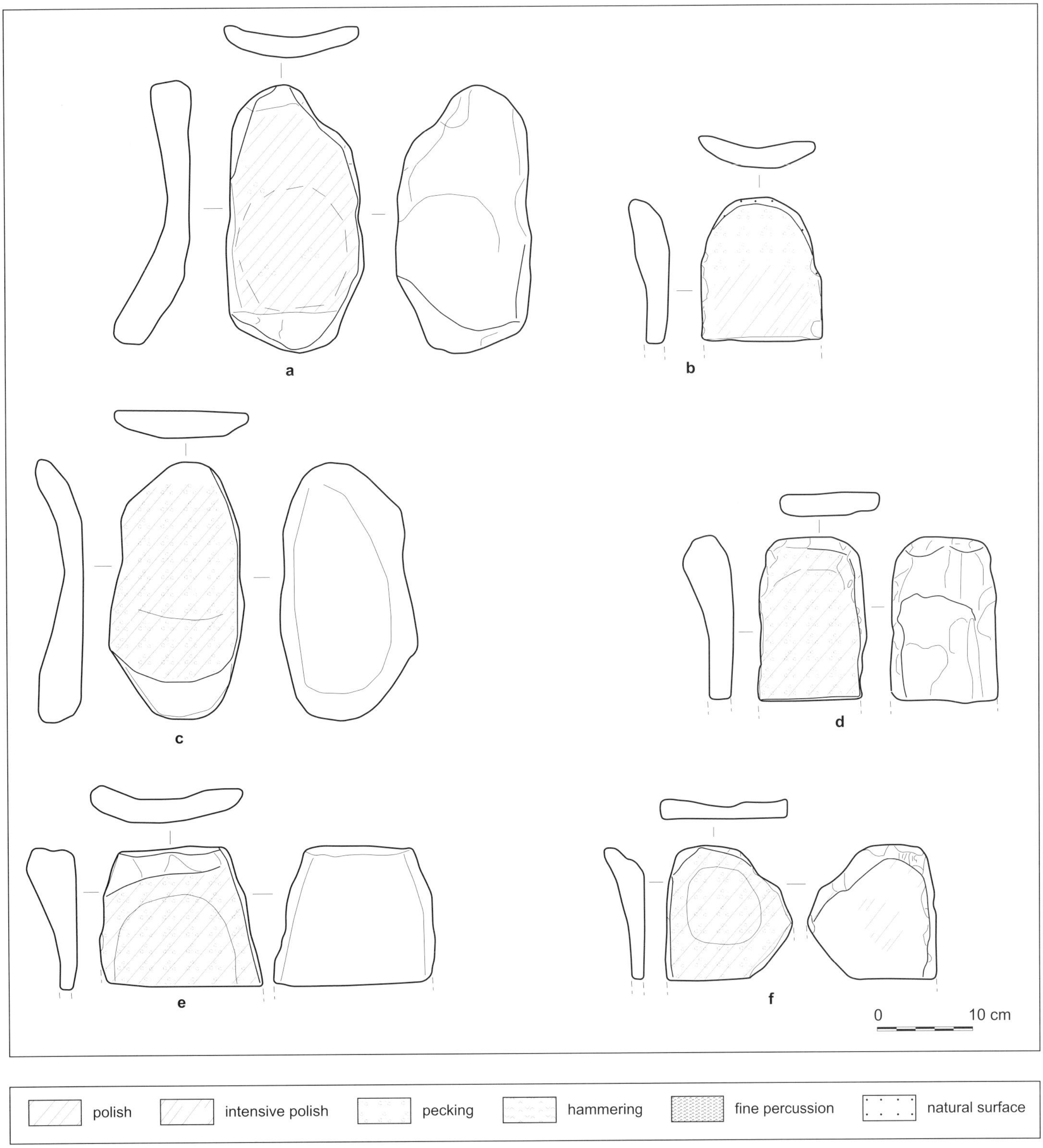

Fig. 8 – *Imiris gora querns:* ***a****, n° 230;* ***b****, n° 163;* ***c****, n° 129;* ***d****, n° 165;* ***e****, n° 130;* ***f****, n° 64.*

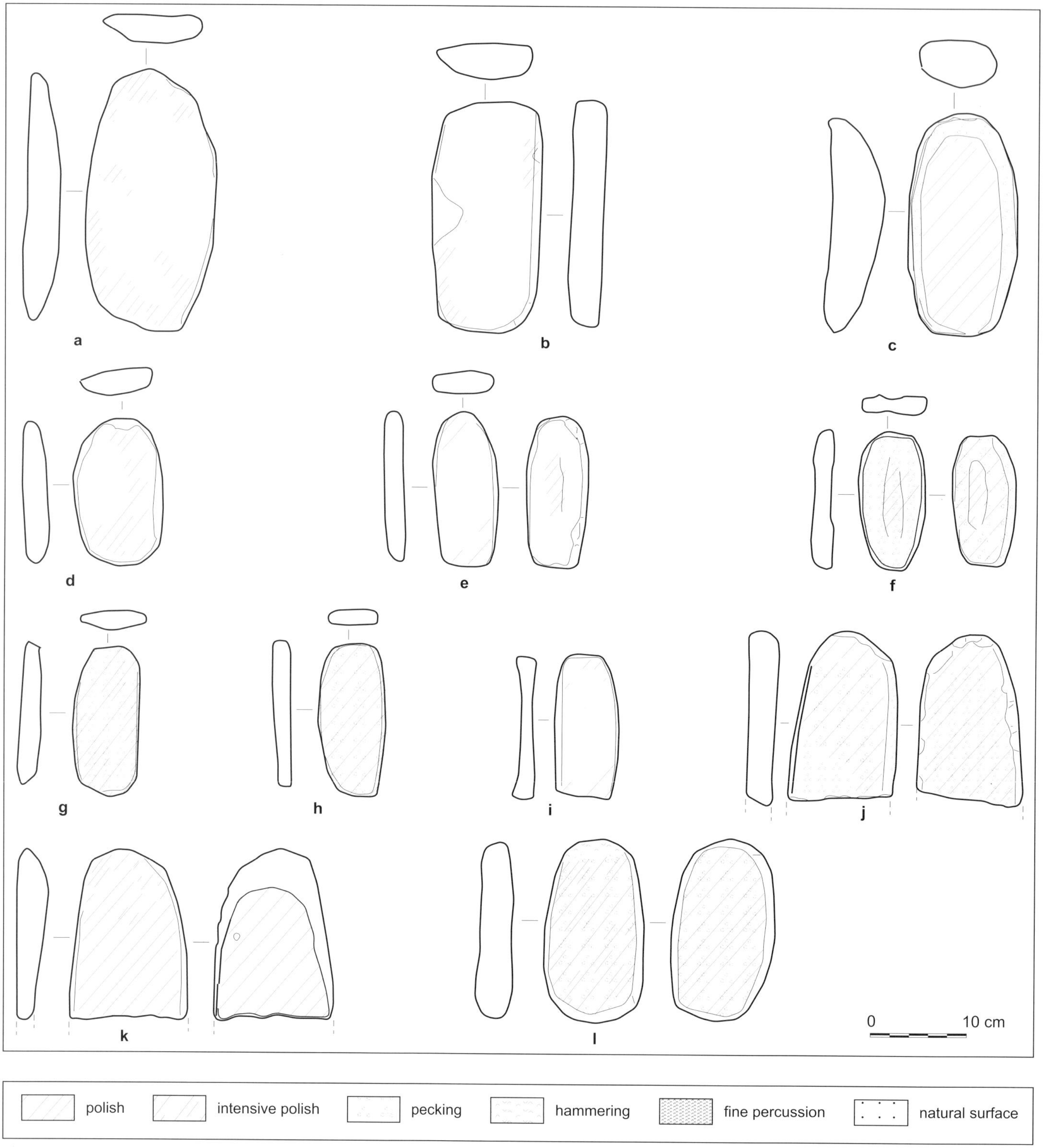

Fig. 9 – *Imiris gora grinders:* ***a****, n° 157;* ***b****, n° 187;* ***c****, n° 226;* ***d****, n° 159;* ***e****, n° 199;* ***f****, n° 198;* ***g****, n° 189;* ***h****, n° 188;* ***i****, n° 166;* ***j****, n° 84;* ***k****, n° 83;* ***l****, n° 151.*

Fig. 10 – *Imiris gora grinding slabs in sand stone:* ***a****, "palette" 160;* ***b****, n° 208;* ***c****, n° 65;* ***d****, n° 75;* ***e****, n° 135.*

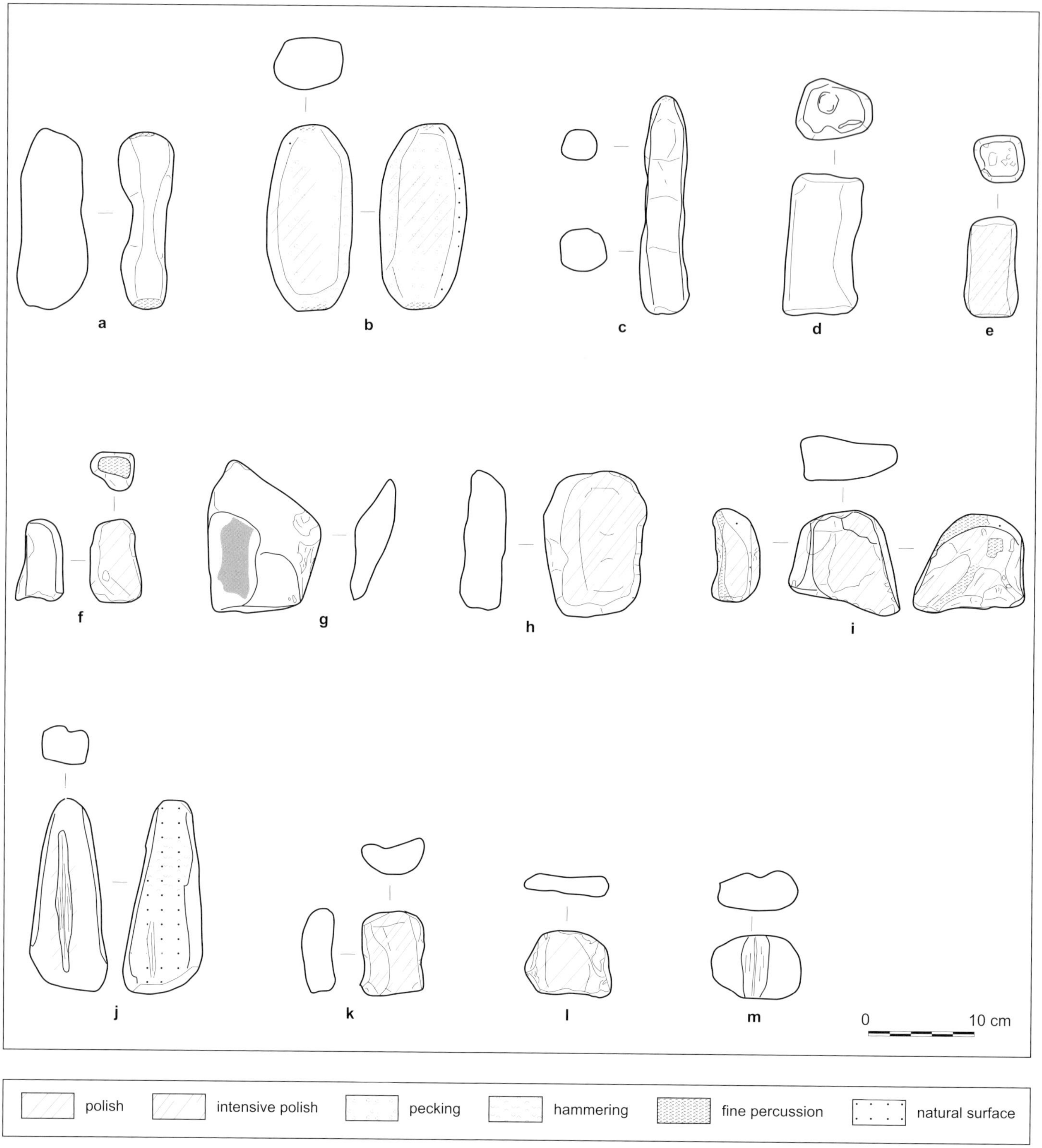

Fig. 11 – *Imiris gora pestles and polishers:* ***a****, pestle 173;* ***b****, pestle 118;* ***c****, pestle 211;* ***d****, pestle 125;* ***e****, pestle 96;* ***f****, pestle 124;* ***g****, "palette" 79;* ***h****, hand polisher 212;* ***i****, hand polisher 115;* ***j****, grooved abrader 127;* ***k****, grooved abrader 93;* ***l****, hand polisher 100.*

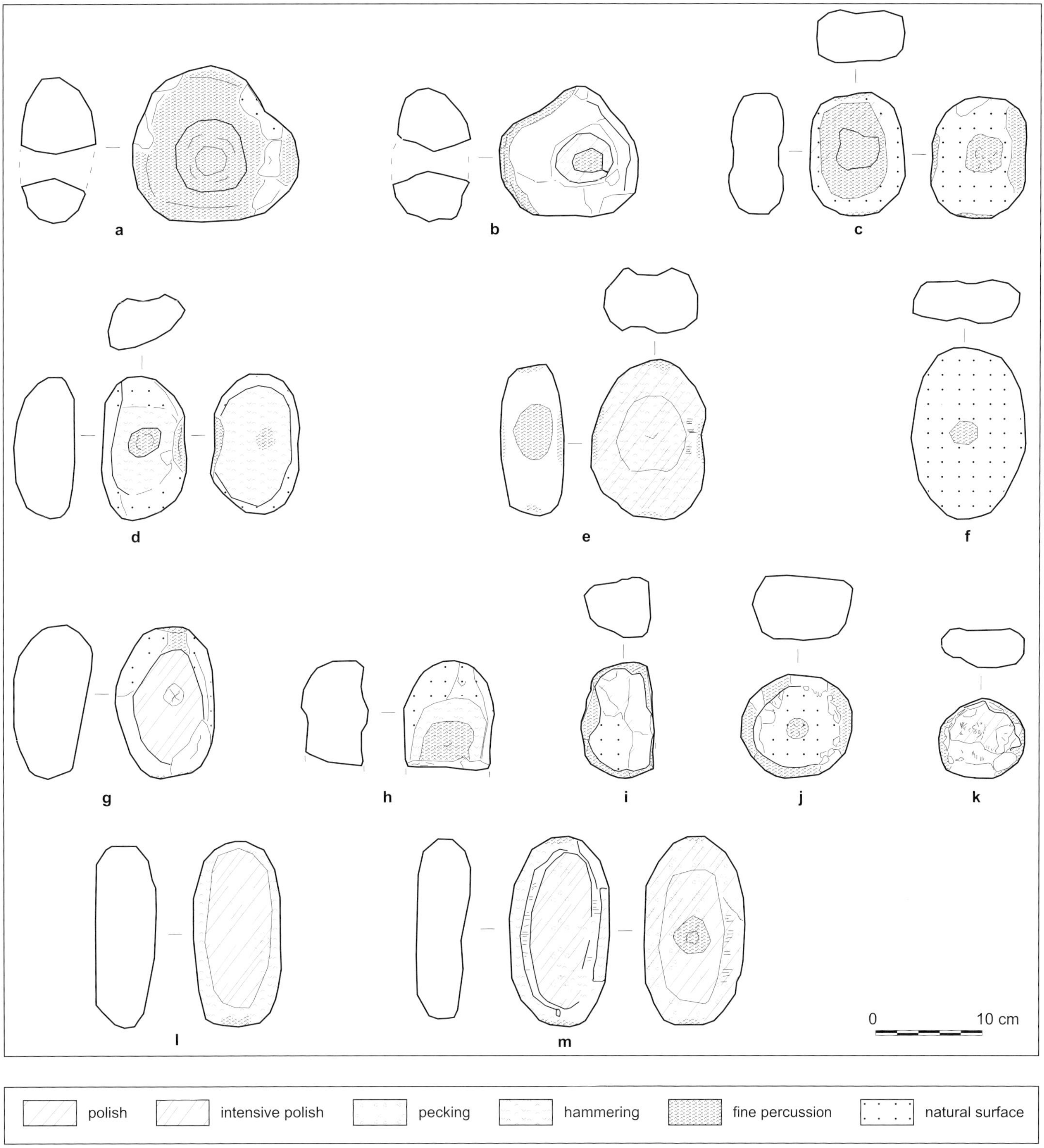

Fig. 12 – *Imiris gora hammer stones and pestles:* ***a****, perforated item 162 basalt;* ***b****, perforated item 97 basalt;* ***c****, crushing cobble 116 basalt;* ***d****, crushing cobble 137;* ***e****, crushing cobble 150;* ***f****, crushing stone 224;* ***g****, hand stone 171 sandstone;* ***h****, crushing cobble 136;* ***i****, hammer stone 69;* ***j****, crushing stone 161;* ***k****, hammer stone 66;* ***l****, hand stone 224;* ***m****, pestle 80 sandstone.*

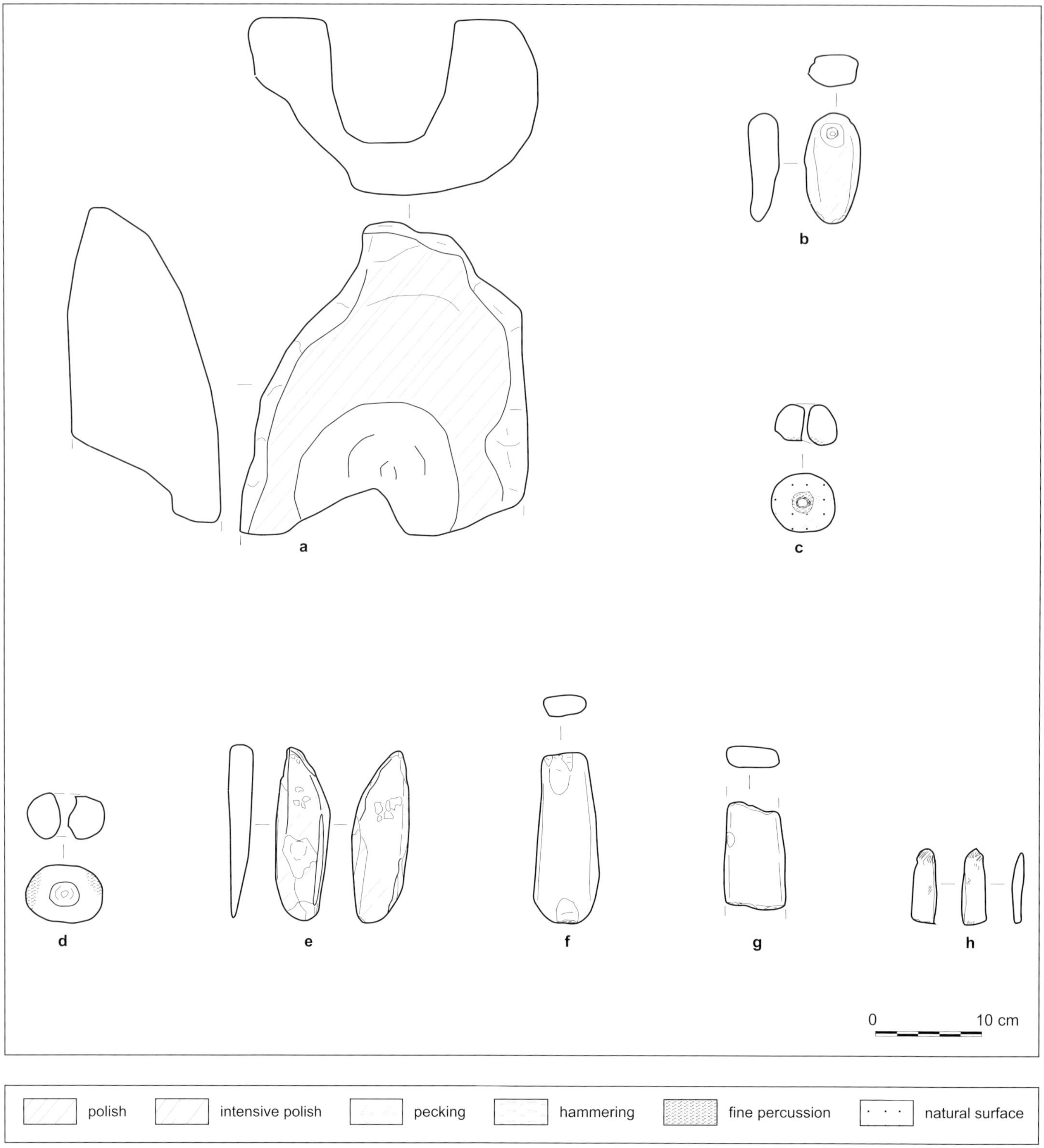

Fig. 13 – *Imiris gora varied items:* ***a****, mortar 233;* ***b****, perforated item 101 limestone;* ***c****, perforated item 102 basalt;* ***d****, perforated item 78 basalt;* ***e****, sharpener 114 limestone;* ***f****, splintered tools 89;* ***g****, splintered tool 90;* ***h****, sharpener 91.*

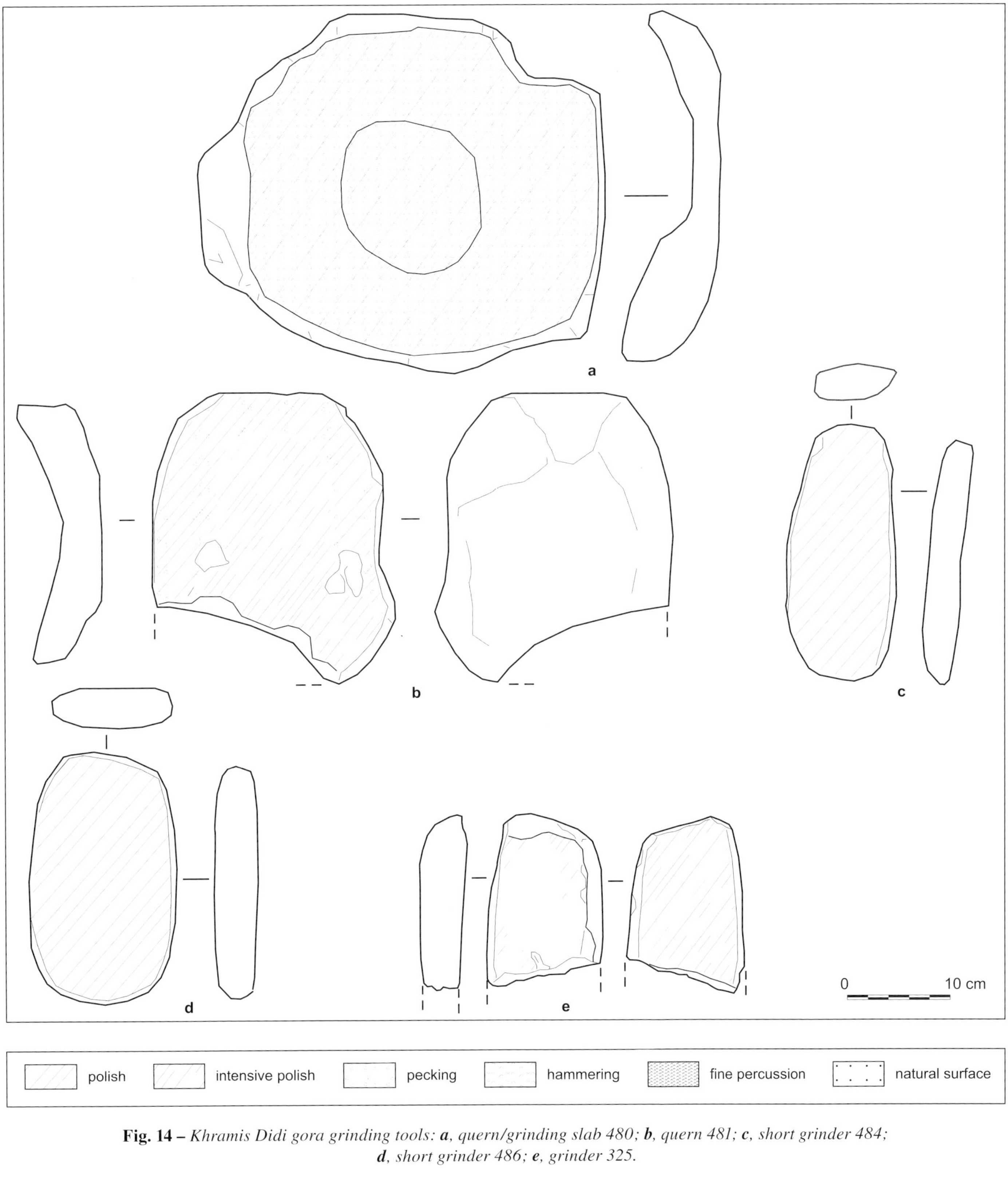

Fig. 14 – *Khramis Didi gora grinding tools:* ***a****, quern/grinding slab 480;* ***b****, quern 481;* ***c****, short grinder 484;* ***d****, short grinder 486;* ***e****, grinder 325.*

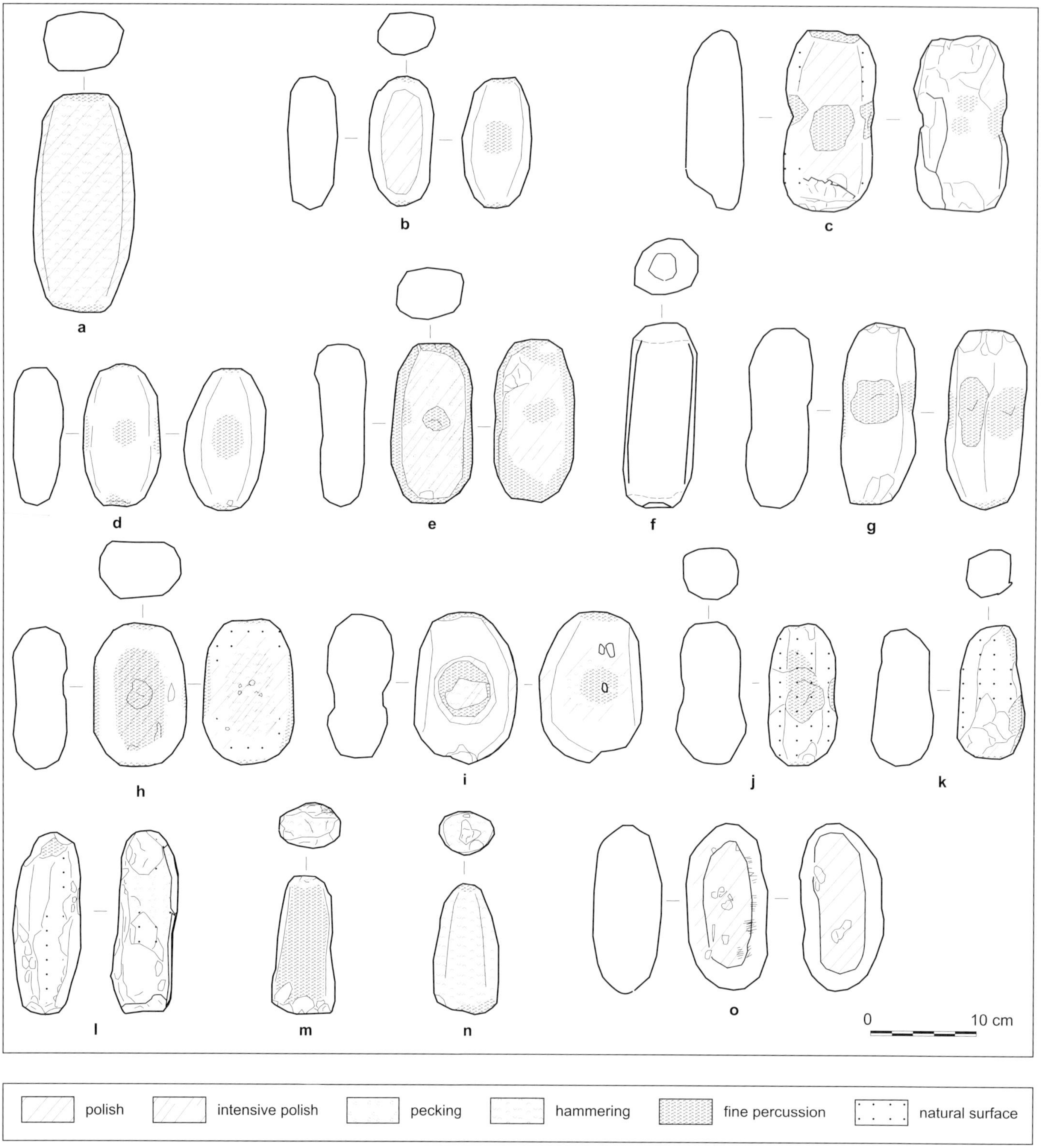

Fig. 15 – *Khramis Didi gora pestles and hand stones:* ***a****, metamorphic rock 499;* ***b****, hand stone in sandstone 422;* ***c****, hammer in sandstone 428;* ***d****, crushing cobble in metamorphic rock 371;* ***e****, crushing stone 490;* ***f****, pestle-hand stone 500;* ***g****, pestle-crushing stone 329;* ***h****, hand stone-crushing stone 324;* ***i****, crushing cobble 350;* ***j****, hammer stone-pestle 400;* ***k****, hammer stone-pestle 405;* ***l****, pestle or reused adze in metamorphic rock 514;* ***m****, pestle 322;* ***n****, pestle 442 in granite;* ***o****, pestle 346.*

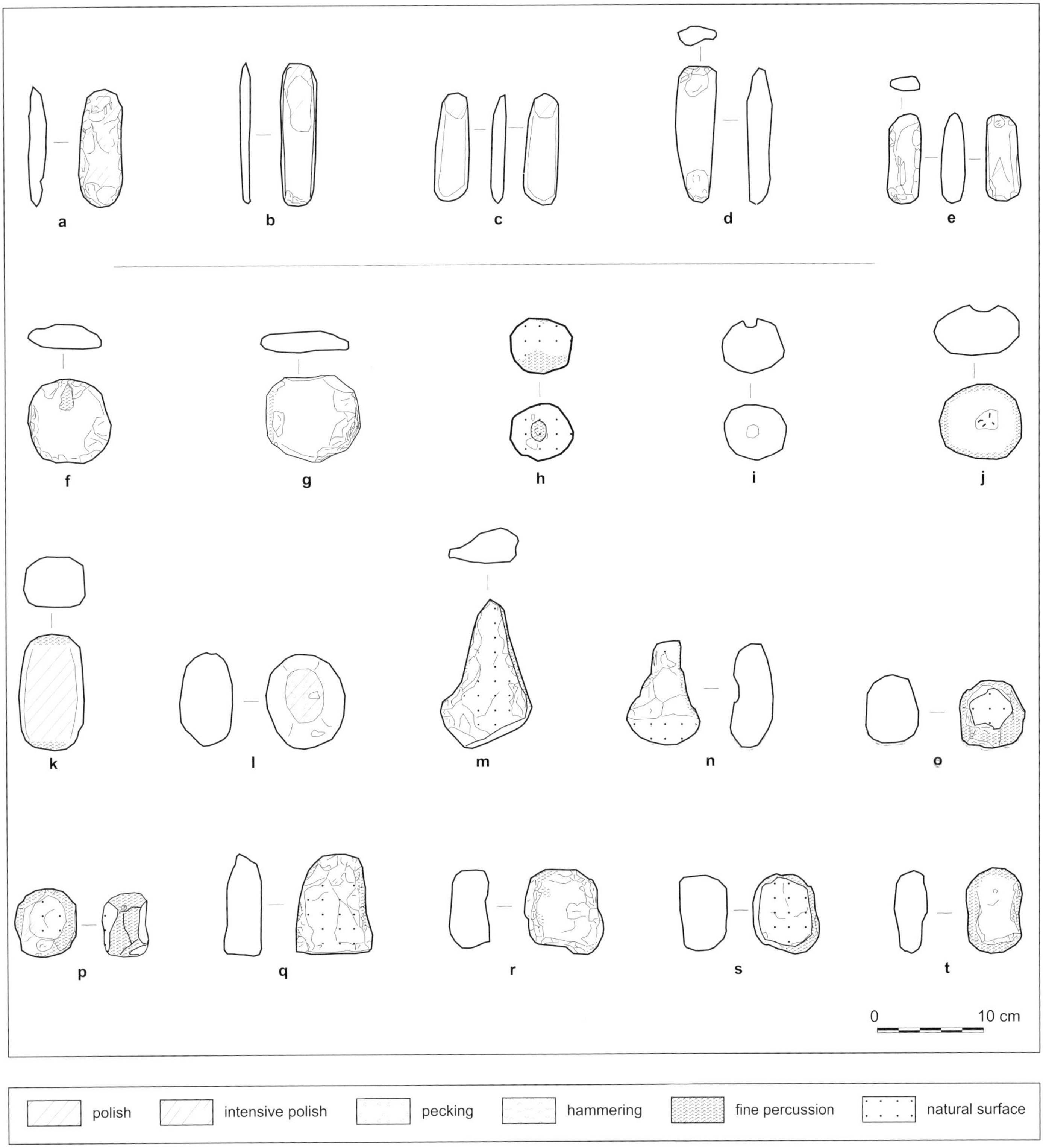

Fig. 16 – *Khramis Didi gora adzes and hammer stones:* ***a****, chisel 380;* ***b****, chisel 383;* ***c****, sharpener in metamorphic rock 351;* ***d****, splintered tool in limestone 303;* ***e****, chisel 368;* ***f****, hammer stone 382;* ***g****, hammer stone 465;* ***h****, crushing stone 319;* ***i****, crushing stone 426;* ***j****, crushing stone 308;* ***k****, hammer in basalt 448;* ***l****, hand stone 375;* ***m****, undetermined flaked cobble 444;* ***n****, pick or hammer stone 358;* ***o****, hammer stone 459;* ***p****, hammer stone 323;* ***q****, undetermined flaked limestone tool 357;* ***r****, hammer stone 359;* ***s****, hammer stone 359;* ***t****, hammer stone 461.*

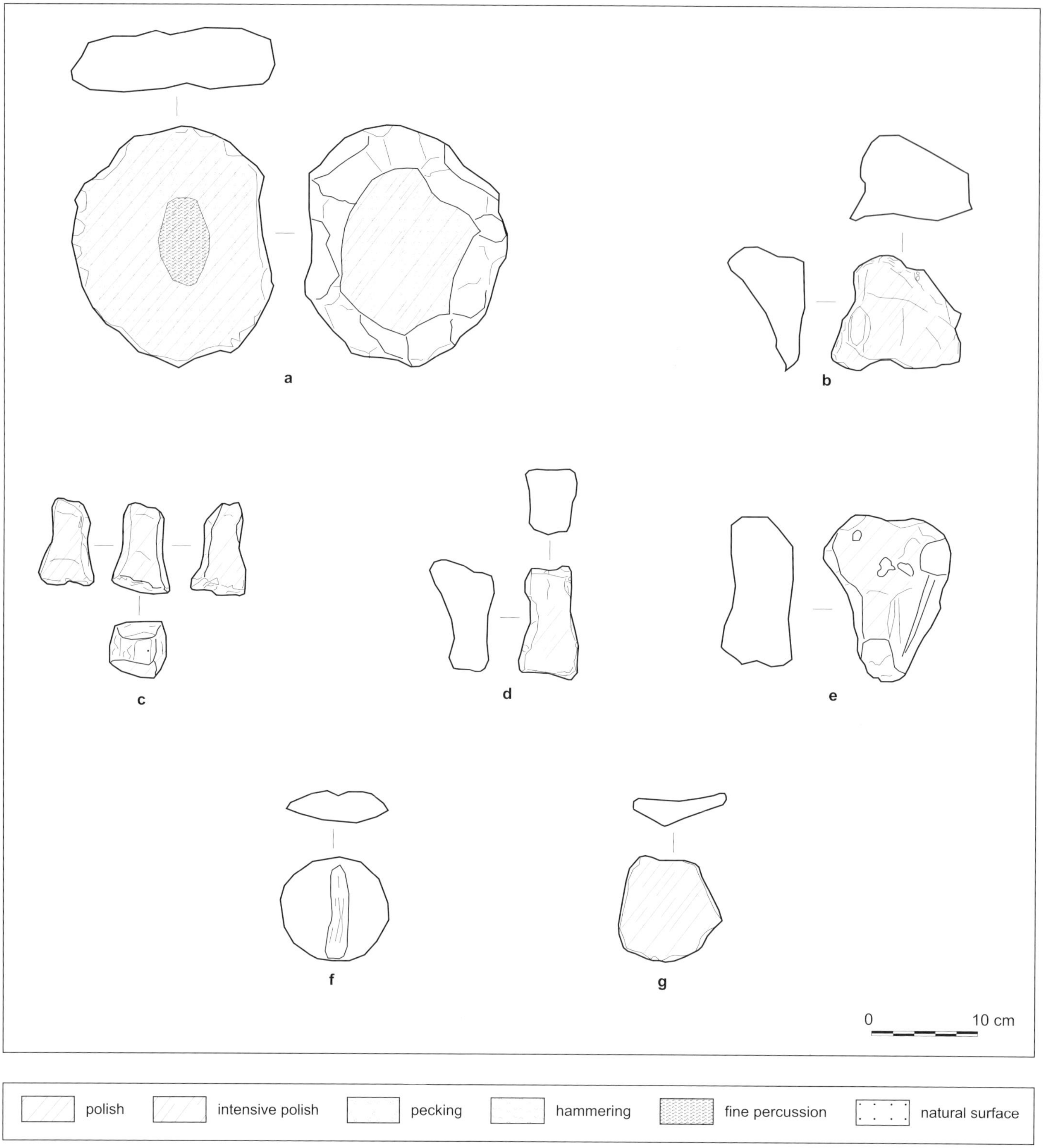

Fig. 17 – *Khramis Didi gora netherstones and abraders:* ***a****, grinding slab 488;* ***b****, hand polisher 386;* ***c****, prismatic hand polisher 376;* ***d****, prismatic hand polisher 456;* ***e****, grooved abrader 367;* ***f****, grooved abrader 397;* ***g****, hand polisher 493.*

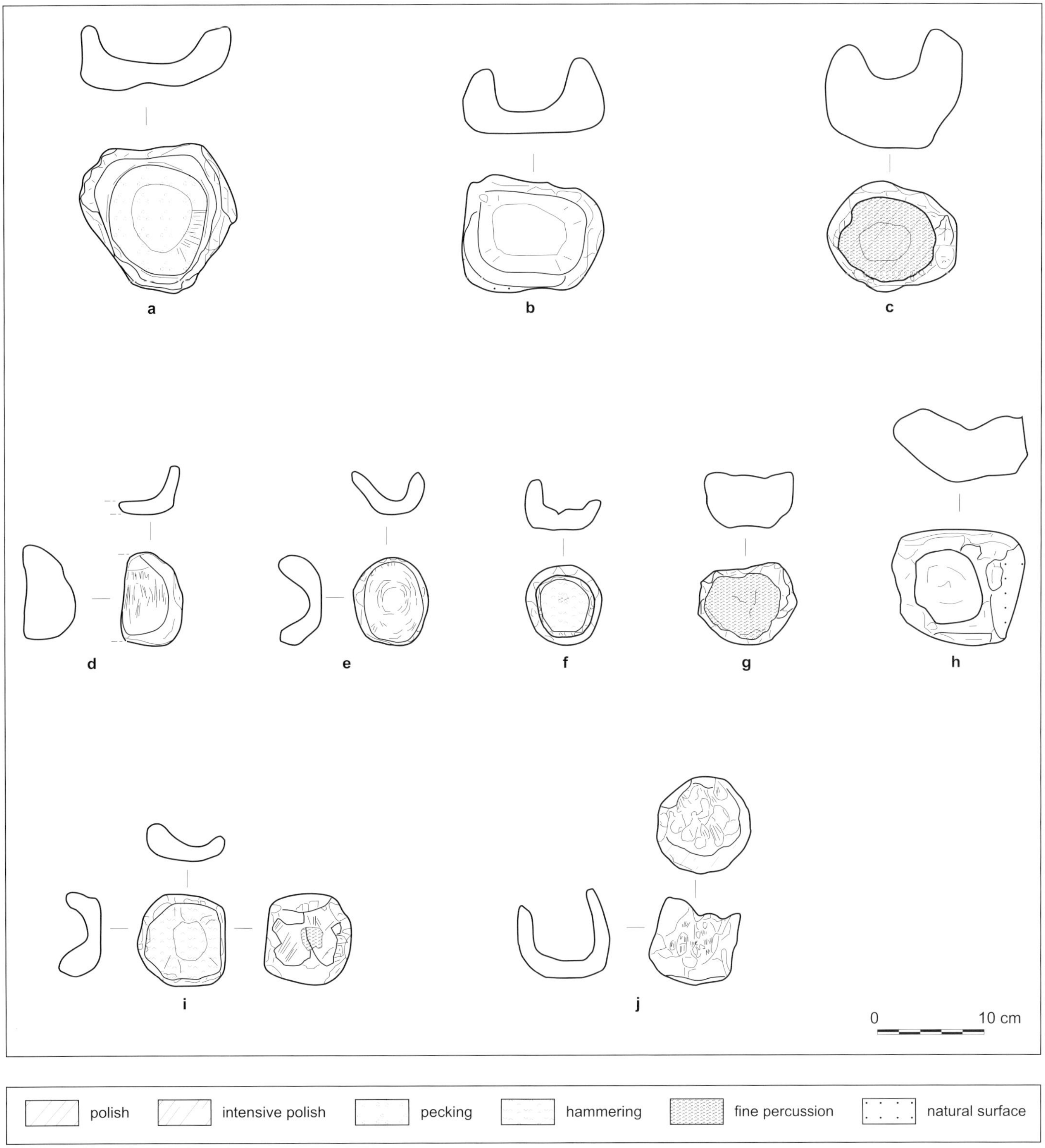

Fig. 18 – *Khramis Didi gora limestone vessel with traces of shaping (scraping, pecking, polishing):* ***a****, n° 407;* ***b****, n° 364;* ***c****, n° 363;* ***d****, n° 406;* ***e****, n° 408;* ***f****, n° 348;* ***g****, n° 385;* ***h****, n° 360;* ***i****, n° 419;* ***j****, n° 335.*

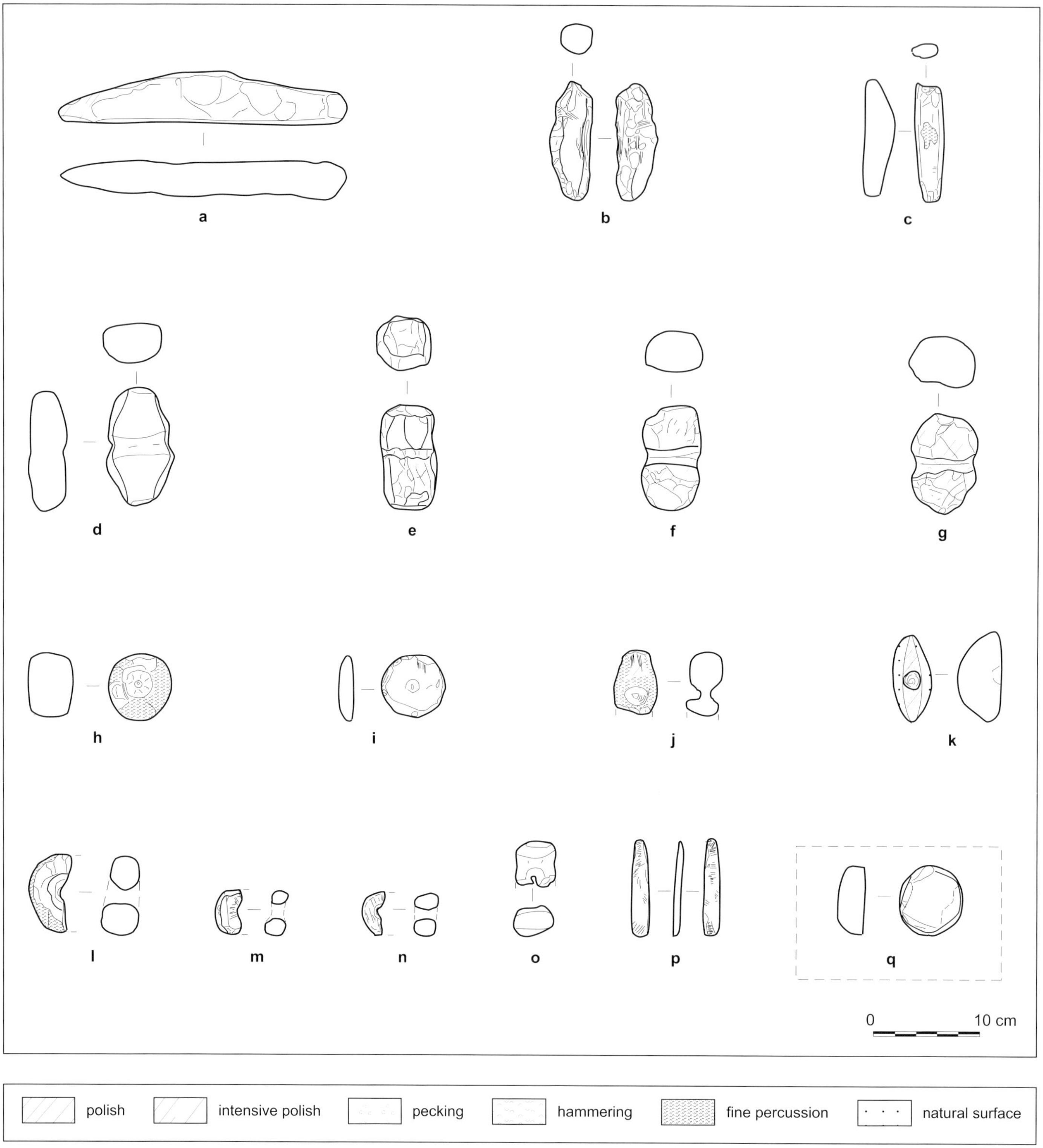

Fig. 19 – *Khramis Didi gora limestone and varied items:* ***a****, pick or sharpener 435;* ***b****, undetermined item 384;* ***c****, splintered tool 333;* ***d****, hammer 429;* ***e****, hammer 349;* ***f****, hammer 526;* ***g****, hammer 468;* ***h****, spheroid perforated tool 470;* ***i****, spindle whorl 469;* ***j****, perforated item 471;* ***k****, perforated sandstone item 317;* ***l****, perforated item 474;* ***m****, perforated cobble 473;* ***n****, perforated cobble 332;* ***o****, hammer in sandstone 336;* ***p****, sharpener in metamorphic rock 352;* ***q****, ochre disc 467.*

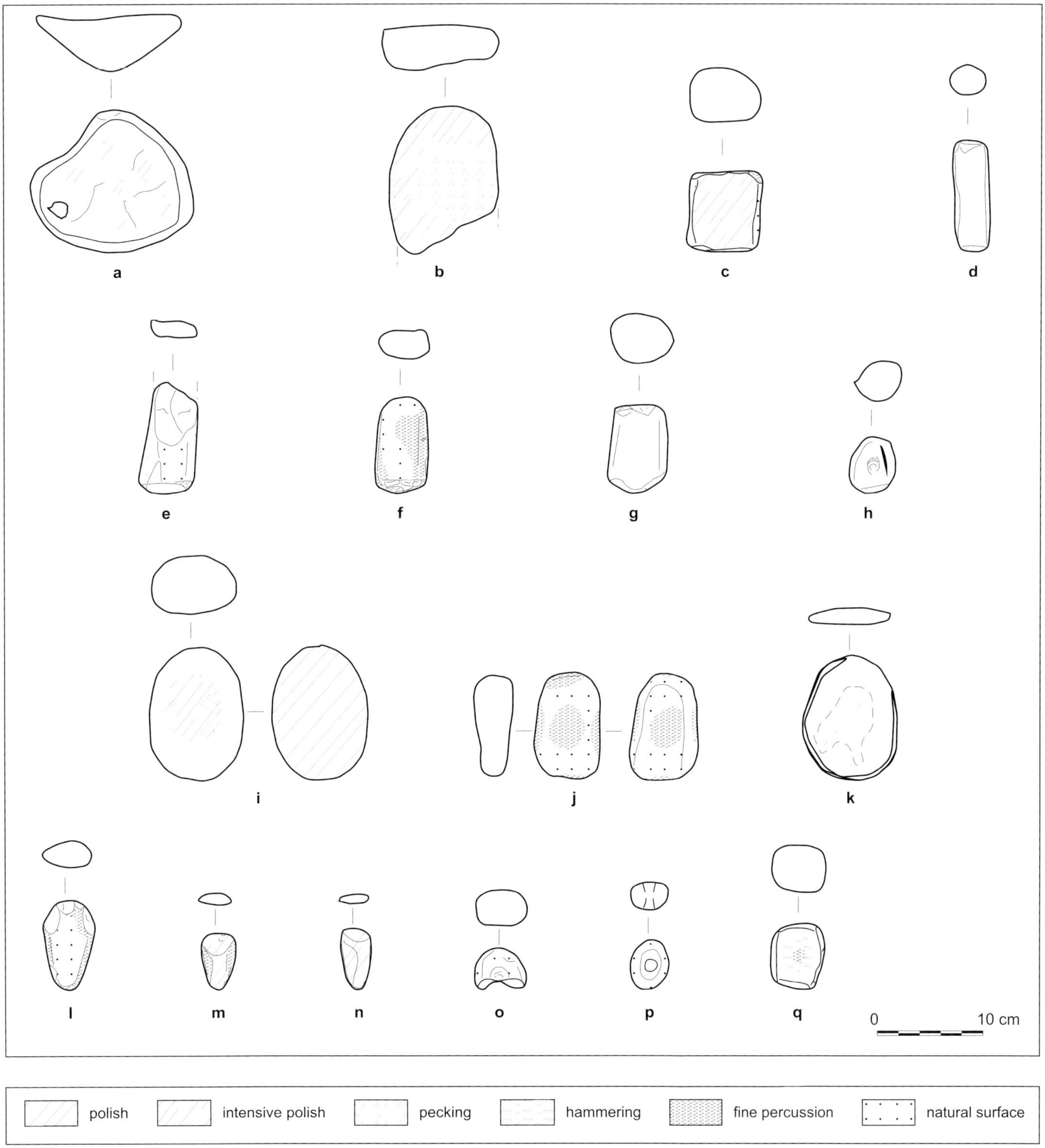

Fig. 20 – *Aruchlo I:* ***a****, mortar 270;* ***b****, short grinder 283;* ***c****, pestle 264;* ***d****, pestle 275;* ***e****, pestle 263;* ***f****, adze-pestle 284;* ***g****, adze-pestle 280;* ***h****, anvil? 282;* ***i****, crushing cobble 278;* ***j****, pestle 261;* ***k****, palette 285;* ***l****, adze 288;* ***m****, adze 286;* ***n****, adze 287;* ***o****, crushing cobble 268;* ***p****, perforated basalt cobble 274;* ***q****, crushing stone 276.*

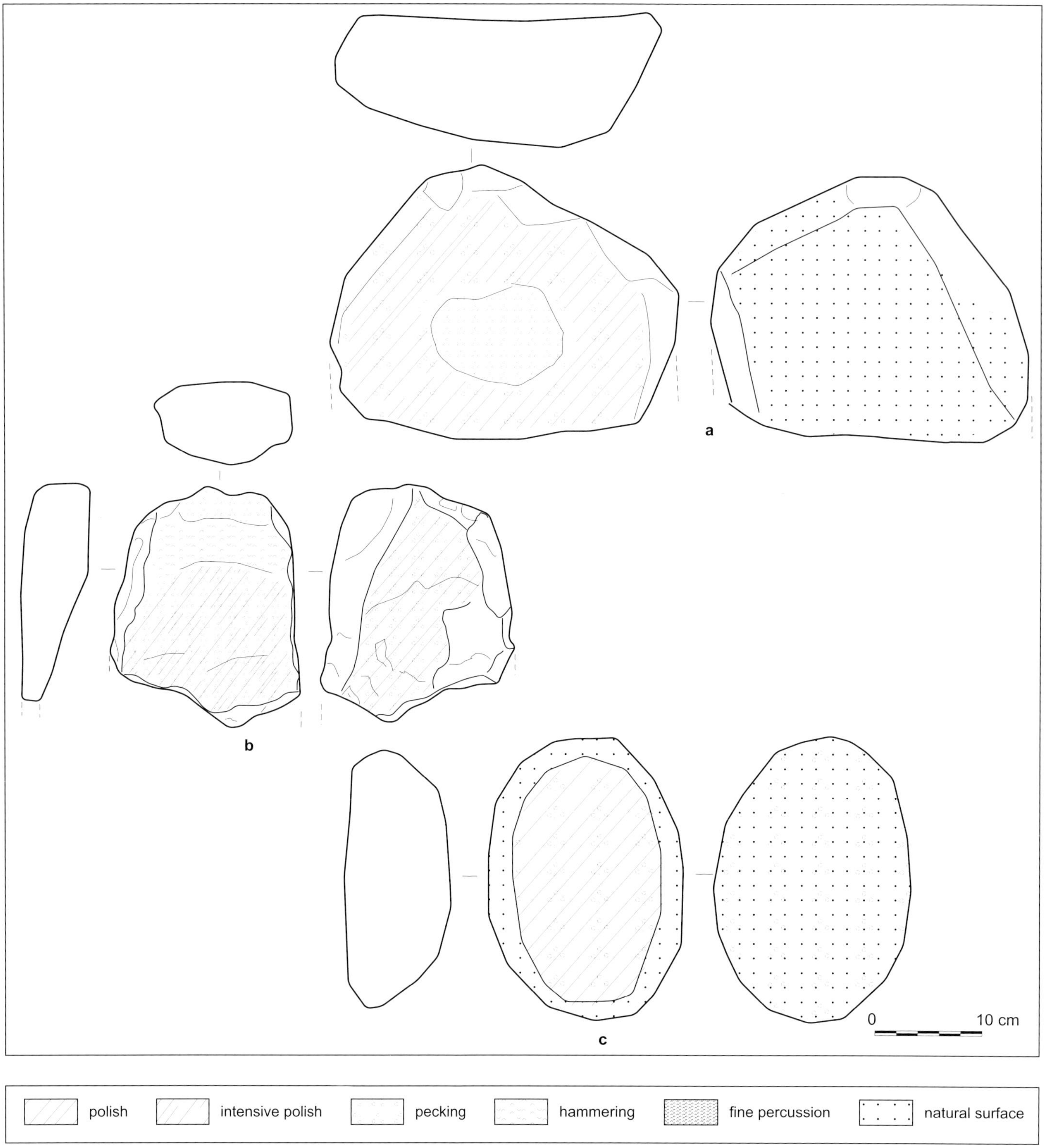

Fig. 21 – *Aruchlo I querns:* ***a****, n° 239;* ***b****, n° 240;* ***c****, n° 241.*

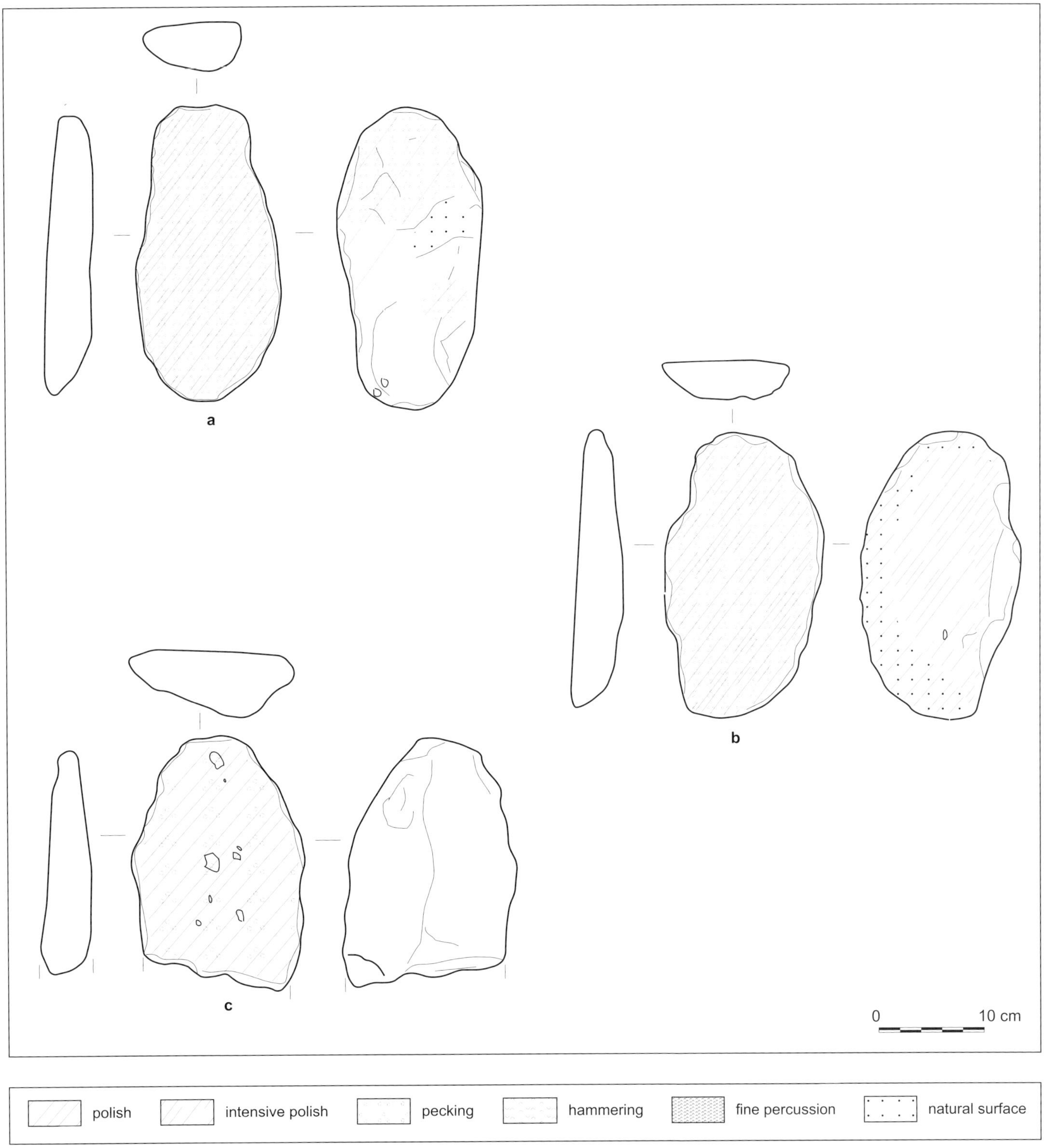

Fig. 22 – *Aruchlo I grinders:* ***a****, n° 243;* ***b****, n° 244;* ***c****, n° 245.*

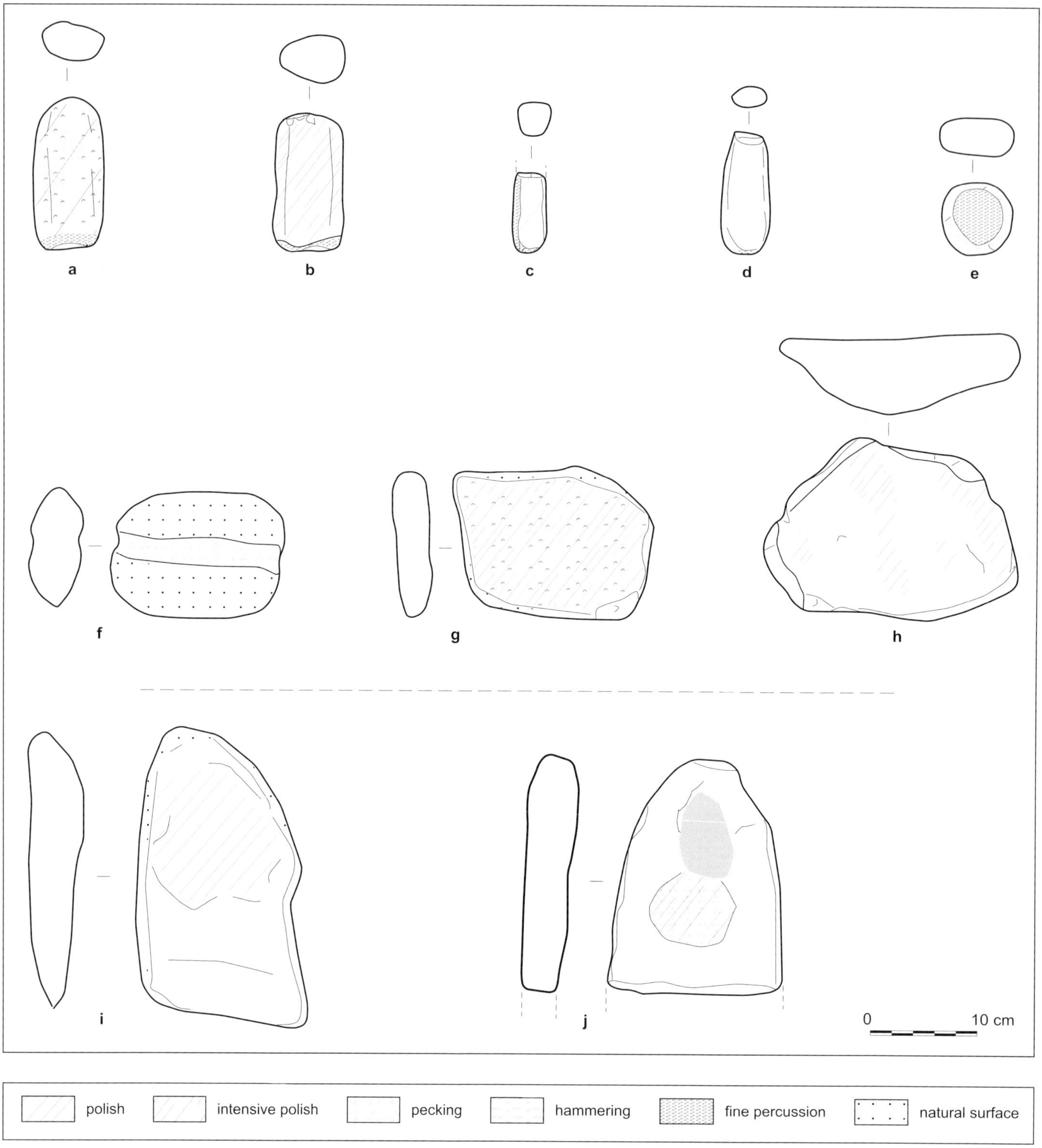

Fig. 23 – *Aruchlo I:* ***a****, pestle 254;* ***b****, pestle 255;* ***c****, pestle 256;* ***d****, splintered tool 251;* ***e****, "broyon" in sandstone 249;* ***f****, undetermined in metamorphic rock 246;* ***g****, "palette" 247;* ***h****, netherstone in metamorphic rock 577;* ***i****, grinding slab 258;* ***j****, quern 257 with traces of colouring.*

BIBLIOGRAPHY

ABIBULLAEV O.A.

1963 Nekotorie itogi izuchenija kholma Kul Tepe [Elements for the study of Kül Tepe. Archaeological researches in Azerbaijan]. *Arkeologicheskie Issledovanija v Azerbajdzhane*: 29-44.

ADAMS J.

2002 *Ground Stone Analysis: A Technological Approach*. Salt Lake City: University of Utah Press, Center for desert archaeology.

ADAMS J., DELGADO S., DUBREUIL L., HAMON C., PLISSON H. and RISCH R.

in press Functional analysis of macro-lithic artefacts. *In*: STERNKE F., EIGELAND L. and COSTA L. (eds.), *Non-flint Raw Material Use in Prehistory. Old Prejudices and New Direction, 15th UISPP congress, Lisbon-September 2006*. Oxford (*BAR Int. Ser.*).

AKKERMANS P.

1989 *Excavations at Tell Sabi Abyad: prehistoric investigations in the Balikh valley, northern Syria*. Oxford (*BAR Int. Ser.* 468).

ALGAZE G, ATAMAN K., INGRAHAM M., MARFOE L., MCDONALD M., MILLER N., SNOW C., STEIN G., VERHAREN B., WATTENMAKER P., WILKINSON T. and YENER A.

1984 *The Chicago Euphrates archaeological, project 1980-1984*. Unpublished interim report.

AMIRKHANOV K.

1987 *Chokhskoe poselenie. Chelovek i ego kul'tura v mezolite i neolite gornogo Dagestana* [*The Chokh settlement. Men and culture in the Mesolithic and Neolithic of mountainous Dagestan*]. Moscou: Nauka.

BADALYAN R., CHATAIGNER C. and KOHL Ph.

2004 Trans-caucasian obsidian: the exploitation of the sources and their distribution. *In*: SAGONA A. (ed.), *A view from the highlands: Trans-Caucasus, eastern Anatolia and north-western Iran. Cultural contacts and diversity, Studies in honour of C.A. Burney*: 437-465. Leuwen: Peeters Press.

BADALYAN R., LOMBARD P., AVESTISYAN P., CHATAIGNER C., CHABOT J., VILA E., HOVSEPYAN R., WILLCOX G. and PESSIN H.

2007 New data on the late prehistory of the southern Caucasus. The excavations at Aratashen (Armenia): preliminary report. *In*: LYONNET B. (éd.), *Les cultures du Caucase (VIe-IIIe siècle avant notre ère). Leurs relations avec le Proche-Orient*: 37-62. Paris: CNRS Éditions (*Éditions Recherches sur les Civilisations*).

CHATAIGNER C.

1995 *La Transcaucasie au Néolithique et au Chalcolithique*. Oxford (*BAR Int. Ser.* 624).

CHIKOVANI G.

1999 *Shida-Kartli V-VIth millennium BC Ph.D* [*Shida-Kartli V-VIth millennium BC Ph.D*]. Tbilisi: Centre for archaeological research publications.

CHUBINISHVILI T. and GOGELIA D.

1981 Kratkij otchet Kvemo-Kartlitskoj arkheologicheskoj ekspeditsii [Report on the archaeological mission of Kwemo-Kartli]. *Polevye arkheologicheskie issledovanija*: 35-42.

CONNOR S. and SAGONA A.

2007 Environment and society in the late prehistory of southern Georgia, Caucasus. *In*: LYONNET B. (éd.), *Les cultures du Caucase (VIe-IIIe siècle avant notre ère). Leurs relations avec le Proche-Orient*: 21-36. Paris: CNRS Éditions (*Éditions Recherches sur les Civilisations*).

DE BEAUNE S.A.

2000 *Pour une archéologie du geste. Broyer, moudre, piler. Des premiers chasseurs aux premiers agriculteurs*. Paris: CNRS Éditions.

DJAVAKHISHVILI A.

1973 *Stroitel'noe delo i arkhitektura poselenii Juzhnogo Kavkaza v V-III- tys. do. n.e.* [*Architecture on the settlements in South Caucasus V-III mill. BC*]. Tbilisi: Metsniereba.

DUBREUIL L.

2004 Long-term trends in Natufian subsistence: a use-wear analysis of ground stone tools. *Journal of Archaeological Science* 31: 1613-1629.

ESIN U., BIÇAKSI E., ÖZBAŞARAN M., BALKAN ATLI N., BERKER D., YAGMUR I. and KORBURT ATLI A.

1991 Salvage excavations at the pre-pottery site of Asikli Höyük in Central Anatolia. *Anatolica* XVII: 123-164.

FULLAGAR R. (ed.)

1998 *A Closer Look. Recent Australian Studies of Stone tool*. Sydney: University of Sydney, Archaeological Computing Laboratory (*Sydney University Archeological Methods Series* 6).

GLONTI L., DJAVAXISHVILI A. and KIGURADZE T.

1975 Antropomorfnye figurki Khramis Didi Gora [The anthropomorphic figurines of Khramisdidi Gora]. *Vestnik Gosudarstvenovo Muzeja Gruzii* 31B: 85-97.

HAMON C.

2006 *Broyage et abrasion au Néolithique ancien. Caractérisation technique et fonctionnelle des outillages en grès du Bassin parisien*. Oxford (*BAR Int. Ser.* 1551).

2007 Modes de subsistance et activités dans le Chalcolithique du Caucase nord : étude fonctionnelle des outils en pierre de la culture de Majkop. *In*: LYONNET B. (éd.), *Les cultures du Caucase (VIe-IIIe siècle avant notre ère). Leurs relations avec le Proche-Orient*: 189-198. Paris: CNRS Éditions (*Éditions Recherches sur les Civilisations*).

2008 Functional analysis of stone grinding and polishing tools from the earliest Neolithic of north-western Europe. *Journal of Archaeological Science* 35: 1502-1520.

HANSEN S., MIRTSKHULAVA G., BASTERT-LAMPRICHS K., BENECKE N., GATSOV I. and NEDELCHEVA P.

2006 Aruchlo 2005-2006. Bericht über die Ausgrabungen in einem neolitischen Siedlungshügel. *Archäologische Mitteilungen aus Iran und Turan* 38: 1-34.

JANELIDZE C.

1975 *Kartlis arkeologiuri expediciebis shedegebi 1965-1971* [*Archaeological expeditions in Georgia 1965-1971*]. Tbilisi: Academy of Sciences.

1984 *On the reconstruction of environment of early farming culture of eastern Georgia, in Human and its environment.* Tbilisi: Metsniereba.

KAVTARADZE G.

1981 *Sakartvelos Eneolit-brinjaos Khanis Arkheologiuri kulturebis kronologia axali Mocamebis Shudze [New data on the chronology of the Eneolithic and Bronze Age in Georgia]*. Tbilisi: Metsniereba.

1999 The importance of metallurgical data for the formation of a central Transcaucasian chronology. *In*: HAUPTMANN H., PERNICKA E., REHREN T. and YALCIN U. (eds.), *The beginnings of metallurgy. Proceedings of the International Conference, Bochum 1995*: 67-101. Bochum: Deutsches Bergbau Museum (*Der Anschnitt* 9).

KIGURADZE T.

1976 *Periodizatsija rannezemledel'cheskoj kul'tury vostochnogo Zakavkazja [Periodisation of the culture of first farmers' settlements of Eastern caucasus]*. Tbilisi: Metsniereba.

1986 *Neolitische siedlungen von Kwemo-Kartli, Georgien*. München: C.H. Beck.

KIGURADZE T. and MENABDE M.

2004 The Neolithic of Georgia. *In*: SAGONA A. (ed.), *A view from the Highlands, Archaeological studies in honour of Charles Burney, Ancient Near East studies*: 245-398. Leuwen: Peeters press (*Supplement* 12).

KORFMANN M.

1982 *Tikiltepe. Die ersten Ansätze prähistoricher Forschung in der Ostlichen Turkei. Istanbuler Mitteilungen. b. 2b.* Tübingen: Verlag Ernst Wasmuth.

KUSHNAREVA K.

1997 *The southern Caucasus in Prehistory. Stages of cultural and socioeconomic development from the 8th to the 2nd millennium BC*, translated by H.N. Michael. Philadelphia: University of Pensylvania (*University Museum Monograph* 99).

KUSHNAREVA K. and CHUBINISHVILI T.

1970 *Drevnie kul'tury Juzhnogo Kavkaza [Old cultures of Southern Caucasus]*. Leningrad: Nauka.

LEROI-GOURHAN A.

1988 *Dictionnaire de la Préhistoire*. Paris: PUF.

LISITSYNA G.N. and PRISCHEPENKO L.V.

1977 *Paleoetnobotanicheskie nakhodki Kavkaza i blizhnego Vostoka. [Paleobotanical data from the Caucasus and the Near East]*. Moscou: Nauka.

LOMBARD P. et CHATAIGNER C.

2004 Le Néolithique et le Chalcolithique de Transcaucasie: l'exemple des bassins de la Kura et de l'Araxe. *In*: GUILAINE J. (éd.), *Aux marges des grands foyers du Néolithique. Périphéries débitrices ou créatrices? Séminaire du collège de France*: 63-84. Paris: Errance (*Collection des Hespérides*).

LORDKIPANIDZE O.

1991 *Archaeologische in Georgien. Von der Altenzeit zum Mittelalter.* Weinheim: VCH, Acta Humanoria.

LUKIN A.L.

1950 Neoliticheskoe selishche Kistrik bliz Gudaut [The Neolithic village of Kistrik near Gudaut]. *Sovetskaja Arkheologija* 12: 247-286.

LYONNET B.

2004 Les cultures anciennes du Caucase du Nord et de l'Ouest (du Néolithique au début de l'Âge du Bronze). *In*: GUILAINE J. (éd.), *Aux marges des grands foyers du Néolithique. Périphéries débitrices ou créatrices? Séminaire du collège de France*: 87-104. Paris: Errance (*Collection des Hespérides*).

LYONNET B. (éd.)

2007 *Les cultures du Caucase (VIe-IIIe siècle avant notre ère). Leurs relations avec le Proche-Orient.* Paris: CNRS Éditions (*Éditions recherches sur les civilisations*).

MASSON V.M. and MERPERT N.Y.

1982 *Eneolit SSSR [The Eneolithic of USSR]*. Moscou: Nauka.

MELLAART J.

1970 *Excavations at Hacilar (I) and (II)*. Ankara / Edinburgh: British Institute of Archaeology / Edinburgh University Press.

1978 *The archaeology of ancient Turkey*. London: Bodley Head.

MENABDE M. and KIGURADZE T.

1981 *Sionis arkeologiuri zeglebi [The archeological site of Sioni]*. Tbilisi: Metsniereba.

NARIMANOV I.G.

1965 Arkheologicheskie issledovanija poselenija Shomu Tepe v 1963 g. [Archeological study of the Shomu Tepe settlement in 1963]. *Arkheologicheskie Issledovanija v Azerbajdzhane*: 45-53.

NEKHAEV A.A.

1990 Eneoliticheskie poselenija Zakuban'ja [An Eneolithic village on Kuban]. *In*: ZHDANOVSKIJ A.M. and MARCHENKO I.I. (eds.), *Drevnie pamyatniki Kubani [The old sites of the Kuban]*: 5-22 and 185-191. Krasnodar: Upravlenye Kul'tury Krasnodarskovo Krajispolkoma, Krasnodarskii Kraevoi Sovet, Vserossijskovo Obshchestva Okhrany Pamyatnikov Istorii I Kul'tury.

NIEBIERIDZE L.

1972 *Dasavlet Amierkavkasiis Neoliti [The Neolithic of Western Caucasus]*. Tbilisi: Metsniereba.

1978 *Darkvetis Mravalpeniani Ekhi [The rock-shelter of Darkveti]*. Tbilisi: Metsniereba.

1986 *Ranniye stupeni razvitja zapadnozakavkazskoj ranezemledelcheskoj kul'tury [The development of agriculture in Western Caucasus]*. Tbilisi: Metsniereba.

NIERLÉ M.-C.

1982 Mureybet et Cheik-Hassan (Syrie): outillage de mouture et de broyage (IX-X^e millénaire). *In: Cahiers de l'Euphrate* 3: 177-216. Paris: ERC.

NISHIAKI Y.

2003 Functional and morphological observations on the Chalcolithic grinding stones from Tell Kosak Shamali. *In*: NISHIAKI Y. and MATSUTANI T. (eds.), *Tell Kosak Shamali, The archeological investigations on the upper Euphrates, Syria, Vol. II, Chalcolithic technology and subsistence*: 121-183. Tokyo: The University Museum (*UMUT Monograph*).

PALMIERI A.

1969 Recenti dati sulla stratigrafia di Arslantepe. *Origini* 3: 7-66.

PROCOPIOU H.

1998 *L'outillage de mouture et de broyage en Crète minoenne*. Paris: Université Paris I. Thèse de doctorat non publiée.

RISCH R.
2002 *Recursos naturales, medios de producción y explotación social. Un análisis económico de la industria lítica de Fuente Álamo (Almería), 2250-1400 antes de nuestra era.* Mainz: P. von Zabern (*Iberia Archaeologica* 3).

ROSENBERG M., NESBITT R., W. REDDING R. and STRASSER T.F.
1995 Hatlan Cemi Tepesi: some preliminary observations concerning early Neolithic subsistence behaviours in eastern Anatolia. *Anatolica* XXI: 1-12.

RUSSISHVILI N.
1990 *Kulturnye rastenja na rannikh poseleniakh Gruzii po paleoetnobotanicheskim issledovanijam.* [*Cultivated species on the ancient sites of Georgia. Paleobotanical studies*]. Tbilissi: Académie des Sciences de Géorgie. Unpublished PhD.

TOROSJAN R.M.
1976 *Rannezemledel'cheskoe poselenie Tekhuta* [*A village of farmers in Tekhut*]. Erevan: Akademii Nauk Armjanskoj SSSR (*Arkheologicheskie Raskopki v Armenii* 14).

TUNCA Ö. et MOLIST M. avec la collaboration de W. CRUELLS
2004 *Tell Amarna, Syrie. 1. La période de Halaf.* Louvain / Paris: Peeters.

VERHOEVEN M.
1999 *An Archaeological Ethnography of a Neolithic Community: Space, Place and Social Relations in the Burnt Village at Tell Sabi Abyad, Syria, Istanbul.* Leiden: Nederlands historisch-archaeologisch Institute Istanbul.

WECHLER K.P.
2001 *Studien zum Neolitikum der östeuropaïschen Steppe.* Mainz: Ph. Von Zabern (*Archaeologie in Eurasien* 12).

WRIGHT K.
1992 A Classification System for Ground Stone Tools from the Prehistoric Levant. *Paléorient* 18,2: 53-81.
2008 Craft Production and the organization of ground stone technologies. *In*: ROWAN Y.M. and EBELING J.R. (eds.), *New approaches to old stones: recent studies of ground stone artefacts*: 130-143. London: Equinox.

FLAKED STONE DISCS OF THE NEOLITHIC AND CHALCOLITHIC PERIODS IN THE SOUTHERN LEVANT

D. ROSENBERG, A. ASSAF, N. GETZOV and A. GOPHER

Abstract: *This paper presents a detailed study of a unique groundstone (henceforth-stone) tool—the flaked stone disc. The study presents the result of an attribute analysis conducted on 703 flaked discs found in stratified contexts during excavations and as surface finds, all from sites in northern Israel. The discs are described in detail and some provisional suggestions regarding their function are discussed.*

Résumé: *Cette étude détaillée présente un outil de pierre particulier: le disque retouché. L'analyse porte sur 703 disques retouchés, provenant aussi bien de contextes stratifiés que de ramassages de surface, tous les sites étant situés dans le nord d'Israël. Ces artefacts sont décrits en détail et des suggestions relatives à leur fonction sont discutées.*

Keywords: *Groundstone, Flaked stone discs, Neolithic, Chalcolithic, Southern Levant.*
Mots-clés: *Pierre travaillée, Disques retouchés, Néolithique, Chalcolithique, Levant Sud.*

Stone discs are round or oval items, usually made on pebbles or cobbles, flakes or fragments of limestone or fine-grained, compact basalt. They have a plano-convex or "flying saucer"-like (tapering of the disc circumference) cross-section depending on whether unifacial or bifacial flaking was used: plano-convex in the former case and "flying-saucer"-like in the case of the later.

A large group of flaked stone discs originating from stratified Neolithic contexts and similar items from surface collections, housed in the Museum of Prehistory of the Hula Valley at Kibbutz Ma'ayan Baruch, northern Israel, were examined. The stratified contexts derive from the sites of Hagoshrim and Beisamoun in the Hula Valley, 'Ein Miri in the Upper Galilee and Nahal Zehora I and II in the Menashe Hills, whereas the surface collections originated from ten sites mostly distributed in the northern part of the Hula Valley (fig. 1; table 1).

The discs seem to appear first in the PPNB, recorded in sites such as Hatanur, Kat, Zug Fukani, Tel Hay / Tel Roeim, Quarantine and Beisamoun. Similar discs were also found at PPNB Ba'ja in Jordan.[1] They continue into the PPNC and were reported from Hagoshrim, Tel Roeim-West, Ain Avazim and Atlit Yam. Discs from Pottery Neolithic (PN) contexts are known from Sha'ar Hagolan and Hamadiyah',[2] Beisamoun,[3] Nahal Zehora I and II, Hagoshrim, Ain Miri and Tel Roeim-West. A single basalt disc was reported from the PN strata at Tel Dan[4] and another one, a relatively large example, was found in the Early Bronze Age stratum at Tel Beit Yerach along with other PN finds,[5] probably suggesting a PN occurrence in the lower strata of the site or in its vicinity, although no diagnostic pottery was found during the excavations. Only a few stone discs, however, were found in Chalcolithic contexts including Hagoshrim, Tel Turmus, and Fazael area in the Jordan Valley.

Discs were also encountered beyond the confines of the southern Levant. A large sample of stone discs was reported from Cayönü Tepesi, Turkey, mainly from the later occupational

1. GEBEL and BIENERT, 1997; GEBEL *et al.*, 1997.

2. Many such discs, including relatively large examples are stored at the Kibbutz Sha'ar Hagolan Museum.
3. ROSENBERG, forthcoming.
4. GOPHER and GREENBERG, 1987: 107-108, fig. 8:1.
5. ROSENBERG, in press 1.

Manuscrit reçu le 10 avril 2008, accepté le 25 novembre 2008

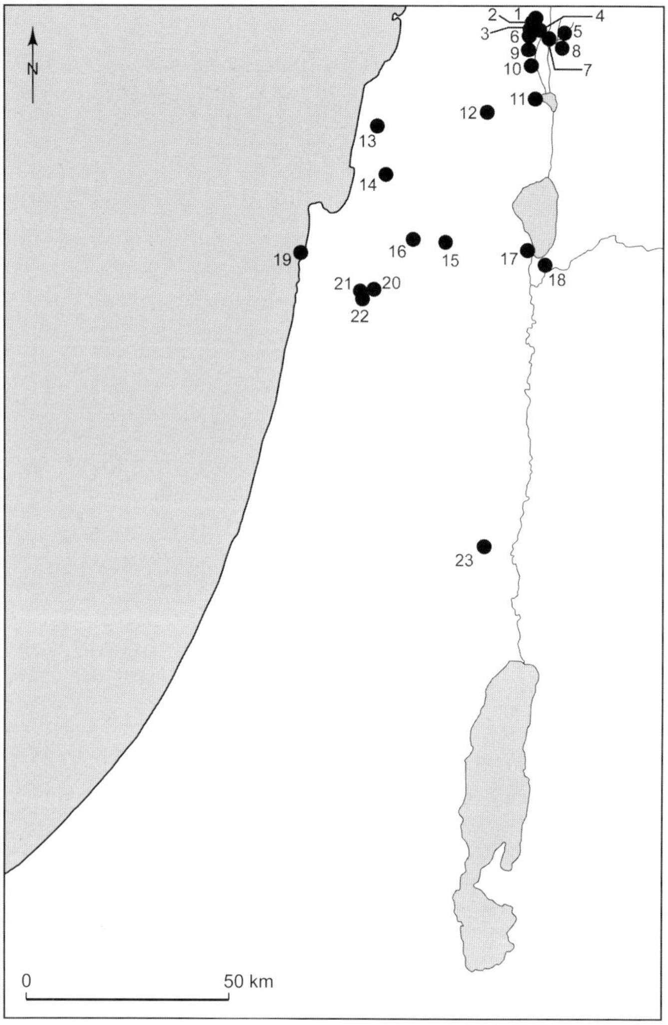

Fig. 1 – *Map showing the position of some of the sites mentioned in the text:* ***1****, Hatanur;* ***2****, Ain Hashomer;* ***3****, Kat;* ***4****, Zug Fukani;* ***5****, Tel Dan;* ***6****, Tel Hay (Tel Roim-West);* ***7****, Hagoshrim;* ***8****, Tel Turmus;* ***9****, Quarantine;* ***10****, Ain Avazim;* ***11****, Beisamoun;* ***12****, Ain Miri;* ***13****, Kabri;* ***14****, Ard el-Samra;* ***15****, Kfar Qana;* ***16****, Yiftahel;* ***17****, Bet Yerach;* ***18****, Sha'ar Hagolan;* ***19****, Atlit Yam;* ***20****, Mishmar Haemeq;* ***21****, Nahal Zehora I;* ***22****, Nahal Zehora II;* ***23****, Fatzael III.*

layers of the site.[6] The occurrence of discs was also noted for the Rouj Basin and were also noted at Fıstıklı Höyük in Syria[7] and Achilleion in Greece.[8] Thus, it appears that discs were widely distributed across the Near East and beyond. In this regard, the fact that the vast majority of discs dealt with here come from the Hula Valley does not mean that the Hula Valley was exceptional, although this possibility must still be considered.

Contextual data regarding stone discs is meagre. Nonetheless, at Hagoshrim discs were found in most of the PPNC stone-containing loci:[9] in various debris accumulations, on floors and living surfaces, in courtyards, pits, in a single grave, and even incorporated in walls. PN Lodian-related[10] loci containing discs at Hagoshrim include debris accumulations, floors and living surfaces and one was found incorporated in a wall. In the later Wadi Rabah-related[11] loci of Hagoshrim, discs were found in debris accumulations, on floors and living surfaces, in a concentration of stone tools[12] and in pits.

The main aim of this paper is to present a descriptive overview of this tool type by means of a detailed attribute analysis and encourage researchers to elaborate and discuss this stone tool in future studies. Further comments with respect to raw material, typology, metric characteristics and technology of production and some tentative comments on their use are also offered.

METHODOLOGY

The current study encompasses 703 flaked stone discs, mostly originating from surface collections with no clear cut chrono-cultural attribution.[13] Items from excavated contexts were assigned to the late PPN, the PN and the PN / Chalcolithic Periods. These include PPNC, Early Pottery Neolithic, Lodian and Wadi Rabah related contexts (table 1).

Parameters such as preservation, raw material, shape and cross-section, colour, as well as size and weight (for whole items) were recorded. Other features considered were production marks, smoothing and smoothened surface shape. Use-wear and adhesives or other features visible to naked eye, were also noted. The measurements and attributes were summarized quantitatively so as to afford a detailed description of these tools.

6. DAVIS, 1982: 118-119, fig. 3.13:12-13.
7. YOHSIZAWA, 2003.
8. WINN and SHIMABUKU, 1989: pl. 9.4: 2-6.
9. ROSENBERG, in press 2.
10. In this study the term "Lodian" relates to all Jericho XI-related assemblages.
11. In this study, Wadi Rabah culture is perceived as a PN entity.
12. Recorded as L. 443.
13. Although PPNB-Chalcolithic attribution may be suggested for all these sites.

Table 1 – *Breakdown of the sites and discs.*

Site	General		Preservation
	N	% of discs in the Studied Assemblage	% Whole discs in the Assemblage
Hatanur (surface)	15	2.13	93.3
Kat (surface)	54	7.68	98.1
Zug-Fukani (surface)	2	0.28	100
Tel Hay/Tel Roim-West (surface)	7	1.00	100
Ain Hashomer (surface)	2	0.28	100
Hagoshrim (surface and unstratified contexts)	201	28.59	98
Hagoshrim-PPNC (Layer VI)	68	9.67	88.2
Hagoshrim-Lodian Related (Layer V)	46	6.54	46
Hagoshrim-Wadi Rabah Related (Layer IV)	21	2.99	100
Beisamoun (surface)	195	27.74	99.5
Beisamoun (early Pottery Neolithic)	12	1.71	100
Ain-Avazim (surface)	2	0.28	0
Tel Turmus (surface)	2	0.28	100
Quarantine (surface)	1	0.14	100
Ain Miri (all strata)	51	7.25	100
Nahal Zehora I (Wadi Rabah)	1	0.14	100
Nahal Zehora II (all strata)	23	3.27	78.2
Total	**703**	**100**	**93.3 ***

*Of the entire studied assemblage.

EXCAVATED SITES

Hagoshrim: The site is situated in the northern outskirts of Kibbutz Hagoshrim and covers approximately 8 ha. Parts of the site were excavated in the 1970s and revisited during the 1990s.[14] The site contained layers belonging to the PPNC (Layer VI), Lodian (Layer V), Wadi Rabah (Layer IV) and Late Chalcolithic periods (Layers III-II).[15] The large stone assemblage retrieved in the course of excavations was studied by one of us.[16] The discs discussed here originated from stratified contexts as well as from mixed and surface contexts.[17]

Beisamoun: This is a complex of several sites, covering an area of some ten ha., extending from the Naphtali Mountains to the west (not far from the Natufian site of Eynan) into the Hula Valley. The site was partly surveyed and excavated during the early 1970s[18] and was revisited recently.[19] The site contained remains dated from the PPNB through the Early PN and Wadi Rabah cultures.[20] The discs presented in this study are mostly of Early PN origin retrieved in the course of a salvage excavation conducted in 2007 (fig. 2), and from surface collection.

Ain Miri: The site is situated near the confluence of the Dishon and Tzivon streams, east of Kibbutz Baram in the Upper Galilee. Survey and excavation at the site revealed finds attributed to the Geometric Kebaran through the PN.[21] Nevertheless, most of the material retrieved from the main area of excavation belongs to the LPPNB-PPNC periods.

Nahal Zehora I: The site is situated in the Menashe Hills and was excavated in 1987-1990. The finds were attributed to the Wadi Rabah culture[22] (fig. 4: 2).

Nahal Zehora II: The site is situated in the Menashe Hills not far from Nahal Zehora I. It was excavated during the 1980-1990s[23] revealing three main PN strata assigned to the Yarmukian, Lodian and Wadi Rabah cultures. The discs included in this study (fig. 3; fig. 4: 1; figs. 5-6) originate from all strata.

14. NOY, 1978: 9 and 1980: 3; GETZOV, 2008.
15. IAA excavation, GETZOV, 2008.
16. ROSENBERG, in press 2.
17. The surface finds consist of four discs found at Hagoshrim-East, separated from the excavated site by an old channel of the nearby Dan stream. The site covers some 3 ha and it is attributed mostly to the Chalcolithic period.
18. LECHEVALLIER, 1978.
19. BOCQUENTIN *et al.*, 2007; ROSENBERG, forthcoming.
20. ROSENBERG *et al.*, 2006; ROSENBERG, forthcoming.
21. SCHYLE and UERPMANN, 2001; SHIMELMITZ *et al.*, 2004.
22. GOPHER, in press.
23. *Ibid.*

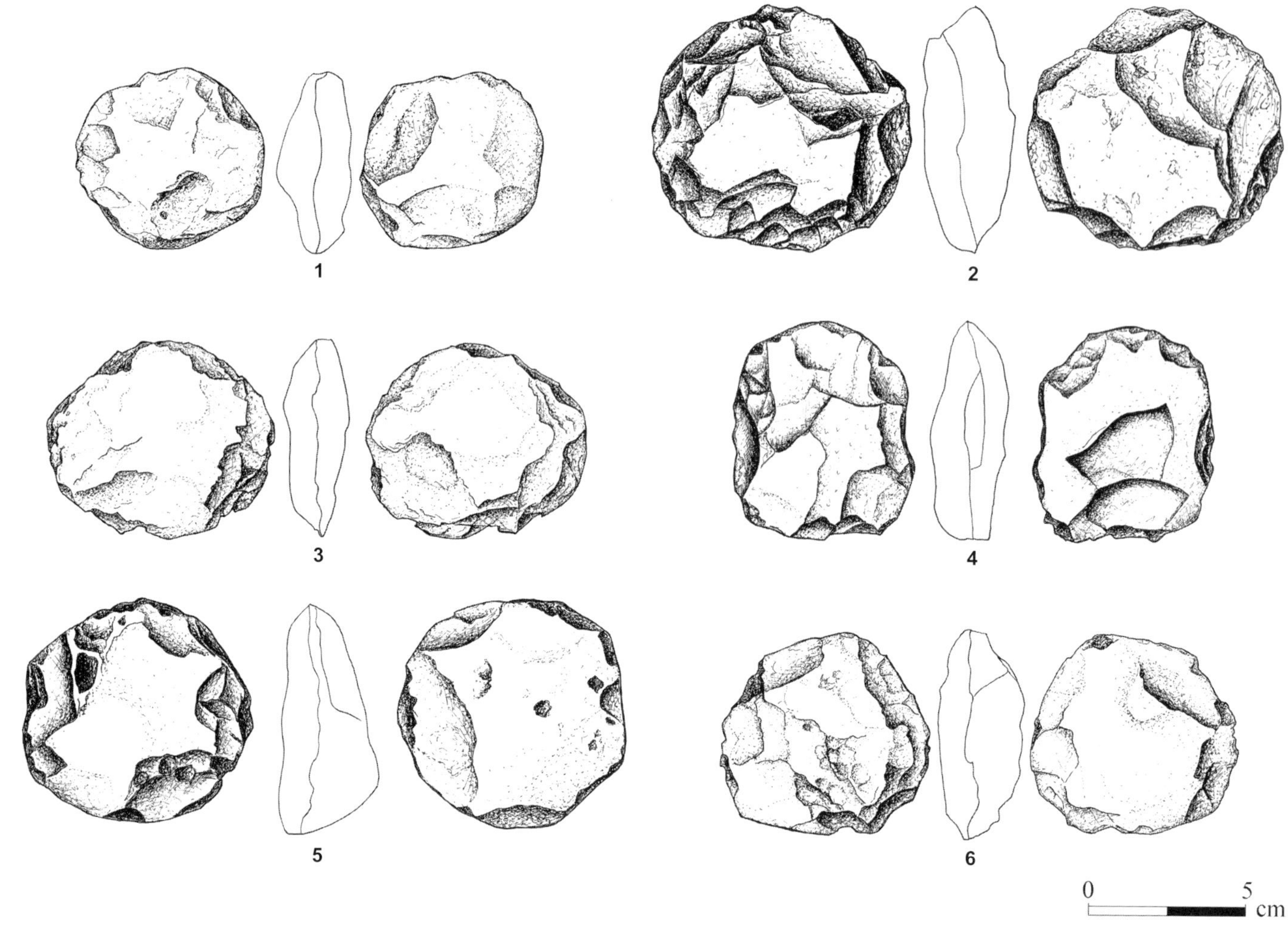

Fig. 2 – *Discs from Early Pottery Neolithic Beisamoun.*

SURFACE-COLLECTED SITES

Hatanur: The site is situated on a small plateau cut by the steep canyon of Nahal Ayun, some 300 m southeast of the channel, and about 50 m above the present gorge. The site surface is covered by dark soil and the finds were scattered over an area of roughly 2 ha.[24] The site was attributed mostly to the PPNB.

Kat: The site is some 1.5 km west of Kfar Yuval, not far from the aforementioned Hatanur site. The estimated site size is *ca* 2 ha. Lechevallier and Dollfus[25] suggested a PPNB attribution for it.

Zug-Fukani: The site is situated about 1 km west of Kfar Yuval and finds were dispersed over an area of about 2 ha. The lithic material collected from the surface until the early 1970s, is characterised by a dominance of PPNB finds.[26]

Tel Hay (Tel Roeim – West): The site is situated at the foot of the eastern slopes of the Ramim Ridge on the northern outskirts of Qiryat Shmone. Site size is unclear but it seems to cover some 3.5 ha.[27] Recently, a salvage excavation took place at the site uncovering finds of the PPNB, PPNC and the PN.[28] During a study of the stone tool assemblage from this

24. LECHEVALLIER et DOLLFUS, 1973.
25. *Ibid.*
26. *Ibid.*
27. SHAKED and MARDER, 1996.
28. D. Nadel, personal communication.

Fig. 3 – *Discs from Pottery Neolithic Nahal Zehora II.*

excavation by one of us (D.R.), several discs were found, originating from all strata.[29]

Ain Hashomer: The site is situated at the foot of the Naphtali Mountains; north of Kfar Gil'adi and near the road leading from Qiryat Shmone to Metula. Some of the lithic material collected from its surface was published in the past,[30] suggesting a PPNB and possibly PN attribution to the site.

Tel Turmus: The site was surveyed and excavated during the 1960s[31] and salvage excavation took place at the site in 1996. Both projects suggest the site should be attributed mainly to the Chalcolithic period, although Early Bronze Age finds were also noted.[32] The two stone discs incorporated in the present study were collected from the site's surface.

Quarantine: The site, located about 3 km south of Qiryat Shmone, on the foot of the Naphtali Mountains. The finds, scattered over an area of about 1 ha, were attributed mainly to the PPNB and to the PN.

Ain-Avazim: The site is located near the highway between the city of Qiryat Shmone and the town of Rosh Pina, some 12 km south of Qiryat Shmone, occupying an estimated area

29. The data is not included in the present study.
30. LECHEVALLIER et DOLLFUS, 1973.
31. DAYAN, 1963: 15.
32. SMITHLINE *et al.*, 1998: 4-5.

Fig. 4 – *Discs:* ***1****, Pottery Neolithic Nahal Zehora II;* ***2****, Nahal Zehora I.*

of 2.5 ha. The site was recently tested[33] and yielded PPNB, PPNC and Chalcolithic material as well as Early Bronze Age finds. The discs for this study originate from the surface collection.

RESULTS

The relative frequencies of discs in the groundstone assemblages cannot be determined, as most of them originate from surface collections.[34] Nevertheless, the quantity of the discs

33. ALON, 2003.

34. This is also true for sites not included in the present study, where discs are only rarely presented.

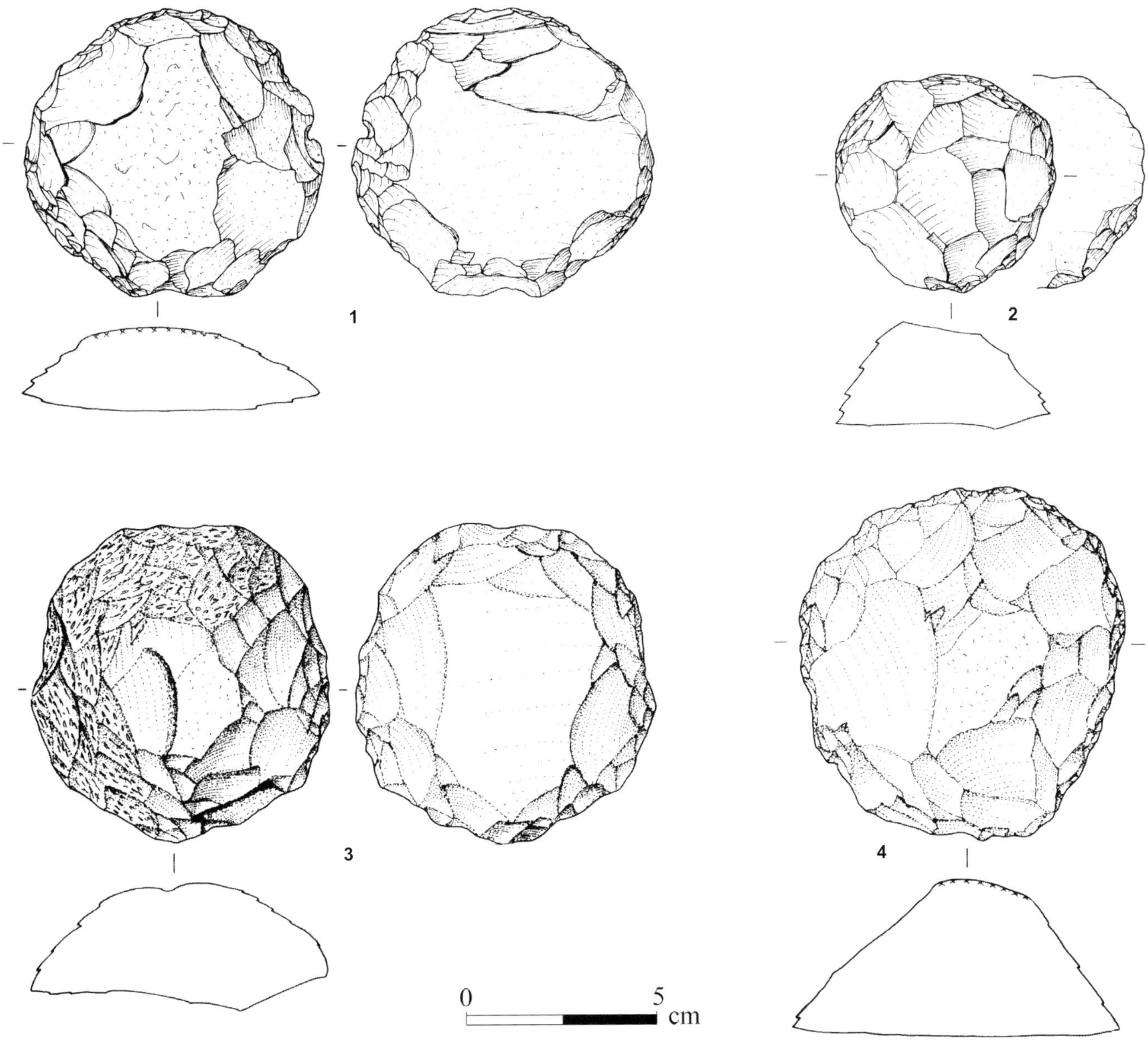

Fig. 5 – *Discs from Pottery Neolithic Nahal Zehora II.*

varies considerably and while some of the sites feature many discs, others have only few (table 1).

At Hagoshrim, the discs comprise 9.4% of the entire stone assemblage (excavation and surface collection combined). The excavated assemblage reveals a decrease in the relative frequencies of discs (table 1), from the PPNC (VI) to the Lodian (V) and the Wadi Rabah (IV) (44.7%, 16.5% and 4.5% of the stone assemblage, respectively). Some 6.4% of the stone assemblage at Early Pottery Neolithic Beisamoun are discs. Although small, a certain decrease in the number of discs over time was also noted at Tel Roeim-West assemblage. Thus, it seems that there was a decline in the frequency of this tool type over time.

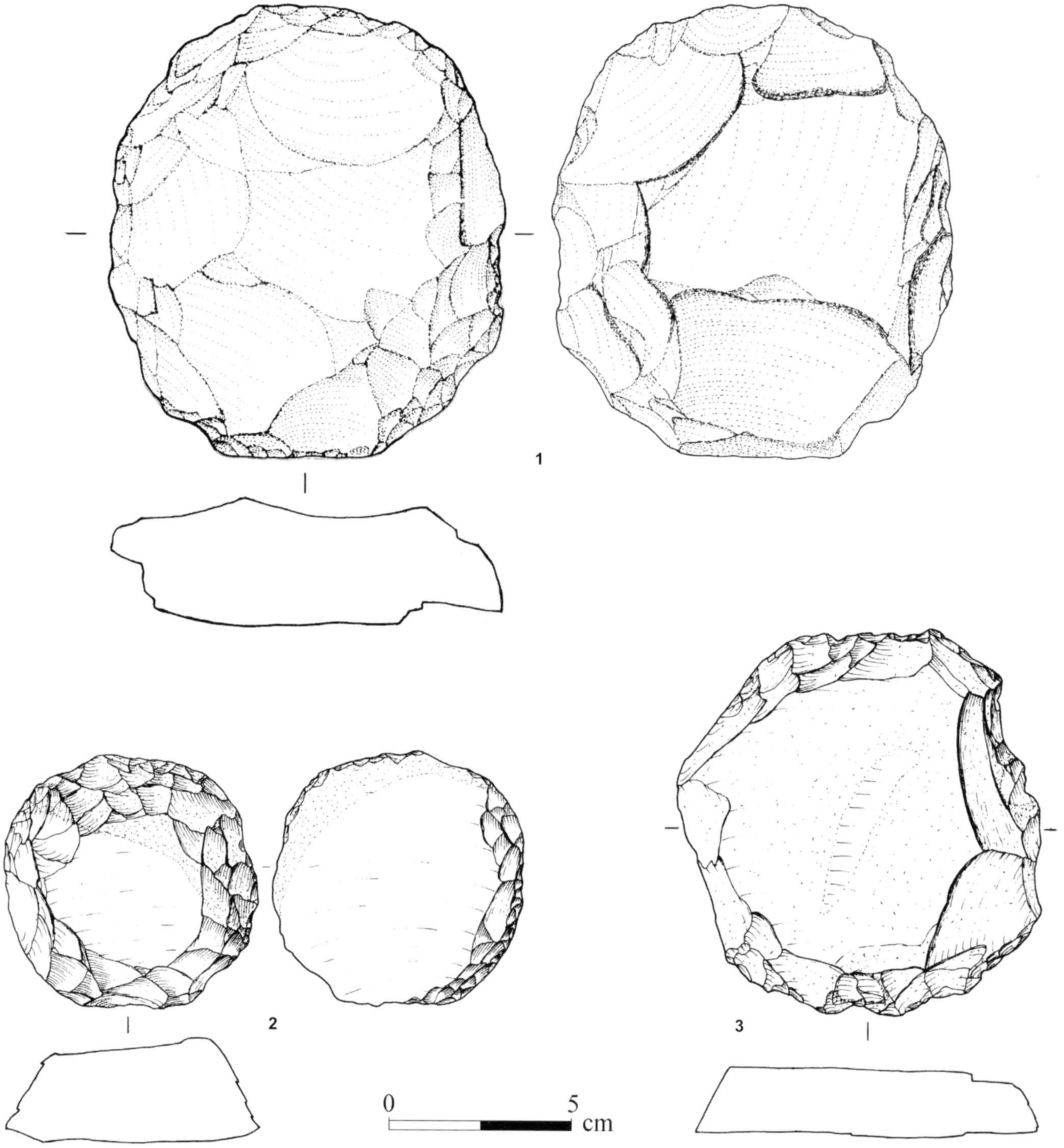

Fig. 6 – *Discs from Pottery Neolithic Nahal Zehora II.*

PRESERVATION

The great majority (93.3%) of the discs are whole with no signs of damage (table 1). This is characteristic of both surface and excavated contexts and thus cannot be attributed to biased collection. It is therefore more likely to be due to the compact nature of the raw material or to how discs were employed. A distinction was made between the scars of production (flaking / chipping) and large scars suggesting fragmentation / breakage. Interestingly, at Hagoshrim, Layers VI and IV comprise mainly whole discs (88.2% and 100%, respectively) while at Layer V, only 46% of the assemblage was found whole. At early PN Beisamoun, all discs are whole (table 1), and this is true for most of the surface collected discs.

RAW MATERIALS

Raw material distribution (table 2) suggests a clear preference for relatively hard limestone (68.71%), followed by fine-grained, compact basalt (29.3%). Other raw materials such as porous basalt and sandstone were only rarely noted (1% each). In most cases, both basalt and limestone were easily accessible, found in proximity to the sites.[35]

The relative distribution of raw materials among the sites suggests a consistent preference for limestone even when basalt was equally accessible.[36] At Hagoshrim for instance, all layers show dominance of limestone (~2:1 limestone to basalt discs), and this accords well with the surface finds of the site. Intriguingly, Hagoshrim is the only site where flaked sandstone discs were found.[37]

Variations in colour are closely correlated to the raw material frequencies. Thus, high frequencies of white (63.1%) and grey (26.1%) discs were found. Rarely, pinkish, red, brown and yellowish disc shades were noted as well. Overall, there seems to have been a preference for light coloured discs (whites and light grey).

TYPOLOGY AND MORPHOLOGY

The typology of discs is based on their morphology. The discs were classified as round (53.06%), oval (42.53%) or other (4.41%), with the latter including heart-like, quasi-triangular and amorphously shaped items (table 3). Although the "other" group is small, it seems that effort was invested in shaping these items and that their morphological characteristics are intentional rather than mere "mistakes" in the production sequence.

Round and oval discs clearly dominate the assemblage and appear in similar frequencies. In most of the sites, a preference for round discs is easily observed. However, the Hagoshrim assemblage is exceptional in this regard, dominated by oval discs in all layers as well as in the surface collection (table 3). Hagoshrim also includes most of the heart-shaped items.

The disc cross-section (table 3) is dictated by the amount nature and orientation of the flaking and to a lesser extent by the nature of the blank (*i.e.* pebble / cobble, flake or fragment). It thus has technological implications and may also be related to the possible function of the discs (cutting, smoothing, scratching or tossing).

The dominant cross-section in the assemblages is biconvex (60.03%), giving the disc a "flying saucer" form. It results from bifacial flaking of the circumference. Only rarely were naturally biconvex blanks used. Plano-convex cross-section was observed in 29.73% of the discs. Other cross-sections present are bi-plano, convex-concave and irregular or unclear (0.85%, 4.84% and 4.55% of the assemblage, respectively). At Hagoshrim Layer VI the number of bi-convex and plano-convex cross-sections is nearly the same, while at Layer V there are more discs with bi-plano cross-sections, and in Layer IV discs with bi-convex cross-sections dominate. The Early PN at Beisamoun shows clear dominance of discs with bi-convex cross-sections.

METRICS

Disc weight ranges from 10 gm to 1 kg (fig. 7) with an average of 168.2 gm (SD = 110.2); most discs weigh between 50 and 350 gm (85.3% of the assemblage). In Hagoshrim, the average mass of the discs remained consistent throughout the three layers (table 4), whereas in the Early PN of Beisamoun, the average mass was considerably greater.

Length was measured only for oval discs, and varied from 4 to 14.2 cm with an average of 7.3 cm (SD = 1.64). The length of most oval discs is in the range of 5 to 10 cm (91.6% of the oval discs). The data (table 4) suggest that hardly any change occurred in the length range, between the late PPN of Hagoshrim and the Early PN of Beisamoun. Hagoshrim Layer IV shows the greatest average length, however.

35. *e.g.*, Beisamoun, Hagoshrim, Kat and Ain Miri

36. *e.g.*, Beisamoun, Hagoshrim and to some extent Ain Miri, where basalt is somewhat less accessible.

37. Sandstone could be found a few kilometers from the site, on the slopes of the Naphtali Mountains, near the city of Qiryat Shmone.

Table 2 – *Breakdown of raw materials by site.*

Site	Fine-grained basalt	Porous basalt	Limestone	Sandstone	Total
Hatanur (surface)	2		13		**15**
Kat (surface)	34	5	15		**54**
Zug-Fukani (surface)	1		1		**2**
Tel Hay / Tel Roim-West (surface)	2		5		**7**
Ain Hashomer (surface)			2		**2**
Hagoshrim (surface and unstratified contexts)	61	2	131	7	**201**
Hagoshrim (Layer VI)	20		48		**68**
Hagoshrim (Layer V)	13		33		**46**
Hagoshrim (Layer IV)	7		14		**21**
Beisamoun (surface)	61		134		**195**
Beisamoun (Early Pottery Neolithic)			12		**12**
Ain-Avazim (surface)			2		**2**
Tel Turmus (surface)	2				**2**
Quarantine (surface)			1		**1**
Ain Miri (all strata)	1		50		**51**
Nahal Zehora I (Wadi Rabah)			1		**1**
Nahal Zehora II (all strata)	2		21		**23**
Total	**206**	**7**	**483**	**7**	**703**
%	**29.3**	**1**	**68.71**	**1**	**100**

Table 3 – *Breakdown of shapes and cross-sections by site.*

Site	Shapes				Cross-sections					
	Round	Oval	Other	Total	Bi-convex	Plano-convex	Bi-plano	Convex-concave	Amorphous-unclear	Total
Hatanur (surface)	12	3		**15**	8	6		1		**15**
Kat (surface)	45	9		**54**	32	21		1		**54**
Zug-Fukani (surface)	2			**2**	1	1				**2**
Tel Hay / Tel Roim-West (surface)	5	2		**7**	3	3		1		**7**
Ain Hashomer (surface)	2			**2**	2					**2**
Hagoshrim (surface and unstratified contexts)	69	121	11	**201**	124	57	3	5	12	**201**
Hagoshrim-PPNC (Layer VI)	13	51	4	**68**	29	27	1	4	7	**68**
Hagoshrim-Lodian Related (Layer V)	19	26	1	**46**	16	25	1		4	**46**
Hagoshrim-Wadi Rabah Related (Layer IV)	4	15	2	**21**	13	5			3	**21**
Beisamoun (surface)	145	41	9	**195**	134	48	1	9	3	**195**
Beisamoun (Early Pottery Neolithic)	7	5		**12**	8	2			2	**12**
Ain-Avazim (surface)			2	**2**				2		**2**
Tel Turmus (surface)	2			**2**	2					**2**
Quarantine (surface)	1			**1**	1					**1**
Ain Miri (all strata)	34	17		**51**	32	9		10		**51**
Nahal Zehora I (Wadi Rabah)		1		**1**	1					**1**
Nahal Zehora II (all strata)	13	8	2	**23**	16	5		1	1	**23**
Total	**373**	**299**	**31**	**703**	**422**	**209**	**6**	**34**	**32**	**703**
%	**53.06**	**42.53**	**4.41**	**100**	**60.03**	**29.73**	**0.85**	**4.84**	**4.55**	**100**

In addition, width values were measured only for oval discs, the range of which is 1.7-13.2 cm, averaging at 6.11 cm (SD = 1.54). Most of the discs fall within the range between 4 and 8 cm (85.1% of the oval discs). There seems to have been considerable uniformity and consistency in the width values, although a somewhat higher average width was recorded in Hagoshrim Layer VI and to lesser extent the Early PN of Beisamoun (table 4).

Diameter values were measured only for round discs, showing a distribution ranging between 3.9 to 11.4 cm, with

Table 4 – *Average thickness for round and oval discs of Hagoshrim and Beisamoun.*

		Site			
		Hagoshrim VI	Hagoshrim V	Hagoshrim IV	Beisamoun (Early Pottery Neolithic)
Weight (gr)	Range	20-440	10-410	10-380	55-305
	Average	121.6	111.19	125	149.5
	SD	74.16	85.3	82.62	75.14
Length (cm) Oval discs	Range	4.3-12.1	4.2-9.1	4.9-9.8	5-8.1
	Average	6.39	6.4	7.19	6.78
	SD	1.57	1.3	1.52	0.95
Width (cm) Oval discs	Range	4.3-9	3-7.9	3-10.6	5.4-7.4
	Average	6.2	5.23	5.51	6.04
	SD	1.47	1.13	1.44	0.81
Diameter (cm) Round discs	Range	4.1-7.7	3.9-9.7	4.4-7.3	5.9-8.6
	Average	6.5	6.54	6.5	7.3
	SD	1.42	1.63	1.4	1
Thickness (cm) Round discs	Average	4.7	2.2	2.1	2.5
	SD	2.46	0.75	0.62	0.46
Thickness (cm) Oval discs	Average	2.05	1.89	2.08	2.08
	SD	0.7	0.55	0.55	0.54

Table 5 – *Distribution of Production Marks.*

Production Marks	N	%
Circumferential bifacial scars only	375	53.34
Circumferential unifacial scars only	142	20.2
Circumferential bifacial scars and body scars	11	1.56
Circumferential bifacial scars, body scars and pecking	11	1.56
Circumferential bifacial scars and pecking	126	17.92
Circumferential unifacial scars and body scars	10	1.42
Circumferential unifacial scars, body scars and pecking	7	1
Circumferential unifacial scars and pecking	21	2.99
Total	703	100

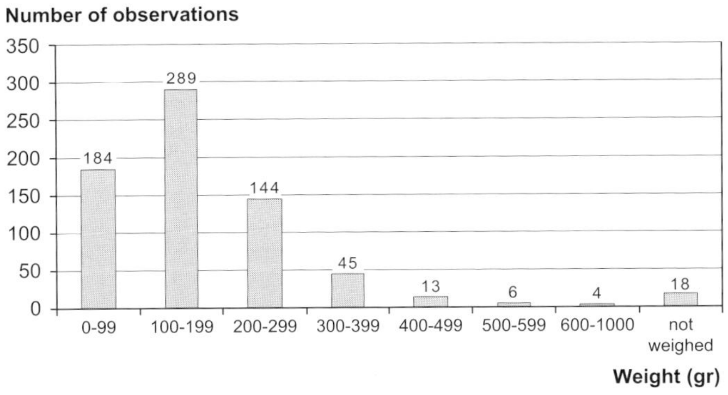

Fig. 7 – *Discs: Number of observations per weight group.*

an average of 7.53 cm (SD = 1.34). Most discs are within the range of 5 to 9 cm (83% of the round discs). In all, the data suggests a high degree of uniformity in diameter throughout the different Layers at Hagoshrim (table 4) and a somewhat higher diameter average in the Early PN of Beisamoun.

Thickness for the entire measured assemblage ranges between 1 and 9.5 cm, with an average of 2.54 cm (SD = 0.97). Most of the discs (87%) were 1.5-3.5 cm thick, suggesting a high degree of uniformity. The thickness of round discs ranges from 1 to 9.5 cm with an average of 2.74 cm (SD = 1.15). Hagoshrim Layer VI features relatively high thickness values for round discs (table 4), whereas in Layers V and IV as well as the Early PN of Beisamoun thicknesses are uniform. The thickness of oval discs, on the other hand, ranges from 1 to 4.4 cm with an average of 2.29 cm (SD = 0.64). Thus, as a whole, the average thickness of both round and oval discs is rather similar. All layers at Hagoshrim and the Early PN of Beisamoun accord well to this uniformity in oval disc thickness (table 4). A difference in thickness between round and oval discs was observed only in Hagoshrim Layer VI, due to the exceptionally thick round discs in the assemblage.

TECHNOLOGY

Most of the blanks from which discs were produced, consisted of pebbles or thick flakes[38] of limestone or basalt. Some discs were shaped on pebble fragments, from which the bulb of percussion was removed by flaking or abrasion. Although it can be said with certainty for only few of the discs, it seems that some preparation took place before flaking. The shaping of the disc was conducted primarily by means of centripetal flaking, directed towards the centre of the disc (table 5) in angles between 45 and 90 degrees, leaving flat scars usually 0.5-1.5 cm in size (figs. 2-4).

Most discs (74.38%) were shaped by bifacial flaking (table 5) of the entire or most of the disc's circumference. Over 50% of the discs were produced by bifacial flaking alone, while 20.2% of the discs were shaped only by unifacial flaking. Some of the unifacial discs may have actually been originally intended as cores or blanks for other tools. Interestingly and contrary to the general trend, Layers VI, V and IV of Hagoshrim demonstrated a consistent preference for unifacially flaked discs (78%, 63% and 81%, respectively), while the discs found at Early PN Beisamoun for example, were all produced using bifacial flaking.

38. As indicated by the bulb of percussion.

Occasionally, bifacial and more rarely unifacial flaking was followed by additional treatments such as thinning,[39] pecking or smoothing (table 5).[40] On many bifacially shaped discs, the two faces did not receive the same amount of care when shaped. In addition, it appears that in some cases the coarse flaking of the circumference was refined by further fine retouch.

Pecking, an important technique in stone tool manufacture, is clearly present and it is more commonly used on basalt. However, many of the discs' surfaces were damaged and roughened by natural agents (rain, running water, etc.) making it hard to distinguish natural deterioration or weathering of the rock from intentional pecking, especially in the surface collected assemblages.

Smoothing may also have concealed pecking signs, although in most cases it is clear that there was no intention of concealing the coarse flaking and retouch of the circumference. On several discs, it seems that smoothing was used for refining parts of the natural surface of the blank rather than the artificially shaped parts; but, for the most part, it is difficult to distinguish natural from intentional smoothening.

Concerning the possible function of the discs, use-wear marks include striations or shallow and narrow grooves. Only in a few cases was wear[41] noticed on the discs. Related residue or traces of attached substances were noted only on a few examples, yet it is not clear whether these[42] are related to the function of the disc or to post-depositional factors.

SUMMARY AND DISCUSSION

Stone discs appeared in the southern Levant sometime during M / LPPNB. Their numbers seem to have diminished following the PPNC[43] and eventually disappeared near the end of the PN. The occurrence of stone discs in Chalcolithic or Early Bronze Age stone assemblages ought to be regarded as provisional for now.

The consistent presence of flaked stone discs in Neolithic sites over a time span of more than two millennia, with only minor changes in their attributes, indicates a stable Neolithic tool type. Stone discs appear to have maintained their formal characteristics over a period of time that has seen significant socio-economic change. Their geographical distribution in Israel suggests that they are mostly a northern phenomenon, appearing from the Hula Valley and the Galilee at least as far south as the Menashe Hills.

Most of the discs examined were whole (93%), both in surface collections and in excavated assemblages. Further north, however, at Çayönü Tepesi a different pattern was noted, where about half of the discs were broken.[44] This may suggest that the northern and southern Levant differed in the functions or discard patterns (deliberate fragmentation?) associated with these items.

Raw materials used for the production of discs include limestone, basalt, sandstone and marble.[45] The general preference for limestone is evident even in sites where basalt is equally prevalent. Thus, in Hagoshrim, where basalt is easily accessible, all layers demonstrated limestone frequencies twice as high as those of basalt. It is likely, therefore, that the qualities of limestone, and to a lesser extent those of fine-grained, compact basalt were appreciated as particularly suitable for the intended use of discs. In addition, the dominance of white and light-grey shades is closely correlated with these high frequencies of limestone in the assemblages, and perhaps had some weight in the choice of raw material.

Most stone discs were either round or oval and of a biconvex cross-section. The round types are dominant in most of the sites, although in Hagoshrim (both surface collected and excavated discs) a clear preference for oval discs was noted. This is the case for both limestone and basalt, suggesting that there is no direct correlation between shape and raw material. The metrical attributes of the discs suggest relative uniformity and consistency through time.

The discs were mostly shaped on flakes and some on pebbles / cobbles. This was done mainly by bifacial, and to a lesser extent by unifacial, flaking of the circumference. Only rarely was any effort made to "camouflage" production scars, suggesting that the fine finishing of the disc was of relatively low priority.

To conclude, this study provides a detailed type-definition of stone discs. The gathered data suggest that most discs demonstrate a high degree of uniformity and follow relatively specific standards in terms of raw material, shape, size and weight. Production modes are generally consistent and may indicate that their functions remained roughly the same throughout

39. As indicated by characteristic thinning flake scars.
40. The two excavated sites of Hagoshrim and Beisamoun showed only rare use of other production techniques, apart of flaking.
41. Naked-eye observation only.
42. Red ochre-like substances and more rarely a whitish material, which may be lime plaster.
43. *i.e.* seen in excavated sites such as Hagoshrim and Tel Roeim-West.
44. DAVIS, 1982: 119.
45. See also BRAIDWOOD and BRAIDWOOD, 1960: 55; GEBEL and BIENERT, 1997; GEBEL *et al.*, 1997.

the two millennia of their existence. The data presented here suggests that discs had a long Neolithic life span, possibly throughout the Levant and beyond.

It was suggested in the past that discs were used as processors / manos[46] or as lids for pottery vessels.[47] It was also suggested that similar items were waste products of initial stages of stone bracelet shaping, such as those found at Ba'ja and other sites in Jordan.[48] Whether these suggestions are valid or not, it seems that the exact function of stone discs still awaits a thorough use-wear study.

The size range of flaked discs and the fact that two main shapes are present (round and oval) could suggest several uses of these tools. We can only speculate and offer a few more suggestions to those noted above:

- Discs were utilized as scrapers, related to the preparation and tenderising of hides, wool, or working wood barks. This function may be relevant specifically to discs shaped by unifacial retouch;
- Discs could have been used in pottery production and design or in applying lime-plaster surfaces;
- Discs may have been thrown objects, used in hunting or as a gaming object.[49] Another option is that they had a function as a mnemonic device used in exchange, each bearing a certain value or "merchant weight".[50] If so, the size, weight and/or raw material characteristics, such as colour or texture, may have been related to the value in the world of commodities,[51] or as part of a barter system. This could be supported by the relative uniformity observed.

Whether these stone discs were used as operational tools of a certain craft or they were some kind of a symbolic device in the economic sphere is left open for now. Reports on stone discs of similar shape from other excavated sites could shed more light on their chronological and geographical distribution as well as on spatial and contextual aspects. A use-wear study of the flaked stone discs is currently underway and we hope it may be of help for better understanding of the exact function of these artefacts.

ACKNOWLEDGMENTS

We would like to thank D. Nadel, Y. Garfinkel and H. Smithline for allowing us to mention data from their yet unpublished excavations. We thank A. Nativ and anonymous reviewers for improving the last version of the text and we are grateful for their helpful comments. Drawings were made by R. Pinhas, A. Avshalomov and I. Rosenberg and photos by P. Shrago and D. Rosenberg.

Danny ROSENBERG
Zinman Institute of Archaeology
University of Haifa
Haifa, 31905 – ISRAEL
aromat@netvision.net.il

Amnon ASSAF
Mayan Baruch Museum of Prehistory of the Hula Valley
Kibbutz Mayan Baruch
ISRAEL

Nimrod GETZOV
Israel Antiquities Authority
POB 586
91104 Jerusalem – ISRAEL
getzov@israntique.org.il

Avi GOPHER
The Sonia and Marco Nadler Institute of archaeology
Tel Aviv University
Tel Aviv – ISRAEL
agopher@post.tau.ac.il

46. DAVIS, 1982: 119.
47. GOPHER and GREENBERG, 1987: 107-108.
48. GEBEL and BIENERT, 1997: 17; GEBEL *et al.*, 1997 and see STARCK, 1988 for a discussion on the rings.
49. These suggestions seem unlikely due to the good preservation of the discs.
50. See also ORRELLE, 1993 and 1996; SCHMANDT-BESSERAT, 1982 and 1986 for similar suggestions for ceramic tokens.
51. *e.g.*, KOPYTOFF, 1988: 68.

BIBLIOGRAPHY

ALON Y.

2003 'En Avazim. *Hadashot Arkheologiyot* 115: 2-3.

BOCQUENTIN F., KHALAILY H., SAMUELIAN N., BARZILAI O., LE DOSSEUR G., KOLSKA-HORWITZ L. and EMERY-BARBIER A.

2007 Renewed excavation of the PPNB site of Beisamoun, Hula Basin. *Neo-Lithics* 2,07: 17-21.

BRAIDWOOD R.J. and BRAIDWOOD L.S.

1960 *Excavations in the Plain of Antioch I, The Earlier Assemblages Phases A-J.* Chicago: University of Chicago Press.

DAVIS M.K.

1982 The Çayönü Ground Stone. *In*: BRAIDWOOD L.S. and BRAIDWOOD R.J. (eds.), *Prehistoric Village Archaeology in Southeastern Turkey*: 73-174. Oxford (*BAR Int. Ser.* 138).

DAYAN Y.

1963 Tel Turmus. *Hadashot Arkheologiyot* 5:15. (in Hebrew).

GEBEL H.G.K. and BIENERT H.D.

1997 The 1997 season at Ba'ja, Southern Jordan. *Neo-Lithics* 3,97: 14-18.

GEBEL H.G.K., BIENERT H.D., KRÄMER T, MÜLLER-NEUHOF B., NEEF R., TIMM J. and WRIGHT K.I.

1997 Ba'ja Hidden in the Petra Mountains. Preliminary report on the 1997 excavation. *In*: GEBEL H.G.K., KAFAFI Z. and ROLLEFSON G.O. (eds.), *The Prehistory of Jordan* II. *Perspective from 1997*: 221-262. Berlin: ex oriente.

GETZOV N.

2008 Ha-Gosherim. *In:* STERN E. (ed.), *The New Encyclopedia of Archaeological Excavations in the Holy Land 5 (Supplementary volume)*: 1759-1761. Jerusalem: Israel Exploration Society.

GOPHER A.

in press *Village Communities of the Pottery-Neolithic Period in the Menashe Hills, Israel. Archaeological Investigations at the Nahal Zehora Sites.* Tel Aviv: Tel Aviv University (*Monograph Series of the Institute of Archaeology*).

GOPHER A. and GREENBERG R.

1987 Pottery Neolithic Levels at Tel Dan. *Journal of the Israel Prehistoric Society* 20: 91-113.

KOPYTOFF I.

1988 The cultural biography of things: Commoditization as process. *In*: APPADURAI A. (ed.), *The Social Life of Things*: 64-91. Cambridge: Cambridge University Press.

LECHEVALLIER M.

1978 *Abou Gosh et Beisamoun. Deux gisements du VIIe millénaire avant l'ère chrétienne en Israël.* Paris: Association Paléorient (*Mémoires et Travaux du Centre de Recherches Préhistoriques Français de Jérusalem* 2).

LECHEVALLIER M. et DOLLFUS G.

1973 Nouveaux sites du VIe millénaire en Haute-Galilée. *Eretz Israel* 11 : 9-21.

NOY T.

1978 Hagoshrim. *Hadashot Arkheologiyot* 67-68: 9

1980 Hagoshrim. *Hadashot Arkheologiyot* 74-75 : 3

ORRELLE E.

1993 *Nahal Zehora I and II: Fifth Millennium Villages of the Wadi Rabah Culture. Quantitative Studies on the Pottery Assemblages.* Tel Aviv: Tel Aviv University. Unpublished M.A. Thesis.

1996 *Pottery discs, plain and perforated: A quantitative study of on pottery assemblage from Nahal Zehora I and II.* Forli. Unpublished paper presented at the 13th UISPP Congress 1996.

ROSENBERG D.

in press 1 The Early Bronze Age stone assemblage. *In*: GREENBERG R. (ed.), *Tel Bet Yerah II: Analysis and Synthesis, Excavations of 1933-1986.* Jerusalem: Israel Antiquities Authority (*IAA Reports*).

in press 2 The Stone Assemblage of Hagoshrim – Continuity and change in the Neolithic of Northern Israel. *In*: GETZOV N. (ed.), *Hagoshrim* (temporary title). Jerusalem: Israel Antiquities Authority (*IAA Reports*).

forthcoming The stone assemblage of the Early Pottery Neolithic occurrence at Beisamoun. *In*: ROSENBERG D. (ed.), *Early Pottery Neolithic of Occurrence at Beisamoun, The Hula Valley, Northern Israel.* Oxford (*BAR Int. ser.*).

ROSENBERG D., ASSAF A., EYAL R. and GOPHER A.

2006 Beisamoun - The Wadi Rabah occurrence. *Journal of the Israel Prehistoric Society* 36: 129-137.

SCHMANDT-BESSERAT D.

1982 The emergence of recording. *American Anthropologist* 84,4: 871-878.

1986 An ancient token system: The precursor to numerals and writing. *Archaeology* 39,6: 32-39.

SCHYLE D. and UERPMANN H.P.

2001 Ein Miri and the Geometric Kebaran of the Northern Levant. *In*: GHLEN B., HEINEN M. und TILLMANN A. (Hrsg.), *Zeiträume, Gedenkschrift für Wolfgang Taute*: 199-217. Bonn: Selbstverlag der Deutschen Gesellschaft für Ur- und Frühgeschichte e. V (*Archäologische Berichte* 14).

SHAKED E. and MARDER O.

1996 Metula Map survey – Prehistoric sites. *Hadashot Arkheologiyot* 106: 7-8.

SHIMELMITZ R., BARKAI R. and GOPHER A.

2004 The Geometric Kebaran microlithic assemblage of Ain Miri, Northern Israel. *Paléorient* 30,2: 127-140.

SMITHLINE H., COVELLO-PARAN K. and MARDER O.

1998 Tel Turmus. *Hadashot Arkheologiyot* 108: 4-5 (in Hebrew).

STARCK J.M.

1988 Stone rings from Baja and Basta – Geographical and chronological implication. *In*: GARRARD A.N. and GEBEL H.G. (eds.), *The Prehistory of Jordan, the state of research in 1986*: 137-174. Oxford (*BAR Int. Ser.* 396-I).

WINN S. and SHIMABUKU D.
1989 Bone and ground stone tools. *In*: GIMBUTAS M., WINN S. and SHIMABUKU D. (eds.), *Achilleion: Neolithic Settlement in Thessaly, Greece, 6400-5600 BC*: 259-272. Los Angeles: University of California Press.

YOHSIZAWA S.
2003 Ground stone artefact. *In*: IWASAKI T. and TSUNEKI A. (eds.), *Archaeology of the Rouj Basin*: 99-117. Tsukuba: Institute of History and Anthropology, University of Tsukuba.

INTERPRETING ANIMAL BONES IN IRAN: CONSIDERING NEW ANIMAL BONE ASSEMBLAGES FROM THREE SITES IN THE QAZVIN PLAIN WITHIN A BROADER GEOGRAPHICAL AND CHRONOLOGICAL PERSPECTIVE

R. YOUNG and H. FAZELI

Abstract: *This paper presents the preliminary analyses and interpretations of animal bone assemblages from three Chalcolithic sites in the Qazvin Plain, Iran. The results of our analyses are then compared to existing analyses from earlier excavations at two of these sites. While there are some similarities between the earlier and later assemblages and analyses, there are also interesting differences. However, issues with dating, excavation and sample collection, as well as sample size from each site meant that analysing each assemblage in detail and linking this analysis to detailed context information has not been possible. Therefore a wider trawl for studies of animal bones from prehistoric sites in a variety of environmental settings in Iran has been carried out in order to determine whether there are broader, longer term trends in human-animal relationships that our Qazvin Plain sites may have been part of. We have looked at animal bone data from a number of prehistoric sites in order to see whether there are any major trends in animal management which might be linked to such things as chronological period, environmental setting, site function and so forth, within which our Qazvin Plain sites can be placed. From the data we consider there are no such trends apparent. Finally, we consider issues of pastoralism in Iran, and are particularly concerned with the suggestion that high numbers of sheep and goat in an assemblage is an indicator of pastoral mobility.*

Résumé : *Sont présentées ici l'analyse préliminaire et des interprétations relatives à des assemblages fauniques provenant de trois sites chalcolithiques de la Plaine de Qazvin, Iran. Les résultats obtenus sont comparés à ceux déjà recueillis sur des collections provenant de fouilles plus anciennes de deux de ces sites. Si l'on note des similitudes entre ces assemblages, des différences peuvent cependant être relevées. Des problèmes restent liés aux datations, aux fouilles, à la nature et à la taille des échantillons de chacun des sites, ce qui n'a pas rendu possible de replacer les résultats de ces analyses dans un contexte sûr. Aussi avons-nous préféré nous tourner vers des collections fauniques de sites localisés en Iran dans des environnements diversifiés afin d'observer si pouvaient se manifester, tant sur la longue durée que sur des aires plus vastes, des tendances dans les relations homme-animal auxquelles pourraient se rattacher celles observées sur les sites retenus de la Plaine de Qazvin ; ainsi en serait-il de la gestion des animaux qui pourrait dépendre aussi bien de la période envisagée, de l'environnement, de la fonction du site. Des données passées en revue, rien ne se dégage réellement. Considérant le pastoralisme en Iran, nous ne pouvons souscrire à la suggestion que des pourcentages importants de moutons et de chèvres dans un assemblage seraient les indices de mobilité et de pastoralisme.*

Keywords: *Chalcolithic, Iran, Qazvin Plain, Animal bones, Interpretive models, Transhumance.*
Mots-clés : *Chalcolithique, Iran, Plaine de Qazvin, Ossements d'animaux, Modèles interprétatifs, Transhumance.*

 Manuscrit reçu le 28 avril 2008, accepté le 25 novembre 2008

Fig. 1 – *Map showing the location of Qazvin Plain and Zageh, Ghabristan and Ismailabad.*

The analysis and interpretation of animal bone assemblages is generally considered to be an excellent mean of learning more about the environment, subsistence patterns and lifeways of past human groups.[1] Since the 1980s the collection of animal bones from excavations (along with soil flotation to recover archaeobotanical remains) has become a key activity on many archaeological sites worldwide and this has resulted in a great deal of data. The quantity and quality of the bone assemblage itself, the quality of the excavation and excavation data, the interpretative stance of the archaeozoologist, the development of a research question, and the availability of models within which to test the data addressing the research question, will all have a major impact on the analysis and interpretation process. Should we disregard all animal bone assemblages that are small, or have been subject to bias (*e.g.* absence of sieving), and only attempt to identify, analyse and interpret large assemblages that are the product of rigorous sampling programs? As we operate in an imperfect world, we argue that we should make the most of all assemblages, and through appropriate analysis and interpretation aim to show that the collection of environmental material can really contribute towards understanding wider archaeological questions.[2] Accordingly, in this paper we want to present our new, preliminary results from three Qazvin Plains sites within a much wider context. Many other animal bone assemblages from Iranian sites are small and often limited, and we hope that by offering a synthesis of results we will be able to distinguish long term trends and developments that reflect the economic, social and cultural events of this region during the late Neolithic and through the Chalcolithic.

We begin by presenting the preliminary results of the archaeozoological analyses of the three Chalcolithic sites of Zageh, Ghabristan and Ismailabad located in the Qazvin Plain, Iran (fig. 1). Zageh and Ghabristan were first excavated by Negahban in the 1970s and subsequently re-excavated in 2001, 2002 and 2003 by a team from the University of Tehran and Iranian Cultural Heritage Organisation (ICHO). Ismailabad was excavated for the first time in 2003 by ICHO and the University of Tehran. All three of these sites have been assigned to the Chalcolithic,[3] which spans a period of more than two millennia in the Tehran and Qazvin Plains from *ca* 5300-4300 BCE (Transitional Chalcolithic), *ca* 4300-4000 BCE (Early Chalcolithic), *ca* 4000-3700 BCE (Middle Chalcolithic) and *ca* 3700-3000 BCE (Late Chalcolithic).[4]

The Chalcolithic is recognized as a period of considerable importance in the northern central plateau, just as it

1. O'CONNOR, 2000; REITZ and WING, 2008.
2. *e.g.*, YOUNG, 2003.
3. FAZELI *et al.*, 2005: 7.
4. FAZELI *et al.*, 2004: table 3.

is throughout the rest of Iran, including the Tehran Plain[5] and Kashan Plain.[6] Defined as a period of great social change, understanding the nature and importance of increasing socio-economic complexity of the Chalcolithic has been achieved through analyses of material culture, including craft specialisation, hierarchical use of space and access to 'exotic' goods.[7] Accordingly, we believe that it should also be possible to detect such major socio-economic changes within the faunal subsistence material recovered from the three sites. In line with work by archaeozoologists such as Zeder[8] who have explored the impact of increasing social complexity on animal bone assemblages (albeit within a complex urban setting), we would expect to see such things as decreased range of animal types, decreased number of wild types, evidence for specialisation in areas such as meat, wool, milk or other secondary products.[9] In order to test this, we undertake various basic analyses, looking in turn at the different proportions of different types and species recovered and identified from each site, and the age at death of sheep and goat.

We then compare our results to those of Mahskour *et al.*[10] where archaeozoological assemblages from two of the three sites explored here were analysed (Zageh and Ghabristan), along with material from the Iron Age site of Sagzabad. This diachronic study aimed to explore "the evolution of faunal spectra in the microregion of Buin Zahra" and the difference between the faunal assemblages from each of the three sites was attributed to "changes in prehistoric subsistence strategies".[11] When our results were considered alongside those of Mashkour *et al.*[12] we found that there were some similarities and also some differences, but neither set of results really told us a great deal about the human choices and social organisation behind the assemblages. Of course this could be attributed to many different factors: small assemblages, limited context information and limited excavations, lack of detailed phasing, and so forth.

In order to maximise the data we do have, we have re-visited a selection of published animal bone analyses from Iranian sites with two aims in mind. Firstly, we wanted to see whether there were any broad trends in the collected data, such as significant changes over time, or major contrasts in assemblages from different environmental zones. Secondly, we wanted to consider the interpretive models that have been applied to these assemblages, and whether these could then be effectively used on other sites, including those of the Qazvin Plain. This does mean that from being a simple presentation of the preliminary results of the animal bone assemblages from three sites, this has become much more of a synthetic presentation, but we hope that the issues relating to interpretation of animal bone assemblages will be of interest to, and relevant to, others working in this region.

THE QAZVIN PLAIN SITES

In order to place the new animal bone material within the context of their sites, this section briefly introduces the three sites, presents their key features and suggested interpretations by their excavators. Figure 1 shows the location of these sites and the position of the Qazvin Plain itself, and the chronology of each site is covered in the table 1.

ZAGEH

Zageh is a 1 m high mound[13] around 2 ha.[14] During the 1970s excavations under the direction of Negahban, more than 1350m^2 were exposed, but only one deep trench was excavated (TTFGX).[15] One of the most important and widely known structures at Zageh is the painted building or decorated shrine, which E.O. Negahban[16] attributed to the late 7th-early 6th millennium BC.[17] The presence of "numerous mountain goat skulls and horns"[18] with the horn bases implanted in the walls of the painted temple may suggest an ideological value for these animals. Eight further trenches were opened in 2001 with the main objectives of locating craft areas on the site and obtaining samples for radiocarbon dating. Trenches A, C3, D and E contained walls, floors, ovens, pits, animal bones, stone tools, and ceramics, giving rise to a general interpretation of

5. CONINGHAM *et al.*, 2006; FAZELI, 2001.
6. MADJIDZADEH, 1978 and 1981; SHAHMIRZADI, 2006.
7. FAZELI, 2001.
8. ZEDER, 1991.
9. SHERRATT, 1983 and 1997.
10. MAHSKOUR *et al.*, 1999: 68.
11. *Ibid.*: 73-74.
12. MASHKOUR *et al.*, 1999.
13. NEGAHBAN, 1977: 34.
14. SHAHMIRZADI, 1979: 183.
15. SHAHMIRZADI, 1980: 14.
16. NEGAHBAN, 1979: 239.
17. See FAZELI *et al.*, 2005 for a discussion of the problems with these dates.
18. NEGAHBAN, 1979: 248.

Table 1 – *Chronology of sites in the order they appear in the text.*

Site name	Date (years BCE)	Cultural Period	Comments	Source
Zageh	5370-5070	Transitional Chalcolithic	Initial occupation	FAZELI *et al.*, 2005: 70-73
	4240-4460	Transitional/Early Chalcolithic	Final occupation	*ibid.*
Ghabristan	4370-4160	Early Chalcolithic	Initial occupation – sequence model	*ibid.*: 68*
	4450-4220	Transitional/Early Chalcolithic	Initial occupation	*ibid.*: 70-73
	3970-3700	Middle Chalcolithic	Final occupation	*ibid.*
Ismailabad	3700-3000	Late Chalcolithic		FAZELI, n.d.
Zageh	5212-4918	Late Neolithic		MASHKOUR *et al.*, 1999: 65, 68
Ghabristan	3782-3361	Early – Late Chalcolithic		*ibid.*: 67-68
Sagzabad	1294-1035	Late Bronze-Iron Age		*ibid.*: 68
Tepe Sarab	6000-5000	Early Ceramic Neolithic		
Umm Dabaghiyah	5800-5300	Neolithic		BÖKÖNY, 1978: 57-58
Tell es-Sawwan	5200-4800	Hassuna-Samarra		*ibid.*
Choga Mami	5000-4800	Samarra		*ibid.*
Tepe Sarab	6000-5500	Early Ceramic Neolithic		DAVIS, 1984: 267
	5500-5000	Late Ceramic Neolithic		*ibid.*
Tepe Siahbid	5500-5000	Late Ceramic Neolithic		*ibid.*
	5800-5000	Early Chalcolithic		*ibid.*
Choga Maran	5800-5000	Early Chalcolithic		*ibid.*
	5000-3600	Middle Chalcolithic		*ibid.*
Tol-e Nurabad	6000-5000	Neolithic		POTTS *et al.*, 2006: 5; MASHKOUR, 2006: 135
	4900-4100	Bakun		*ibid.*
	3900-3500	Lapui		*ibid.*
	3400-2800	Banesh		*ibid.*
	2100-1800	Kaftari		*ibid.*
Tol-e Spid	3900-3500	Lapui		*ibid.*
	3400-2800	Banesh		*ibid.*
	2100-1800	Kaftari		*ibid.*
Tuwah Khoshkeh	5000-4000	Middle Chalcolithic		ABDI *et al.*, 2002: 43

* Sequence model developed by Bronk Ramsay in OxCal v3.8 program

domestic areas by the excavators.[19] Copper tools and spindle whorls were recovered, but from only one context (11) in Trench A.

No ovens, walls or floors were uncovered in Trench K, but the remains of three kilns were found, along with the highest concentration of sherds (accounting for 31.8% of the total ceramic assemblage). In contexts related to the kilns fragments of finished, unfinished and deformed figurines were recovered, together with spindle whorls and unworked clay lumps for pottery production. Pigment was found inside a bowl suggesting that it may have been connected to pottery painting. These artefacts are similar to those found by M. Shahmirzadi in the 'workshop area' at Zageh.[20] In addition, large stone bowls with pounding marks on their interior walls were recovered, although the exact function of these bowls is unknown. Cores showing signs of being heated have also been found, suggesting lithic tool production may also have taken place here. Textile production is indicated by the presence of several hundred spindle whorls.[21]

In summary, the structures, features and artefacts of Zageh show that there were domestic areas, possibly a temple or other public building, and that there is considerable evidence for craft activity, including workshop areas indicating craft specialisation in pottery and textiles. Zageh has been interpreted as a site with evidence for social complexity and a social hierarchy.[22]

19. FAZELI *et al.*, 2005; FAZELI and ALIARI, 2007.
20. SHAHMIRZADI, 1977.
21. MATTHEWS and FAZELI, 2004: 63.
22. *Ibid.*: 63-64.

GHABRISTAN

Ghabristan mound covers an area approximately 1.6 ha[23] and is less than half a metre in height. Like Zageh, Ghabristan was excavated under the direction of Negahban in the early 1970s. A number of large horizontal areas were opened during the three seasons (1970, 1971, 1972), with at least two of these reaching depths of 4 to 5 m in limited soundings. In 2002 eleven trenches were opened in different areas across the site, and of these, five trenches contained only 1st millennium BCE graves and natural clay deposits. Two trenches in the southern area of the site contained only Late Chalcolithic materials, consisting primarily of the remains of a kiln and finds related to craft production.[24] Four trenches in the central area of the site provided dates from the Early Chalcolithic through to the Late Chalcolithic, and these are thought to show evidence for industrial activity.[25] This interpretation is based largely on evidence from Trench L34, from which various features such as walls, ovens, pits and floors were recorded, along with finds such as bones and bone tools, stone tools, copper tools and copper slag.

The focus of the 2003 excavation was to further explore the different functions represented at Ghabristan, and to spatially define the residential and workshop areas. In order to achieve this, a large 10 x 10 m trench (Trench 1) was opened at the edge of Negahban's Trench C. Extensive soil erosion[26] and disturbance by the earlier excavations slowed down the excavation, and only Late Chalcolithic layers were reached by the end of this season. One complete and one partial kiln were uncovered, along with bone and stone tools, and some copper slags.

Ghabristan, like Zageh, is clearly a village with an agricultural subsistence base. There is also considerable evidence for workshops, both for pottery production and copper working. The copper slag from both the 1970s and the 2000s excavations remain unquantified, making an assessment of the scale (industrial *vs* domestic) difficult. There is also at least one building of different style and scale to others, which may be a temple, public building or private elite residence, thus introducing the possibility of an emerging hierarchy alongside craft specialization.[27] In terms of structures, features and artifacts, Ghabristan differs from Zageh mainly in the evidence for a metal working industry here, as well as dedicated workshops for pottery production. We would therefore expect that the animal bone assemblages would be essentially similar to the Zageh assemblage.

23. NEGAHBAN, 1977: 27.
24. FAZELI and ABBASNEGAD, 2007.
25. FAZELI *et al.*, 2005: 46.
26. Also discussed by MADJIDZADEH, 1977: 46-47.
27. NEGAHBAN, 1977: 37.

ISMAILABAD

Ismailabad is a 10m high mound covering an area approximately 300 m², and is located in the hilly flanks of the southern foothills of the Alborz Mountains.[28] No radiocarbon dates have been calculated for this site, but the ceramics recovered through excavation and surface collection have been classified as Late Chalcolithic, *i.e. ca* 3500-3000 BCE.[29] One trench of 2 x 5 m was opened in 2003, and reached natural at a depth of 4.2 m indicating that human activity here took place on top of a natural mound. Sixteen contexts were recorded, and the features uncovered included stone walls, pits, and ovens or hearths. In addition to the characteristic Late Chalcolithic pottery, a large number of stone tools were recovered, which the excavators found surprising when compared to Ghabristan, where very few lithics were recorded.[30]

To date, Ismailabad is the first example of a Chalcolithic site identified and recorded in the southern foothills of the Alborz, bordering the Qazvin Plain. Today, the foothills are used for grazing sheep and goats rather than agriculture, and all other Chalcolithic sites in the region have been located in the plain itself where modern agriculture is practiced. Because of this marginal location, the excavators suggested that Ismailabad may be a pastoral site, possibly a transhumant site. Some of the reasons put forward to support this interpretation include its location in an agriculturally marginal area and the presence of a far higher than expected number of stone tools of a particular type.[31] The majority of lithic finds at Ismailabad were flakes and blades, and this is thought to indicate tools used primarily for non-agricultural tasks.[32] One of the main ceramic types identified here is Burnished Grey Ware, which has far greater affinity with the earliest Kura-Araxes Transcaucasian ceramics than with other Chalcolithic assemblages in the Qazvin or even Tehran Plains, and the primary economic base for the Kura-Araxes settlements has been defined as mobile pastoralism.[33]

With regards architecture on the mound, remnants of irregular stone walls were recorded, which may have formed part of a circular stone foundation.[34] No *pisé* or mud walls were uncovered at Ismailabad, which contrasts with the walls of the larger settlements at sites such as Zageh and Ghabristan. This is of course a very tentative interpretation and requires

28. FAZELI and AJOURLU, 2007.
29. FAZELI, 2007.
30. FAZELI and AJOURLU, 2007.
31. *Ibid.*
32. *Ibid.*
33. ROTHMAN, 2003.
34. FAZELI and AJOURLU, 2007.

further excavation and analysis to develop or refute this model of Ismailabad as a small mound used periodically by mobile pastoral or transhumant groups. Analysing the animal bones from Ismailabad and comparing the results to those from Zageh and Ghabristan may well be of use in understanding more about the function of Ismailabad, and suggesting whether or not it is a transhumant or mobile pastoral site. As there is evidence here for lithic production and no other craft activity, and as the main ceramic type and structures are in contrast to those at either Zageh or Ghabristan, we might well expect that the animal bones would reflect this contrast.

THE ANIMAL BONE ASSEMBLAGES FROM THE QAZVIN PLAIN SITES

Animal bones were collected from each site during the recent University of Tehran/ICHO excavations, and this section summarizes the preliminary analyses of these assemblages. The bones were collected by hand, and while the absence of regular sieving of soil from each context does mean that the recovery of fish, reptile and small mammal bones, is far less likely than the recovery of larger mammal bones, we do have an assemblage of the latter with which to explore various subsistence related issues for the three sites.[35] While the lack of sieving has undoubtedly introduced bias towards larger animals in this assemblage, we believe that undertaking some analysis is worthwhile.[36] Future field work could certainly involve a modified collection strategy and the inclusion of an archaeozoologist during the planning phases and excavation of any site, while future analytical work on the current assemblages could include different tests to estimate the degree of bias present.[37]

A total of 9230 bones and bone fragments were collected from all contexts at the three sites. The majority of these bones were very fragmented, showing signs of breakage before and during deposition, as well as some new excavation damage. Of these 9230 bones, 1482 were identified to body part and genus or species level; table 2 shows the percentage of bones identified for each site.

During the analysis of the animal bone assemblages from the three sites, the research objectives outlined in the introduction to this paper were used as guidance. We were therefore interested in characterising the assemblages from each site

35. HESSE and WAPNISH, 1985; REITZ and WING, 2008: 147.
36. PAYNE, 1975.
37. *Ibid.*: 14, 16-17.

Table 2 – *Species identified (%) for all sites in the order they appear in the text (with numbers available in brackets).*

Species	Sites											
	Zageh	Ghabristan	Ismailabad	Zageh (1999)	Ghabristan (1999)	Sagzabad	Sarab (ECN)	Sarab (LCN)	Siahbid (LCN)	Siahbid & Choga Maran	Choga Maran (MC)	Choga Maran (EBA)
Domestic caprine	70 (392)	53 (257)	60 (260)	62.1 (1548)	66 (3927)	38.1 (918)	89.3 (138.5~)	94.7 (152)	98 (49)	78.8 (123)	81.2 (119)	79.3 (92)
Wild caprine	11 (60)	10 (51)	6 (27)	12.5 (311)	5.7 (340)	8.0 (193)						
Cattle	6 (33)	16 (81)	17 (75)	4.1 (15)	10 (587)	16.3 (392)	1.2 (2)	3.1 (5)		11.5 (18)	17.4 (25.5~)	14.2 (16.5~)
Pig	1.6 (9)	5 (24)	2 (9)	0.3 (6)	2.5 (150)	4.2 (100)			2 (1)	9.6 (15)	1.3 (2)	6.4 (7.5~)
Dog	1.4 (8)	0.6 (3)	2 (10)	0.7 (18)	0.6 (38)	0.7 (16)	P	P	P		P	
Equid	3 (16)	7 (35)	2 (7)	2.8 (70)	6.8 (407)	24.0 (576)		0.6 (1)		P	P	P
Gazelle	1 (6)	1 (5)	5 (20)	16.4 (408)	7.2 (429)	7.9 (191)	9.3 (14.5~)	4.6 (7.5~)	P			
Deer	1.6 (9)	3 (17)	3 (14)		0.2 (12)			P				
Camelid	0.5 (3)	0.2 (1)	0.5 (2)	0.1 (2)		0.3 (8)						
Felidae	0.2 (1)		0.7 (3)	0.1 (2)	0.1 (1)	0.1 (2)						
Rat	0.2 (1)						P		P			
Rabbit / hare		0.2 (1)										
Fowl/bird	1.8 (10)	3 (17)	2 (9)	0.4 (9)	0.2 (13)	0.1 (2)	P					

Species	Sites											
	Umm Dabaghiyah	Tell es-Sawwan	Choga Mami	Tepe Sarab (Bokonyi)	Tol-e Nurabad (Kaftari)	Tol-e Nurabad (Banesh)	Tol-e Nurabad (Lapui)	Tol-e Nurabad (Bakun)	Tol-e Nurabad (Neo)	Tol-e Spid (Kaftari)	Tol-e Spid (Banesh)	Tol-e Spid (Lapui)	Tuwah Khoshkeh
Domestic caprine	8.97	88.74	88.74	62.8	95.3 (281)	91.9 (102)	80 (16)	90.6 (48)	93.6 (427)	88.8 (293)	64.6 (53)	84.5 (142)	85
Wild caprine													
Cattle	1.03	0.07	0.07		4.1 (12)	8.1 (9)	20 (4)	9.4 (5)	3.9 (18)	10.9 (36)	28 (23)	13.7 (23)	5
Pig	1.04	0.07	0.07	0.2							7.3 (6)	1.8 (3)	
Dog	0.52	1.4	1.4	1					0.2 (1)				
Equid	68.35	2.44	2.44	0.01									
Gazelle	15.99	4	4	11.5	0.3 (1)								
Deer	4.10*	3.28*	3.28*	24.5*									10**
Camelid													
Felidae													
Rat													
Rabbit / hare													
Fowl/bird													

P = Davis denotes presence of certain types but gives no figure ~ 0.5 refers to single metapodial condyles (DAVIS, 1984: 267)
* Bökönyi (1978) gives this figure for 'other wild types'. ** In ABDI *et al.* (2002: 64), the authors have grouped together gazelle, deer, pig in this figure.

in turn, discerning the major trends, and comparing each of the three sites with the others in order to determine whether they were essentially similar in nature, or whether there were appreciable differences between the animal bone assemblages. In order to address these questions we have determined the species present and the relative quantity of each species or type and calculated age at death for caprines.

SPECIES IDENTIFIED

Table 2 gives the percentage of bones identified for each site, according to type or species. It is clear that domesticated caprines (*Ovis aries* and *Capra hircus*) are the dominant species at all three sites, although the percentages do vary between the sites. Wild caprines and then bovines (cattle) are the next most significant types at all the sites. All identifications and discriminations between different types were made on the basis of both morphological and metrical data.[38] The equid group will be the subject of a separate, detailed investigation. Pig (*Sus* sp.), gazelle (*Gazella subgutturosa*) and deer (*Cervus* sp.) have been identified in all assemblages. The pig may be domestic on the basis of metrical comparisons[39] although the small numbers make this difficult to ascertain with any certainty. The bird category includes all types recovered.

The percentage of wild types in the assemblage for each site is: Zageh 20%; Ghabristan 25%; Ismailabad 19%. This suggests that wild animals were an important dietary element in the Qazvin Plain, and that this remained relatively constant over the time span covered by these three sites, and also between the different types of sites. Perhaps the most interesting trend apparent in the percentage of species data is the difference between Zageh and the other two sites with regard the proportions of both domestic caprines and cattle. At Zageh caprines comprise 70% of the assemblage, and cattle only 6%, while at Ghabristan caprines comprise 53% and cattle 16%, and at Ismailabad caprines comprise 60% and cattle 17% of the assemblage.

AGE AT DEATH

The survivorship representations for caprines (fig. 2) show some interesting contrasts, although the small sample numbers mean that any interpretations must remain very tentative.

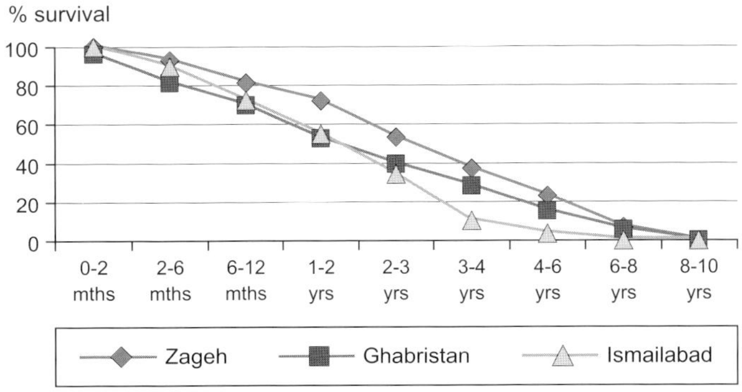

Fig. 2 – *Survivorship curve for caprines from Zageh, Ghabristan and Ismailabad.*

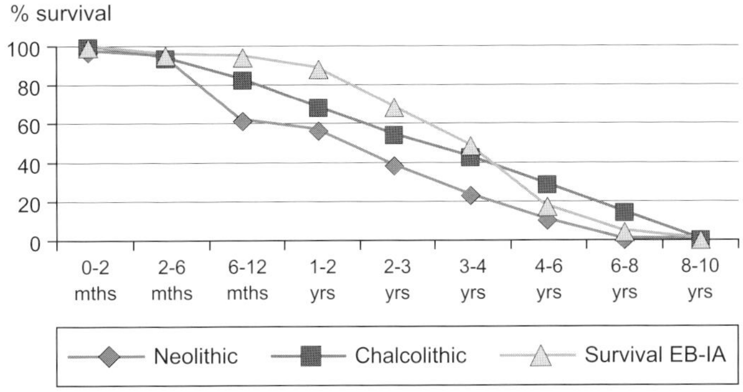

Fig. 3 – *Survivorship curve for caprines from Kermanshah sites (DAVIS, 1984).*

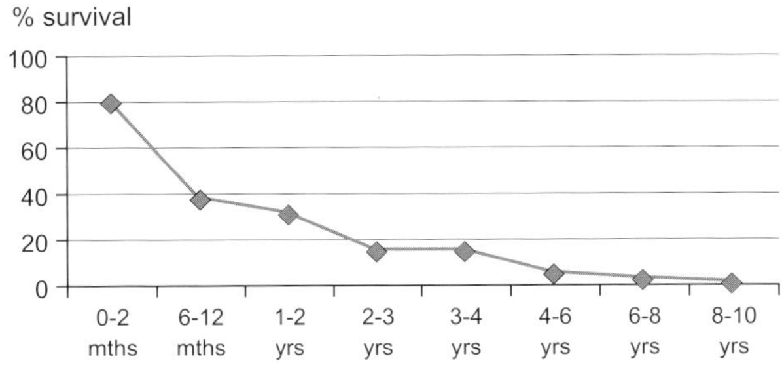

Fig. 4 – *Survivorship curve for caprines from Tuwah Khoshkeh (ABDI et al., 2002).*

The Zageh figures and curve show that nearly three quarters (72%) of the animals survived to two or more years, compared to Ghabristan and Ismailabad where around half the animals (53% and 55% respectively) survived beyond two years. Nearly 40% of animals at Zageh survived to four years or older, and

38. BOESSNECK, 1969; PAYNE, 1969 and 1985; ROWLEY-CONWY, 1998.
39. PAYNE and BULL, 1988; ROWLEY-CONWY, 1995.

nearly 30% at Ghabristan, while at Ismailabad, only 10% of animals survived four years of age. When these survivorship representations are compared to Payne's models of kill-off patterns and survival curves for milk, meat and wool production in modern Turkish animals,[40] it is possible to suggest that in general terms, the survivorship patterns at Zageh and Ghabristan represent more diversified herd management strategies, utilising both meat and wool, in contrast to a somewhat more specialized focus at Ismailabad, perhaps more oriented towards meat production.

SHEEP TO GOAT RATIOS

Sheep outnumber goats at all three sites (table 3), although Ghabristan appears to have fewer goats in relation to sheep than either Zageh or Ismailabad. At Zageh and Ismailabad the ratios of goat to sheep are high, suggesting that the numbers of each in live herds may have been similar. At Ghabristan the ratio of goat to sheep suggests that sheep would have outnumbered goats around three times. There are many possible factors behind the choice of sheep or goat that might shape their numbers within a herd, and these include environmental conditions, wool production and rates of reproduction.[41] Again, given the very small number of bones that have been identified as either sheep or goat mean that only very tentative suggestions can be drawn from this analysis at this point.

ANIMAL BONE ANALYSES FROM THE EARLIER EXCAVATIONS AT ZAGEH AND GHABRISTAN COMPARED TO ANIMAL BONE ANALYSES FROM THE MORE RECENT EXCAVATIONS

The identifications and analyses carried out by Mashkour *et al.*[42] show that for Zageh, Ghabristan and Sagzabad, domestic caprines dominate the assemblages. Mashkour *et al.* noted that wild animals were well represented at each site,[43] and the total proportion of wild animals in each assemblage is: Zageh, 33%, Ghabristan 20%, and Sagzabad 36%. Mashkour *et al.* also noted that there was an increase in cattle and equid numbers from Zageh to Ghabristan and again to Sagzabad.[44] In contrast, caprine numbers are relatively constant between Zageh and Ghabristan (62% and 66% respectively), but are notably less in Sagzabad (38%). Mashkour *et al.*[45] report using morphological and metrical observations to distinguish wild and domestic species.

When we compare this with the recent analyses, we see that while numbers of cattle increased from Zageh to Ghabristan, there is no further increase from Ghabristan to Ismailabad. Equids offer a further contrast—whereas Mashkour *et al.* are able to trace a clear increase in equid numbers from Zageh through to Sagzabad, we see an increase in equid numbers from Zageh to Ghabristan, then a notable decrease between Ghabristan and Ismailabad.

There are also interesting differences with respect to other categories of wild mammals. Mashkour *et al.* observe that:

> "Wild mammals are represented especially by wild ovicaprids and gazelle in Zageh, whereas in Qabrestan and Sagzabad there seems to have been more hunting as expressed by increases in the number of bones from equids and wild boar."[46]

The recent analyses show that gazelle comprised only 1% of the assemblage at Zageh and Ghabristan, but makes up 5% of the assemblage at Ismailabad. Further, while the quantity of wild animals in the more recent assemblages remains relatively constant, the difference in the types of animals may well be the result of human agency, *i.e.* changes in hunting fashions or preferences. Mashkour stresses that there is significant difference between the assemblages from the three sites[47] and that rather than representing a move from a diverse faunal base in the Neolithic to a more restricted faunal base in the Iron Age, in fact the reverse is true. The assemblage from Sagzabad, the Iron Age site, has a general increase in wild or hunted species, and far less reliance on domesticated caprines than the two earlier sites.[48] No age at death information is given for the Zageh, Ghabristan or Sagzabad.

In terms of understanding how animal bone assemblages and the human choices they represent might be interpreted with regard to the type of site, the chronological period of the site and the environmental setting of the site, we felt it was necessary to do more than simply look at similarities and differences between the 1970s and 2000s excavations. We wanted

40. PAYNE, 1973.
41. REDDING, 1984; ZEDER, 1991.
42. MASHKOUR *et al.*, 1999.
43. *Ibid.*: 73.
44. *Ibid.*: 73-74.
45. *Ibid.*: 70.
46. *Ibid.*: 73.
47. *Ibid.*: 74
48. *Ibid.*

Table 3 – *The sites in chronological order with location, environment, type of site, and faunal analyses.*

Site name	Date	Location	Environmental Setting	Type of site	% wild	% caprine	Sheep: goat ratios	Caprine: cattle ratios
Tepe Sarab (ECN)	6000-5000	Kermanshah Valley, Western Iran	Narrow high valley	Small settlement – Neolithic	36.01	62.8	1:6.14	
Tepe Sarab (ECN) (DAVIS, 1984)	6000-5000	Kermanshah Valley, Western Iran	Narrow high valley	Small mound – settlement – Neolithic	9.3	89.3	1:3	1:0.01
Tol-e Nurabad (Neolithic)	6000-5000	Mamsani District, Fars province	River plain in high valley	Small settlement	6.3	93.6		1:0.42
Umm Dabaghiyah (ECN)	5800-5300	Jezira, between Tigris and Euphrates, south of Mosul	Hilly, treeless steppe, arid, saline	Small mound – settlement – Neolithic (hunting economy – marginal?)	88.44	8.97	1:0.51	
Tepe Siahbid	5800-5000	Kermanshah Valley, Western Iran	Narrow high valley	Small mound – settlement	-	78.8*	1:1*	1:0.14*
Choga Maran (EC)	5800-5000	Kermanshah Valley, Western Iran	River plain in high valley	Small mound – settlement – Chalcolithic	-	78.8*	1:1*	1:0.14*
Tepe Sarab (LCN) (DAVIS, 1984)	5500-5000	Kermanshah Valley, Western Iran	Narrow high valley	Small mound – settlement – Neolithic	5.2	94.7	1:10	1:0.03
Tepe Siahbid (LCN)	5500-5000	Kermanshah Valley, Western Iran	Narrow high valley	Small mound – settlement – Neolithic	-	98	1:3	-
Zageh (FAZELI and ALIERI, 2007))	5370-4460	Qazvin Plain	Plain	Mound – small village – Transitional – Early Chalcolithic	20	70	1:0.83	1:0.08
Zageh (MASHKOUR *et al.*, 1999)	5212-4561	Qazvin Plain	Plain	Mound – small village – Late Neolithic	32.3	62.1	1:0.95	1:0.06
Tell es-Sawwan (Hassuna-Sammara)	5200-4800	Left bank of Tigris, south of Samarra	River flood plain	Small settlement – Hassuna-Samarra	9.72	88.74	1:1.94	
Choga Mami (Samarra)	5000-4800	Iraq-Iran border, foothills of Zagros mountains	Foothills	Small settlement – Samarra	10.68	77.71	1:7.33	
Choga Maran (MC)	5000-3600	Kermanshah Valley, Western Iran	River plain in high valley	Small mound – settlement – Chalcolithic	-	81.2	1:03	1:0.21
Tuweh Khoshkeh	4900-4000	West central Zagros mountains	Foothills	Pastoral campsite	10	88	1:0.7	
Tol-e Nurabad (Bakun)	4900-4100	Mamsani District, Fars province	River plain in high valley	Small settlement	9.4	90.6		1:0.1
Ghabristan (FAZELI and ABBASNEGAD, 2007)	4450-3700	Qazvin Plain	Plain	Mound – small village – Early – Middle Chalcolithic	24.6	53	1:0.34	1:0.3
Tol-e Nurabad (Lapui)	3900-3500	Mamsani District, Fars province	River plain in high valley	Small settlement	20	80		1:0.25
Tol-e Spid (Lapui)	3900-3500	Mamsani District, Fars province	River plain in high valley	Small settlement	15.5	84.5		1:0.016
Ghabristan (MASHKOUR *et al.*, 1999)	3782-2876	Qazvin Plain	Plain	Mound – small village – Chalcolithic	19.93[1]	66.3	1:1.01	1:15
Ismailabad	3700-3000	Qazvin Plain, Alborz foothills	Foothills	Mound – seasonal settlement?	19.4	60	1: 0.70	1:.028
Tol-e Nurabad (Banesh)	3400-2800	Mamsani District, Fars province	River plain in high valley	Small settlement	8.1	91.9		1:0.08
Tol-e Spid (Banesh)	3400-2800	Mamsani District, Fars province	River plain in high valley	Small settlement	35.3	64.6		1:0.43
Choga Maran (EBA)	2800-2500	Kermanshah Valley, Western Iran	River plain in high valley	Small mound – settlement	-	79.3	1:1	1:0.17
Tol-e Nurabad (Kaftari)	2100-1800	Mamsani District, Fars province	River plain in high valley	Small settlement	4.4	95.3		1:0.04
Tol-e Spid (Kaftari)	2100-1800	Mamsani District, Fars province	River plain in high valley	Small settlement	10.9	88.8		1:0.12
Sagzabad	1294-863	Qazvin Plain	Plain	Mound – small village – Late Bronze – Iron Age	36.4[1]	38.1	1:1.14	1:0.43

[1] The figure given by MASHKOUR *et al.* (1999) for domestic equids has been subtracted from this total.
* Davis gives both sites / periods together.

to work out whether any differences between the assemblages from Zageh and Ghabristan were due to issues associated with small samples, differences in analytical approaches, or differences resulting from different areas of each site being excavated and sampled. However, the limited context information available for the animal bones from the earlier analyses means that there are restrictions on the sorts of comparisons regarding overall site function, and the different function of the different excavated areas within each site that we can make. As this is a widespread issue for many bone reports, and as we want to make use of the material that we do have, in order to consider the sites of Zageh, Ghabristan and Ismailabad within a wider range of Iranian prehistoric sites, we have undertaken a brief and far from exhaustive examination of published animal bone analyses from Iranian excavations.

ANALYSES OF ANIMAL BONE ASSEMBLAGES FROM SELECTED PREHISTORIC SITES IN IRAN

In order to place the material from the Qazvin Plain within a wider context, environmentally, chronologically, and in terms of attributed site function, we consider here analyses and interpretations from prehistoric sites (Neolithic through to Chalcolithic) from a range of environmental zones across Iran. The sites are: Umm Dabaghiyah, Choga Mami, Tepe Sarab and Tell es-Sawwan in western Iran and Iraq;[49] Tepe Sarab, Siahbid and Choga Maran in the Kermanshah valley of western Iran;[50] Tuweh Khoskeh in the west central Zagros;[51] and Tol-e Nurabad and Tol-e Spid in the Mamasani region of Fars province.[52] Note that the Tepe Sarab assemblages have resulted from different excavations (Bököny's material is from the Braidwood excavations and Davis's material is from the Levine excavations) and have been analysed separately.

UMM DABAGHIYAH, TELL ES-SAWWAN, CHOGA MAMI, TEPE SARAB

Four Neolithic sites in Southwest Asia were analysed by S. Bököny:[53] Umm Dabaghiyah, located in the Jezira region of what is now Iraq, in a dry, saline steppe environment and dated to between 5800-5300 BCE;[54] Tell es-Sawwan, located in a riverine oasis on the Tigris River to the south of Samarra and dated from the end of the 6th to the beginning of the 5th millennium BCE;[55] Choga Mami, located in the dry foothills of the Zagros Mountains and dated to the beginning of the 5th millennium BCE;[56] and Tepe Sarab, located in the high narrow Kermanshah Valley in mountains of Western Iran, and dated to the 6th millennium BCE.[57]

Domestic caprines dominate the faunal assemblages at three of these sites, ranging from 63% at Tepe Sarab, to 89% at Tell es-Sawwan and 78% at Choga Mami. In terms of chronology, this shows an increase from Tepe Sarab (6th millennium BCE) to Tell es-Sawwan (5200-4800 BCE), but a decline to Choga Mami (5200-4800 BCE). Cattle are of little significance at any of these sites—Bököny does not think cattle were present at all at Tepe Sarab, and at the other two comprise less than 2% of the total assemblage.[58]

Overall, the proportion of wild types recovered at each site ranges from 37% at Tepe Sarab, to 89% at Umm Dabaghiyah, 10% at Tell-es Sawwan, and 18% at Choga Mami. The assemblage at Umm Dabaghiyah is clearly dominated by wild animals, particularly equids and gazelle. Bököny says that this shows the relative importance of hunting rather than animal husbandry.[59] The dominance of onager at Umm Dabaghiyah was unique in this region at the time of analysis. However, work in the Marv Dasht region of Fars has resulted in the recovery and analysis of animal bone material from the ceramic Neolithic site of Tall-i Mushki dated to 5th-6th millennium BCE, where wild equid bones (*E. hemionus*) dominated the assemblage, leading to an interpretation of a marginal economy based on equid hunting.[60]

Although he does not give age at death data, Bököny reports that the ratio of immature individuals to mature individuals is high among domesticates at all of the sites. He explains this as a product of animals being kept primarily for meat, and keeping only selected breeding stock through winter.[61] Bököny gives a summary of the ratios of sheep to goat at these sites (but not the actual data): at Umm Dabaghiyah, sheep were more frequent than goats (66 to 34%); while at Tell es-Sawwan, Choga

49. BÖKÖNY, 1978: 58.
50. DAVIS, 1984: 267, 269.
51. ABDI *et al.*, 2002: 65.
52. MASHKOUR, 2006: 135.
53. BÖKÖNY, 1978.
54. *Ibid.*: 57.
55. *Ibid.*
56. *Ibid.*
57. *Ibid.*
58. *Ibid.*: 59.
59. *Ibid.*: 58.
60. ALIZADEH *et al.*, 2004; PAYNE, 1991: 161.
61. BÖKÖNY, 1978: 61.

Mami and Tepe Sarab goats outnumbered sheep (66 to 34%, 88 to 12%, and 86 to 14% respectively).[62] This might indicate deliberate selection of sheep for wool at Umm Dabaghiyah where hunted onager would provide most of the meat, while the more 'typical' herding assemblages at the other three sites may be incorporating higher numbers of goats for reproductive reasons linked to their primary use for meat. Overall, Bököny explains differences in the animals hunted or kept at each of these four sites in terms of the environment of each site, rather than attributing this to cultural factors or human agency.[63]

TEPE SARAB, TEPE SIAHBID AND CHOGA MARAN

A series of excavations were carried out in the Kermanshah Valley and adjacent Mahidasht Plain in western Iran during the late 1970s.[64] These sites are small mounds dated from the Ceramic Neolithic through to historical periods, and the mammal bones from each were recovered and identified. The main aim of the archaeozoological study was to determine the domestic and wild proportions of each assemblage and then consider explanations for the husbandry of caprines and pigs. Here we review material recovered from six different periods from three sites: Tepe Sarab (*ca* 6000-5500 BCE), Tepe Siahbid (*ca* 5500-5000 BCE and *ca* 5000-4100 BCE) and Choga Maran (*ca* 5000-4100 BCE, *ca* 4100-2900 BCE and *ca* 2200-1900 BCE).[65]

S.J.M. Davis points out a broad trend from the earliest to latest periods analysed, showing a decrease in the number of species represented from a relatively wide range to a far narrower range, which he says is the result of a cultural change from hunting economies to husbanding economies in this region.[66] In the Neolithic phases at Sarab and Siahbid caprines comprise 89-98% of the assemblages, while in the Chalcolithic to Early Bronze Age phases at Siabhid and Choga Maran they comprise 79-81% of the assemblages. This decline in the amount of caprines in the assemblages is accompanied by a corresponding increase in the amount of cattle. Pig numbers are also significant in both the Siahbid and Choga Maran Early Chalcolithic and the Choga Maran Early Bronze Age assemblages. Davis suggests that the sheep and goat from these sites were "mostly if not all domestic"[67] and that by the Early Neolithic at Tepe Sarab, subsistence was thus based primarily on domestic caprines rather than any significant hunting of deer, gazelle or onager.

The survivorship representations for these three sites is again based on relatively small sample sizes, and so remains very tentative, but when the data is combined by period (fig. 3) it suggests that in the Neolithic periods, only 40% of caprines survived beyond two-three years, which Davis suggests is compatible with meat production.[68] In the Chalcolithic and Early Bronze Age periods there is greater representation of older animals, and Davis suggests that this could indicate a move towards animals exploited for reasons other than meat, such as wool and possibly milk.[69] However, neither the Chalcolithic nor the Early Bronze-Iron Age survivorship curves show a focus on slaughtering young (male) animals that would be expected with a milk production model.[70] This would of course require sex data (which is unavailable) in addition to age at death to explore further, but we can consider ratios of sheep to goat in this context (table 3). In the Neolithic periods (Sarab ECN, Sarab LCN and Siahbid LCN) there are between three to ten times more goats than sheep. In the Early Chalcolithic and Early Bronze Age, there are now equal numbers of goats and sheep at both Siahbid and Choga Maran—while the ratios for Middle Chalcolithic Choga Maran show around three times as many sheep as goat. The greater proportions of sheep would support a wool production model in the later periods, while the greater proportions of goats in the earlier periods would support a meat producing model.

TOL-E NURABAD AND TOL-E SPID

The two tell sites of Tol-e Nurabad and Tol-e Spid are located in the Mamasani District of Fars Province, Southwest Iran. They were the subject of small-scale excavations during 2003 which permitted the retrieval of animal bone assemblages and the percentage of animal bones by type and period is shown in table 2.[71] Material from Tol-e Nurabad spans the Ceramic Neolithic (6th to mid 5th millennia BCE) right through to the late 2nd millennium BCE.[72] At Tol-e Spid, material remains from the early to mid 4th millennium BCE to the mid 2nd millennium BCE have been recovered.[73] In the

62. *Ibid.*: 59.
63. *Ibid.*: 61.
64. DAVIS, 1984: 265.
65. *Ibid.*: 267.
66. *Ibid.*: 269.
67. *Ibid.*
68. *Ibid.*: 274-275.
69. *Ibid.*: 275.
70. PAYNE, 1973.
71. MASHKOUR, 2006: 135; POTTS *et al.*, 2006: 5.
72. MASHKOUR, 2006: 135.
73. *Ibid.*

analysis of these animal bone assemblages the main aim "is to present the diachronic evolution of animal exploitation in the Mamasani region throughout the succession of cultural phases attested there".[74] Mashkour emphasizes the smallness of the assemblages, and the preliminary nature of analysis at the time of publication, yet this work still provides useful material in our broad comparison of animal bone assemblages.

Although no age at death data is yet available, Mashkour[75] says that the sheep/goat ratio is almost equal at both sites and all periods, and also that there is little evidence for wild caprines at either site in any period. At both sites the overwhelming dominance of caprines in all periods is quite clear, with cattle the next most important species, although there is some variations in cattle numbers from period to period. This variation in cattle numbers may be linked to "chronological differences" although further analysis is considered necessary to substantiate and develop this idea.[76] Interestingly, there is an almost total absence of wild types at all levels of these sites.[77] When Tol-e Nurabad and Tol-e Spid are compared to other sites in the region, they stand out because of the very high proportion of caprine remains and the very limited variation within the assemblage.[78] At other sites in the region gazelle and equid hunting were clearly important, including the extreme Tall-i Mushki (6th-5th millennium BCE), where equid remains totally dominated the assemblage.[79] On the basis of the animal assemblages from Tol-e Spid and Tol-e Nurabad, Mashkour suggests that these sites had a meat based economy exploiting caprines and cattle, and that these assemblages show a clear micro-regional level economy.[80]

TUWAH KHOSHKEH

Tuwah Khoshkeh is a 5th millennium BCE site in West Central Zagros mountains excavated in 2000.[81] The excavators noted the site during survey, "and because of its configuration and marginal location, made a note of it as a possible mobile pastoralist campsite"[82] where the configuration appears to refer to the site as a low rise. Further analyses of a range of material culture and the animal bone assemblage have led to confirmation of the initial interpretation of Tuwah Khoshkeh as a mobile pastoral site. One of the most striking features of the animal bone assemblage is the dominance of caprine remains (some 85%), which is all treated as domesticates and has been explained thus: "The imposing percentage of Ovi-Caprine bone can be considered a pastoral signature and forms a particular feature of the Tuwah Khoshkeh faunal assemblage."[83] Cattle accounted for only 5% of the assemblage, while 10% of bones identified belonged to wild mammals such as boar, gazelle and deer.[84]

In terms of age at death, Abdi *et al.*[85] have attempted to "characterize the economic pattern of the people of Tuwah Khoshkeh through the study of the age at death of the Ovi-Caprines" yet they point out that they are doing this using a sample of 49 teeth from 12 individuals from the Middle Chalcolithic period, which spans several centuries at least.[86] This draws attention to the problem of using small samples; however, as this type of sample issue is far from unique in archaeology in Iran (or indeed many other parts of the world), it is still reported here, bearing in mind its potential limitations.

The survivorship representation shows that 60% of animals were slaughtered between 6-12 months, with 14% surviving beyond two-three years of age. Tuwheh Khoshkeh resembles Payne's milk production model,[87] with a high proportion of animals being slaughtered before reaching 12 months, though the relatively low numbers of animals older than three years may well suggest a focus on meat as well. Indeed, Abdi *et al.*[88] suggest "a mixed exploitation for meat and secondary products, more for dairy products than fibre products, can be suggested." The high mortality of very young animals could perhaps be linked to seasonal occupation of the site, during periods when animals are less than twelve months old. Abdi *et al.*[89] also compare the kill-off pattern at Tuwah Khoshkeh with that of nearby Chogha Gavaneh (reported in passing within the Abdi *et al.* paper), described by the excavators as a major sedentary site with evidence for long occupation, and this has some interesting implications in terms of interpretative models and frameworks. They claim that there are appreciable differences between the kill-off patterns, although as sample size for Chogah Gavaneh is also small (n = 38) this analysis should be treated with some caution. The Chogah Gavaneh

74. MASHKOUR, 2006: 135.
75. *Ibid.*: 136-137.
76. *Ibid.*: 136.
77. *Ibid.*: 138.
78. *Ibid.*
79. ALIZADEH *et al.*, 2004.
80. MASHKOUR, 2006: 138.
81. ABDI *et al.*, 2002: 43.
82. *Ibid.*
83. *Ibid.*: 64.
84. *Ibid.*
85. *Ibid.*: 65.
86. *Ibid.*
87. PAYNE, 1973.
88. ABDI *et al.*, 2002: 67.
89. *Ibid.*: 69.

assemblage shows more diversified animal exploitation, while the Tuwah Khoshkeh assemblage shows a more specialized function. Fewer very young animals (under six months) were noted at Chogah Gavaneh, along with a higher number in the old age groups (six-ten years). The authors suggest that the kill-off pattern at Chogah Gavaneh shows less focus on dairy products than at Tuwah Khoshkeh, and a greater emphasis on wool.[90]

Abdi *et al.*[91] point out that the sample size and other difficulties mean that there is little real possibility of being able to determine seasonal occupation at Tuwah Khoshkeh through the analysis of the faunal assemblage. Also, they point out that a model for prehistoric nomadic or seasonally occupied campsites are lacking, and without a clear understanding of the sorts of material culture markers for different lifeways and the sorts of social relationships and interactions these things represent, archaeologists are not able to interpret sites as the product of mobile groups with any confidence. Ethnographic accounts exist, but they tend towards qualitative descriptions of seasonal pastoral activity, while Abdi *et al.*[92] point out the need for long term and statistical observation which we support.

DISCUSSION

This section looks at certain trends in the data to see if they vary mostly with environment, time or socio-economic context. Table 3 presents the sites discussed in chronological order and their location, with a summary of their environmental setting, cultural affiliation and various bone indices. We will consider the assemblages in terms of the dominant species, the proportions of wild and domestic species, age at death, sheep to goat ratios, and caprine to cattle ratios.

SPECIES PRESENT

Caprines dominate the assemblages at all sites and in all periods with the single exception of Umm Dabaghiyah, where equids dominate. The range of animal species found in an assemblage is generally considered to be a key analytical element for archaezoologists, and pastoral economies tend to focus on a narrower range of species, indicating greater specialisation. This may also be linked to wider social and cultural changes, such as the domestication of animals and a shift to sedentism, as well as increasing urbanisation.[93] However, the presence of a range of species, both domestic and wild, might also indicate an opportunistic approach to subsistence, as suggested by M. Zeder after analysis of the animal bone assemblage at the Syrian site of Umm Qseir.[94]

Tepe Sarab is one of earliest sites dates in this paper, and the two faunal analyses that have been carried out on this material[95] have identified between 63% and 90% of the bone assemblages as caprine. Only two assemblages examined here have caprine proportions accounting for less than 50% of the assemblages: Umm Dabaghiyah (9%) and Sagzabad (38%); Umm Dabaghiyah is another early site, and Sagzabad is the latest. In between these two sites proportions of caprines range from 98% (Tepe Siahbid 5500-5000 BCE) to 53% (Ghabristan 4450-3700 BCE), and there is no simple increase in domestic caprines over time, accompanied by a restriction in hunting or in the range of other animals exploited at sites. Umm Dabaghiyah is located in what is described as a very dry, saline steppe environment between the Tigris and Euphrates rivers,[96] and this distinctive environment may be a contributing factor to the unusual exploitation of one particular type of wild animal. Sagazabad on the other hand, is located in the Qazvin Plain, a relatively fertile and benign environment, and eminently suited to caprine husbandry as shown by the figures from the other sites located in this area.

At both Tol-e Nurabad and Tol-e Spid caprine figures fall and rise; in the Banesh period at Tol-e Spid (3400-2800 BCE) caprine figures decline to 65% and wild animals increase to 35%, while in the Banesh period at Tol-e Nurabad caprines comprise 92% of the assemblage and wild animals only 8%. As these figures are for the same broad period, and as these sites are situated in the same environmental location, it would seem that there must be site-specific factors causing the decline in caprine numbers at Tol-e Spid during the Banesh period.

With regard to diversity of the species present at these sites, both Zageh (2000s excavations) and Ismailabad have some nine different types of animals present in quantities greater than 1%; Ghabristan (2000s excavation) has eight; Choga Mami has seven; Ghabristan (1970s excavations), Sagzabad, and Umm Dabaghiyah all have six. As both Zageh and Ghabristan have evidence for craft specialization and increasing complexity[97]

90. *Ibid.*
91. *Ibid.*: 67.
92. *Ibid.*

93. ZEDER, 1991; O'CONNOR, 2000.
94. ZEDER, 1994.
95. BÖKÖNYI, 1978; DAVIS, 1984.
96. BÖKÖNYI, 1978.
97. FAZELI and ABBASNEGAD, 2007; FAZELI and ALIARI, 2007.

we might expect that the animal bone assemblages would show a decrease in the range of animals present as discussed by Zeder.[98] While this might be true at larger urban sites, this is not evident at the sites reviewed here, for example at Zageh there is a greater range of species present than at sites of earlier periods (see table 2 although this is obscured by Davis's reporting of many species simply as present, rather than quantifying them). The range of species present does not decrease noticeably from Zageh to later sites such as Choga Mami (seven types recorded) and Ghabristan (six or eight types recorded). It is likely that at these sites the occupants were probably practicing small scale, mixed farming keeping a range of animals in order to spread risk, management and so forth,[99] and regularly hunting to supplement the domesticated animals. Again, we would argue that there is no simple correlation between time and range of species, and it is necessary to look at other factors in order to understand differences between sites.

Sites with very high numbers of goats compared to sheep are Tepe Sarab (Early Ceramic Neolithic), Tepe Siahbid (Late Ceramic Neolithic), Tell es-Sawwan (Hassuna-Samarra), Choga Mami (Samarra), Ghabristan (1970s excavation, Chalcolithic), Choga Maran (Early Bronze Age), and Sagzabad (Iron Age). A similar number of sites from all the periods covered here have a greater number of sheep than goats. Sites with a high ratio of goat to sheep are located across the whole range of environmental settings from narrow high valleys, foothills, river flood plains and plains; while sites with a low ratio of goats to sheep are also found in similarly diverse settings. Ghabristan (1970s excavation, Chalcolithic) has the highest caprine to cattle ratio at 1:1.15. Of the other sites, Tol-e Nurabad (Neolithic) and Tol-e Spid (Banesh) have the next highest ratios of caprine to cattle of 1:0.43, and Sagzabad has 1:0.42, while Tepe Sarab (Early Ceramic Neolithic) and Tol-e Spid (Lapui) have the lowest caprine to cattle ratios at 1:0.01. Davis's analysis and interpretations of the three Kermanshah sites do show a shift from a wide range to a narrower range of species from the Neolithic to the Chalcolithic and the Early Bronze Age. However, when considered with other sites, this trend is not sustained across a range of sites. Again, there is no simple relationship that changes over time, and there is no obvious link between environment and proportions of domesticates or any particular species in these assemblages.

98. ZEDER, 1991.
99. HALSTEAD, 1996: 33.

RELATIVE PROPORTIONS OF WILD AND DOMESTICATED TYPES

Of the twenty-six periods reviewed from these sites, five have more than 30% wild types in their assemblages. These range from Tepe Sarab (ECN) through to Sagzabad (IA), and are from sites located in diverse environmental settings. A further nine sites have between 10% and 30% wild types in their assemblages, and many of these form a chronological cluster in the 5th and 4th millennia BCE, which equates to the Transitional and Early Chalcolithic periods. These sites are located in a range of environmental settings, and also vary with regard percentage of caprines, and sheep / goat and caprine / cattle ratios. In order to determine whether the presence of relatively high proportions of wild animal bones in these assemblages is a potential indicator of social and economic change, it would be necessary to explore this in conjunction this to other archaeological data from these sites. Many bone reports are often divorced from their wider archaeological context, and this is a big problem when trying to link bones to past human activity and decision making.

Umm Dabaghiyah is a clear outlier in terms of the proportion of wild animals, and this suggests that the economy at this site was determined by factors not at work at the other sites considered here. The site of Tal-e Mushki (near Shiraz) is not considered here, but also has a very high proportion of wild equid remains, which has been attributed to hunting and is believed to indicate the marginality of this site compared to other sites from the same period and region.[100]

AGE AT DEATH

The survivorship representations described above (see also figs. 2-4) show that different sites are exploiting domesticates in different ways. At Zageh and Ghabristan the survivorship patterns indicate diversified strategies, making use of milk, wool and meat, while at Ismailabad it is possible to see a more specialized management strategy, where only 10% of animals survive beyond three-four years. This may indicate an emphasis on meat rather than other products. The survivorship curve from Tuwheh Khoshkeh suggests an emphasis on milk and milk products, with a high proportion of animals being slaughtered before reaching 12 months. Just over 10% of animals survive beyond three-four years, and this indicates the

100. MASHKOUR, 2006.

importance of meat as well. Abdi *et al.*[101] interpret the age at death data as showing exploitation of caprines for dairy products and meat.

Davis presented age at death data according to chronological period rather than site, and the survivorship data shows that in each period, different economic strategies were in place. In the Neolithic, the data supports an interpretation of animals kept primarily for meat, while in the Chalcolithic and Early Bronze Age to Iron Age there are older animals, which may indicate animals being exploited for wool. This diversification in later periods fits the secondary products model, in terms of timing and age at death profiles.[102]

Although the analysis of age at death has been restricted by a frequent lack of clear information in the bone reports, or consistency in the way it has been reported, it is possible to argue that yet again there is no simple change from meat economies in the Neolithic to diversified economies in the Chalcolithic and later periods. This may be evident at some sites or groups of sites (*e.g.*, Davis's Kermanshah sites) but it is not evident at all of the sites considered here. Environmental setting also seems to have very little to do with the type of economy practiced.

PASTORAL MODELS

Given the suggested preliminary interpretation of Ismailabad as a transhumant site, we thought that it would be interesting to include here a brief consideration of how transhumance might be recognized archeologically, primarily through the analysis of animal bone assemblages. This then allows us to compare Ismailabad and Tuweh Khoshkeh to available models, and also considers whether bone assemblages from economically and socially distinctive sites are themselves distinctive. Transhumance as a subsistence approach and lifeway has long been studied in anthropology and archaeology, in many different parts of the world[103] but in Iran the majority of recent work on transhumance has been carried out by Mashkour and co-workers.[104] This work has taken the form of both traditional archaeozoological analyses following excavation at sites such as Tuwah Khoshkeh,[105] and innovative isotope analyses, particularly drawing on material connected to the Bakhtiari nomads in the Zagros Mountains.[106] The potential for isotope analysis to shed light on the movement, diet and palaeoenvironmental conditions of both people and animals is immense,[107] but it is important to remember that such analyses are not absolute, clear-cut solutions to all questions about mobility and diet in the past. With regard to the seasonal movement of animals, Mashkour's results to date show encouraging results on modern specimens, but we are still awaiting results on archaeological assemblages.

Rather than summarise past work on mobile pastoralism here, we want to consider whether it is possible to distinguish transhumant sites on the basis of animal bone assemblages by looking at the two suggested transhumant sites (Ismailabad and Tuwah Khoshkeh) within the animal bone spectrum of all the sites considered here. This requires some idea of what we might expect as an archaeological signature of transhumance within a bone assemblage, and as has been noted previously,[108] considerably more work has been carried out exploring nomadism or mobile pastoralism in archaeology, than actual transhumance. In his discussion of identification of archaeological pastoral campsites, Cribb says that sites such as Tepe Sarab

> "have long been suspected as pastoral campsites... the difficulty involved in positively identifying them as such lies not so much in the lack of archaeological data but on the absence hitherto of any firm defining criteria."[109]

While Cribb's own research has provided such criteria for pastoral campsites, it is still largely lacking for transhumant sites. With regard to animal bones associated with pastoral campsites, Cribb notes that caprines dominate assemblages from such sites, and furthermore, that age at death analysis shows a high proportion of adults in the assemblage of the site of Tepe Tula'i.[110]

Mashkour[111] also says "the imposing percentage of Ovi-Caprine bone can be considered a pastoral signature and forms a particular feature of the Tuwah Khoshkeh faunal assemblage." However, table 3 shows that many sites from different periods and in different geographical locations have very high percentages of sheep and goat in their animal bone assemblages —thirteen of the twenty-six periods/sites considered here have 80% or more caprine bones. Are they all to be considered

101. ABDI *et al.*, 2002: 67.
102. DAVIS, 1984: 274-275.
103. *e.g.*, BARTH, 1956; CHANG and TOURTELLOTTE, 1993; CRIBB, 1991; HALSTEAD, 1990; KUZMINA, 1995; LEVY, 1992; PARKES, 1987; YOUNG, 2003.
104. *e.g.*, MASHKOUR *et al.*, 2005; MASHKOUR, 2003.
105. ABDI *et al.*, 2002.
106. MASHKOUR, 2003.
107. *e.g.* MONTGOMERY *et al.*, 2005; RICHARDS *et al.*, 2003.
108. YOUNG, 2003.
109. CRIBB, 1991: 216-218.
110. *Ibid.*: 215.
111. ABDI *et al.*, 2002: 64.

pastoral or transhumant sites? Obviously the answer is no, and the purpose of this is not to argue against the interpretation of Tuwah Khoshkeh as a transhumant or pastoral site, but to agree with the overall finding of the work there, which is based on more than faunal remains alone. For example, both Abdi *et al.*[112] at Tuwah Khoshkeh and Fazeli and Ajourlu[113] at Ismailabad have noted that among the lithic assemblages flakes and blades were present in very high quantities, and it is thought that such tool types were not used for agricultural tasks, rather associated with butchery and so forth. The architecture uncovered at Tuwah Khoshkeh comprises stone walls, made by placing large rocks in lower courses with smaller rocks above them and placing cobbles in between these rocks with an absence of any mortar,[114] and the structures revealed have been interpreted as part of a nomadic settlement using modern analogy.[115] The remnants of stone walls recorded at Ismailabad may also be part of circular foundations for temporary structures.[116] Relying on other categories of material culture may also prove difficult, as Cribb notes in his analysis of household goods in modern nomadic pastoral campsites that

> "while there may be a bias towards lighter, more portable materials, the range of objects that may be observed in a nomad camp is comparable to that observable in Near Eastern villages."[117]

Alizadeh *et al.* note the difficulties of using ethnographic analogies to identify mobile pastoral groups in archaeology.[118] They therefore combine the analysis of landscape settlement patterns with the excavation of the site of Dar Khazineh in the Upper Khuzestan Plain in order to identify archaeological evidence for pastoral mobility. The excavation uncovered occupation evidence in the form of *pisé* walls, postholes and shallow hearths, plus a stone pavement and three grave pits.[119] Grave artefacts included rubbing stones, grinders, a miniature bitumen vessel and a copper pin, and occupation artefacts included small amounts of pottery and chipped stone, and animal bones and plant remains. Of the animal bones less than 200 fragments were identifiable. Both domestic and wild types were noted, and "many of the caprine bones are from juvenile or newborn animals, possibly indicating occupation of Dar Khazineh in the spring months."[120] Less than 30 seeds were recovered and identified and of these, barley was the only cultivated type.[121] The excavators offer this interpretation:

> "From this evidence, we infer that that in the later 5th millennium BCE Dar Khazineh was an intermittently and probably seasonally occupied location along a major wadi system draining the foothills of the Zagros Mountains."[122]

The landscape and geoarchaeological elements of this study place the site in its wider environmental setting, which contributes to this interpretation. Yet it is clear that the evidence from Dar Khazineh which has been used to provide this interpretation of mobile, pastoral activity is rather different to the material from either Tuwah Khoshkeh or Ismailabad which has been used to develop an interpretation of these as sites occupied by mobile pastoral groups. This further reinforces the need for work on developing holistic models that will allow us to assess site and landscape evidence for mobility.

This leads us to one of the big gaps in both archaeozoological and archaeological studies in Iran and surrounding areas: the study of purely transhumant groups in order to develop the type of criteria Cribb has for nomadic pastoralists. This requires careful definition, where transhumant groups are not viewed as a type of mobile pastoralists, but are people with a particular type of economic base involving a permanent home base, the practice of agriculture alongside their animal husbandry, and the seasonal movement of flocks and at least part of the village from the home base to another base.[123] The home base is likely to be a permanent structure, probably indistinguishable from permanently occupied structures; the second base may be anywhere in a spectrum from permanent structure to temporary camp site, depending on the nature of transhumance for a given group. A possible solution to this may be the sustained study of modern transhumant groups in Iran in order to observe such things as material culture, spatial organization within different occupation sites, and likely animal bone assemblages at different points during a seasonal round. It would then also be necessary to have similar data from an entirely sedentary group in order to provide an effective control.

We would also suggest that trying to differentiate between sedentary and mobile herds themselves might be a slightly specious aim. When we consider the role of mobile herders

112. *Ibid.*: 62.
113. FAZELI and AJOURLU, 2007.
114. ABDI *et al.*, 2002: 51, 70.
115. *Ibid.*: 69.
116. FAZELI and AJOURLU, 2007.
117. CRIBB, 1991: 69.
118. ALIZADEH *et al.*, 2004: 71.
119. *Ibid.*: 73.
120. *Ibid.*: 74.
121. *Ibid.*
122. *Ibid.*
123. JONES, 2005; YOUNG, 2003.

in modern Iran and the recent past, it is hard to separate some form of mobility, whether seasonal, year round, long distance, or very short distance, from the husbandry of animals, particularly sheep and goat. When we talk about pastoralism and transhumance, are we talking about the movement of animals under the control of selected members of the community, or are we talking about the involvement of whole communities?[124] Isotope analysis may well be an important key to understanding the movement of herds, but until it can be carried out on human populations, we will only know about the seasonal movement of animals. This means that we still need to develop the whole archaeological package, or model, for distinguishing mobile pastoralism from sedentary groups and sites.

CONCLUSION

Taking an overview of sites covering a time period of some three millennia has allowed us to look for long term trends or patterns occurring in animal bone assemblages. Although differences in collection strategies, published informations and details mean that we are certainly obscuring many subtle trends and variations, we have shown that there is no gradual change from early periods with high proportions of wild species and a wide range of types, to later periods with few wild species present and a much narrower range of types. Caprines are the most numerous animals at almost all of the sites considered here from the earliest periods through to the latest periods, while some of the later sites such as Ghabristan have a far greater range of types (both wild and domesticated) than sites with Neolithic phases such as Tepe Sarab or Tol-e Nurabad. Nor are there any clear trends in the faunal assemblages that can be linked to the environmental location of these sites. There are similarities and differences between sites in narrow high valleys, on plains and in foothills, and no obvious correlations in the data we have considered here.

That we are unable to do more than consider gross trends is perhaps attributable to two factors: firstly, the tendency to remove animal bone assemblages from the rest of an archaeological site; and secondly, the absence of good models for animal bone assemblages that allow us to move beyond description and begin to explore issues of explanation. Greater communication and the integration of specialists would go a long way to overcoming the first problem, and long-term dedicated ethnographic research the second. Drawing on large amounts of data from big urban sites, the work of Zeder[125] has shown that detailed context information, intra-site comparison, clear research questions and relevant models can produce very exciting results. This overview shows that there are many issues in the interpretation of Iranian animal bones that need to be addressed if we are to produce archaeozoological reports that offer insight into past human-animal relationships, and perhaps most importantly, convince excavators of the continuing need for time and energy, and cost, while in the field and during post-excavation stages in order to produce these bone reports.

ACKNOWLEDGEMENTS

The authors would like to thank the British Academy for their generous support which allowed R. Young to visit Tehran, to carry out the animal bone identifications and the British Institute of Persian Studies for use of their accommodation and library in Tehran. We would also like to express our gratitude to the staff of the Iranian Cultural and Heritage Organization, Tehran; The Institute of Archaeology, University of Tehran; students from the Department of Archaeology, University of Tehran, in particular Mr B. Ajourlu and Ms Leila Ibrahim; The School of Archaeology and Ancient History, University of Leicester. We are also grateful for the support of Dr M. Azarnoush; Dr S.M. Mousavi Kouhpar; Dr V. Sarkosh Curtis; Mr N. Pazouki; Pr R. Coningham, Durham University; Dr D. Kennet, Durham University; Mr R. Sabin, Curator of Mammals, Natural History Museum, London; Pr G. Barker, Cambridge University; Pr T. Insoll, Manchester University. We would also like to express our grateful thanks to the reviewers for their extremely thoughtful and useful comments and suggestions.

Ruth YOUNG
School of Archaeology and Ancient History
University of Leicester
Leicester
ENGLAND
rly3@le.ac.uk

Hassan FAZELI
Director, Iranian center for Archaeological Research
Tehran
IRAN
hfazelin@ut.ac.ir

124. *e.g.* BAR-YOSEF and KHAZANOV, 1992; HALSTEAD, 1996; PARKES, 1987; YOUNG, 2003; YOUNG *et al.*, 2008.

125. ZEDER, 1984 and 1991.

BIBLIOGRAPHY

ABDI K., NOKANDEH G., AZADI A., BIGLARI F., HEYDARI S., FARMANI D., REZALI A. and MASHKOUR M.

2002 Tuwah Khohkeh: A Middle Chalcolithic Mobile Pastoralist Camp-site in the Islamabad Plain, West central Zagros Mountains, Iran. *Iran* XL: 43-74.

ALIZADEH A., KOUCHOUKOS N., WILKINSON T.J., BAUER A.M. and MASHKOUR M.

2004 Human-Environment Interactions on the Upper Khuzestan Plains, Southwest Iran. Recent Investigations. *Paléorient* 30,1: 69-88.

BAR-YOSEF O. and KHAZANOV A.

1992 Introduction. *In*: BAR-YOSEF O. and KHAZANOV A. (eds.), *Pastoralism in the Levant. Archaeological Materials in Anthropological Perspectives*: 1-9. Madison, Wisconsin: Prehistory Press.

BARTH F.

1956 Ecologic Relationships of Ethnic Groups in Swat, North Pakistan. *American Anthropologist* 58,6: 1079-1089.

BOESSNECK J.

1969 Osteological differences between sheep (*Ovis aries Linné*) and goats (*Capra hircus Linné*). *In*: BROTHWELL D. and HIGGS E.S. (eds.), *Science in Archaeology*: 331-358. London: Thames and Hudson.

BÖKÖNYI S.

1978 Environmental and cultural differences as reflected in the animal bone samples from five Early Neolithic sites in Southwest Asia. *In*: MEADOW R.H. and ZEDER M.A. (eds.), *Approaches to Faunal Analysis in the Middle East*: 57-62. Harvard: Peabody Museum of Archaeology and Ethnology.

CHANG C. and TOURTELLOTTE P.A.

1993 Ethnoarchaeological Survey of Pastoral Transhumance Sites in the Grevena Region, Greece. *Journal of Field Archaeology* 20: 249-264.

CONINGHAM R.A.E., FAZELI H., YOUNG R.L., GILLMORE G.K., KARIMIAN H., MAGHSOUDI M., DONOHUE R.E. and BATT C.M.

2006 Socio-economic transformations: Settlement Survey in the Tehran Plain and Excavations at Tepe Pardis. *Iran* XLIV: 33-62.

CRIBB R.

1991 *Nomads in Archaeology*. Cambridge: Cambridge University Press.

DAVIS S.J.M.

1984 The advent of milk and wool production in western Iran: some speculations. *In*: CLUTTON-BROCK J. and GRIGSON C. (eds.), *Animals and Archaeology: 3. Early Herders and their Flocks*: 265-278. Oxford (*BAR Int. Ser.* 202).

FAZELI H.

2001 *An investigation of Craft Specialisation and Cultural Complexity of Neolithic and Chalcolithic Periods in the Tehran Plain*. Unpublished PhD thesis, University of Bradford.

2007 *The Archaeology of the Qazvin Plain*. Tehran: University of Tehran Press.

FAZELI H. and ABBASNEGAD R.

2007 Prehistoric Chronology of Tepe Ghabristan. *In*: FAZELI H. (ed.), *The Archaeology of the Qazvin Plain*: 79-120. Tehran: University of Tehran Press.

FAZELI H. and AJOURLU B.

2007 Tepe Ismailabad. *In*: FAZELI H. (ed.), *The Archaeology of the Qazvin Plain*: 149-210. Tehran: University of Tehran Press.

FAZELI H. and ALIARI A.

2007 Chronology of Tepe Zageh. *In*: FAZELI H. (ed.), *The Archaeology of the Qazvin Plain*: 29-77. Tehran: University of Tehran Press.

FAZELI H., CONINGHAM R. and BATT C.M.

2004 Cheshmeh-Ali Revisited; Towards an Absolute Dating of the Late Neolithic and Chalcolithic of Iran's Tehran Plain. *Iran* XLII: 13-23.

FAZELI H., WONG E. and POTTS D.T.

2005 The Qazvin Plain revisited: A Reappraisal of the Chronology of Northwestern Central Plateau, Iran, in the 6th to the 4th Millennium BC. *Ancient Near Eastern Studies* 42: 3-82.

HALSTEAD P.

1990 Present to Past in the Pindhos: diversification and specialization in mountain economies. *Rivista di Studi Liguri* LVI 1-4: 61-80.

1996 Pastoralism or Household Herding? Problems of Scale and Specialization in Early Greek animal Husbandry. *World Archaeology* 28,1: 20-42.

HESSE B. and WAPNISH P.

1985 *Animal Bone Archaeology*. Washington: Taraxacum.

JONES S.

2005 Transhumance Re-Examined. *Journal of the Royal Anthropological Institute* 11,2: 357-359.

KUZMINA E.E.

1995 The cultural Connections between the Shepherds of the Steppes and south Central Asia, Afghanistan and India in the Bronze Age. *In*: ALLCHIN R. and ALLCHIN B. (eds.), *South Asian Archaeology 1995*: 279-289. New Delhi-Oxford: Mohan Primlani-IBH Publishing Co.

LEVY T.E.

1992 Transhumance, Subsistence and Social Evolution. *In*: BAR-YOSEF O. and KHAZANOV A. (eds.), *Pastoralism in the Levant. Archaeological Materials in Anthropological Perspectives*: 65-82. Madison, Wisconsin: Prehistory Press.

MADJIDZADEH Y.

1977 Excavations in Tepe Ghabristan: The First Two Seasons, 1970 and 1971. *Marlik* 2: 45-62.

1978 Corrections of the internal chronology for the Sialk III period on the basis of the pottery sequence at Tepe Ghabristan. *Iran* XVI: 93-101.

1981 Sialk III and the Pottery Sequence at Tepe Ghabristan. The Coherence of the Cultures of the Central Iranian Plateau. *Iran* XIX: 141- 146.

MASHKOUR M.

2003 Tracing ancient 'nomads': isotopic research on the origins of vertical 'transhumance' in the Zagros region. *Nomadic Peoples* 7,2: 36-47.

2006 Faunal remains from Tol-e Nurabad and Tol-e Spid. *In*: POTTS D.T. and ROUSTAEI K. (eds.), *The Mamasani Archaeological Project Stage One. A report on the first two seasons of the ICAR – University of Sydney expedition to the Mamasani District, Fars Province, Iran*: 135-146. Tehran: Iranian Center for Archaeological Research.

MASHKOUR M., BOCHERENS H. and MOUSSA I.

2005 Long distance movement of sheep and goats of Bakhtiari nomads tracked with intra-tooth variation of stable isotopes ($_{13}C$ and $_{18}O$). *In*: DAVIES J., FABIS M., MAINLAND I., RICHARDS M. and THOMAS R. (eds.), *Diet and Health in Past Animal Populations: Current Research and Future Directions*: 113-124. Oxford: Oxbow Books.

MASHKOUR M., FONTUGNE M. and HATTE C.

1999 Investigation on the evolution of subsistence economy in the Qazvin Plain (Iran) from the Neolithic to the Bronze Age. *Antiquity* 73: 65-76.

MATTHEWS R. and FAZELI H.

2004 Copper and Complexity: Iran and Mesopotamia in the fourth millennium B.C. *Iran* XLII: 61-75.

MONTGOMERY J., EVANS J.A., POWESLAND D. and ROBERTS C.A.

2005 Continuity in Anglo-Saxon England? Isotope evidence for mobility, subsistence practice, and status at West Heslerton. *American Journal of Physical Anthropology* 126: 123-138.

NEGAHBAN E.O.

1977 Preliminary report of Qazvin Expedition: excavation of Zaghe, Qabristan, Sagzabad, 1971-1972. *Marlik* 2: 26-44.

1979 A Brief Report on the Painted Building of Zaghe. *Paléorient* 5: 239-250.

O'CONNOR T.P.

2000 *The Archaeology of Animal Bones*. Stroud: Sutton Publishing.

PARKES P.

1987 Livestock Symbolism and Pastoral ideology among the Kafirs of the Hindu Kush. *Man* 22,4: 637-660.

PAYNE S.

1969 A metrical distinction between sheep and goat metacarpals. *In*: UCKO P.J. and DIMBLEBY G.W. (eds.), *The Domestication and Exploitation of Plants and Animals*: 295-305. London: Duckworth.

1973 Kill-off patterns in sheep and goats: the mandibles from Asvan Kale. *Anatolian Studies* 23: 281-303.

1975 Partial recovery and sample bias. *In*: CLASON A.T. (ed.), *Archaeozoological Studies*: 7-17. Amsterdam: North-Holland Publishing Company.

1985 Morphological distinctions between the mandibular teeth of young sheep, *Ovis*, and goats, *Capra*. *Journal of Archaeological Science* 12: 139-147.

1991 Early Holocene Equids from Tall-i-Mushki (Iran) and Can Hasan III (Turkey). *In*: MEADOW R.H. and UERPMANN H.-P. (eds.), *Equids in the Ancient World. Volume II*: 132-177. Wiesbaden: Dr. Ludwig Reichert Verlag.

PAYNE S. and BULL G.

1988 Components of variation in measurements of pig bones and teeth, and the use of measurements to distinguish wild from domestic pig remains. *Archaeozoologia* II: 27-65.

POTTS D.T., ROUSTAEI K., WEEKS L.R. and PETRIE C.A.

2006 The Mamasani District and the Archaeology of Southwestern Iran. *In*: POTTS D.T. and ROUSTAEI K. (eds.), *The Mamasani Archaeological Project Stage One. A report on the first two seasons of the ICAR – University of Sydney expedition to the Mamasani District, Fars Province, Iran*: 1-16. Tehran: Iranian Center for Archaeological Research.

REDDING R.W.

1984 Theoretical determinants of a herder's decisions: modeling variation in the sheep/goat ratio. *In*: CLUTTON-BROCK J. and GRIGSON C. (eds.), *Animals and Archaeology 3. Early Herders and their Flocks*: 223-241. Oxford (*BAR Int. Ser.* 202).

REITZ E.J. and WING E.S.

2008 *Zooarchaeology*. Cambridge: Cambridge University Press.

RICHARDS M.P., PEARSON J.A., MOLLESON T.I., RUSSELL N. and MARTIN L.

2003 Palaeodietary evidence from Neolithic Çatalhöyük, Turkey. *Journal of Archaeological Science* 30: 67-76.

ROTHMAN M.S.

2003 Ripples in the stream: Trans Caucasian – Anatolian interaction in the Murat/Euphrates Basin at the beginning of the Third Millennium BC. *In*: SMITH A. and RUBINSON K. (eds.), *Archaeology in the Borderlands: Investigations in Caucasia and Beyond*: 94-109. Los Angeles: UCLA Institute of Archaeology.

ROWLEY-CONWY P.

1995 Wild or domestic? On the evidence for the earliest domestic cattle and pigs in South Scandinavia and Iberia. *International Journal of Osteoarchaeology* 5: 115-126.

1998 Improved separation of Neolithic metapodials of sheep (*Ovis*) and goats (*Capra*) from Arene Candide Cave, Liguria, Italy. *Journal of Archaeological Science* 25: 251-258.

SHAHMIRZADI S.M.

1977 The Excavation of Sagzabad Mound, Qazvin Plain, Iran 1970-71. *Marlik* 2: 67-79.

1979 A Specialized Housebuilder in an Iranian Village of the 6th Millennium BC. *Paléorient* 5: 183-192.

1980 Tepe Zageh and the problem of the fugitive painted pottery. *Survey and Excavation*: 13-23.

2006 *Sialk. The Oldest Fortified Village of Iran*. Tehran: Iranian Centre of Archaeological Research.

SHERRATT A.

1983 The Secondary Exploitation of Animals in the Old World. *World Archaeology* 15,1: 90-104.

1997 *Economy and Society in Prehistoric Europe*. Edinburgh: Edinburgh University Press.

YOUNG R.

2003 *Agriculture and Pastoralism in the Late Bronze and Early Iron Age, North West Frontier Province, Pakistan*. Oxford: Archaeopress (*BAR Int. Ser.* 1124).

YOUNG R., CONINGHAM R., ALI I. and ALI T.
2008 The archaeological visibility of transhumance tested using faunal material from NWFP, Pakistan. *In*: RAVEN E.M. and POSSEHL G.L. (eds.), *South Asian Archaeology 1999*: 203-210. Groningen: Egbert Forsten Publishing.

ZEDER M.A.
1984 Meat distribution at the highland Iranian urban centre of Tal-e Malyan. *In*: CLUTTON-BROCK J. and GRIGSON C. (eds.), *Animals and Archaeology 3. Early Herders and their Flocks*: 279-307. Oxford (*BAR Int. Ser.* 202).
1991 *Feeding Cities: specialized animal economy in the ancient Near East*. Washington DC: Smithsonian Institution Press.
1994 After the revolution: post-Neolithic subsistence in southern Mesopotamia. *American Anthropologist* 96: 97-126.

RESOURCE EXPLOITATION OF THE UPPER KHABUR BASIN (NE SYRIA) DURING THE 3RD MILLENNIUM BC

K. DECKERS and S. RIEHL

Abstract: *A dataset of 908 tells, 1823 radial lines and streams in the Upper Khabur Basin was produced by digitizing CORONA satellite images. Several analyses are applied to this database in order to understand the hydrological and agrarian contexts of tells that were occupied mainly in the mid 3rd millennium BC. A strong alignment of tell sites along wadis is observed, which may explain the building of settlements upwards. The analysis of radial lines with exaggerated 3D-SRTM data indicates that, in most cases, they are unlikely to have been used as an irrigation system, but rather may be hollow ways with sometimes fade-out points delineating the end of cultivation zones. This is supported by a comparison between cultivation zones based on hollow ways and calculations of the necessary amount of land starting from the tell sizes. Resource exploitation calculations indicate that the area was not overexploited during the mid 3rd millennium BC and that tribute ca 2300 BC may have been a reason for land overexploitation, perhaps playing a role in settlement disruptions between 2200 and 1900 BC.*

Résumé: *Une base de données comprenant 908 tells et 1823 lignes radiales et cours d'eau a été réalisée pour le bassin du Haut Khabur, en digitalisant des images-satellites CORONA. Plusieurs analyses ont été menées à partir de cette base de données afin de comprendre les contextes hydrologiques et agraires des tells répertoriés, tells qui furent principalement occupés au milieu du IIIe millénaire avant J.-C. Les sites sont majoritairement localisés le long des wadis, ce qui peut expliquer pourquoi les installations étaient établies sur des hauteurs. L'étude des lignes radiales, à partir de données 3D-SRTM exagérées, indique que, dans la plupart des cas, elles ne constituaient probablement pas un système d'irrigation, mais qu'il s'agissait plutôt de « chemins creux », parfois sans issue, dont le maillage délimite des zones cultivées. Cette hypothèse est appuyée par une comparaison entre l'étendue des zones cultivées (en se basant sur les lignes radiales) et la quantité de terre agricole nécessaire, calculée à partir de la taille des sites. Le calcul des ressources nécessaires indique que le Haut Khabur n'était pas surexploité au milieu du IIIe millénaire avant J.-C. Cependant, le tribut dû aux environs de 2300 avant J.-C. peut avoir été la cause d'un déséquilibre dans les ressources disponibles, jouant alors un rôle dans l'abandon de certains sites entre 2200 et 1900 avant J.-C.*

Keywords: *Site location, Resource exploitation, 3rd millennium BC, Early Bronze Age, Radial lines, Hollow ways, Tribute.*
Mots-clés: *Localisation des sites, Stratégie d'exploitation des ressources, IIIe millénaire avant J.-C., Bronze ancien, Lignes radiales, « Chemins creux », Tribut.*

The Upper Khabur in northeastern Syria is an undulating plain just south of the Anti-Taurus Mountains and north of the Jebel Sinjar (fig. 1). Several wadis within this area have a north-south direction and drain to the Wadi-el-Radd, which is a tributary of the Khabur, after which the name of the region is given. From west to east are the Wadi Aweij, the Wadi Khanzir, the Wadi Jaghjagh, the Wadi Jarrah, the Wadi Kuneizir, and the Wadi Rumeilan (fig. 1). Although the Jaghjagh and Jarrah would be permanent streams in the absence of damming and irrigation, they presently run dry during the summer. The Khanzir only flows during the peak of the rainy season. Evaporation in the Wadi-el Radd is so great that only in times

Manuscrit reçu le 2 octobre 2008, accepté le 25 novembre 2008

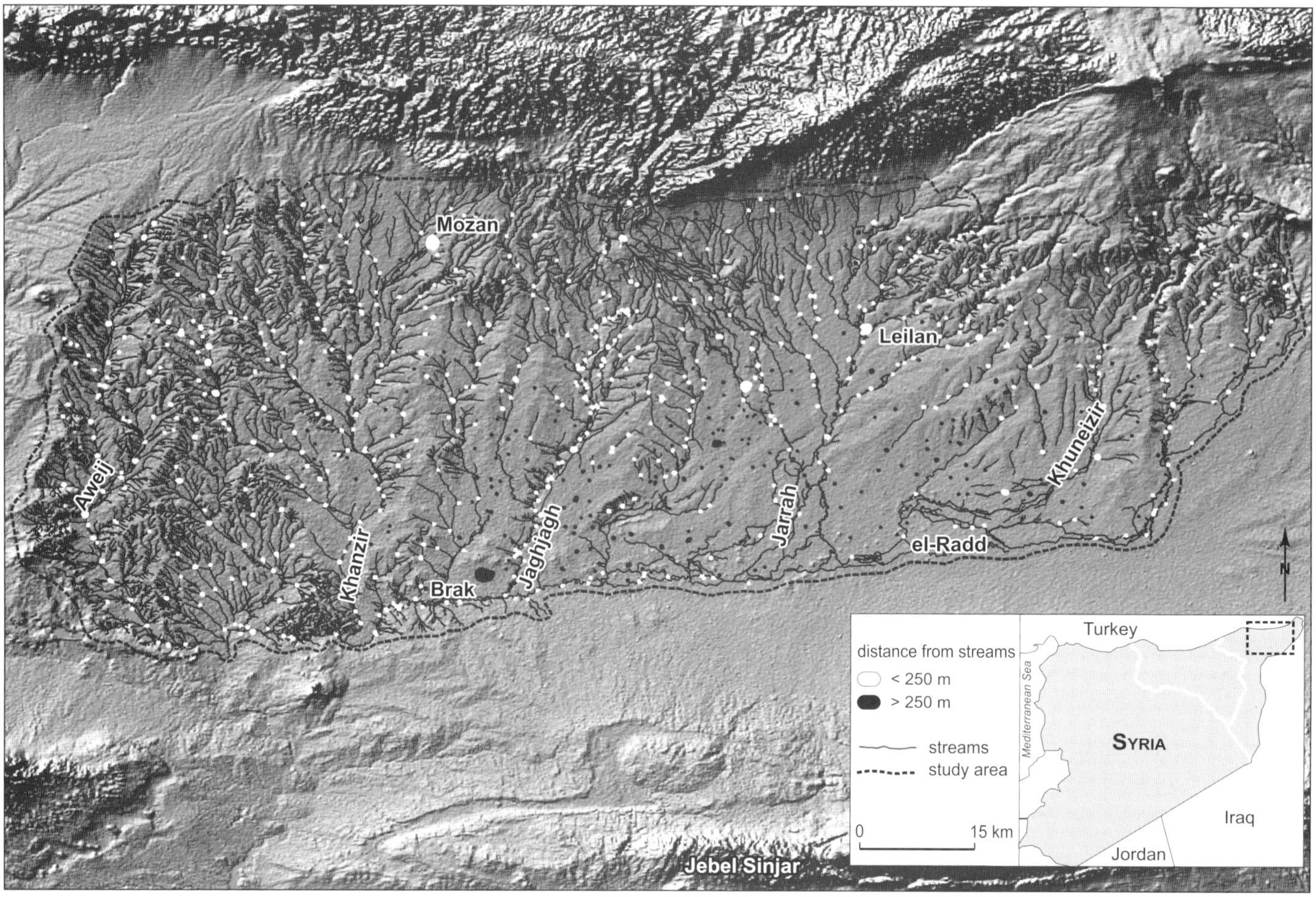

Fig. 1 – *Map of the Upper Khabur area with locational analysis of tells for their distance from streams.*

of flooding does any amount of water find its way west to the Khabur.[1] Although the climate has clearly been fluctuating between wet and dry phases, most evidence[2] suggests that the Upper Khabur basin was better watered than today during the mid-Holocene.

Summers in the Khabur area are very hot and dry today, with the most rain falling between October and April when snowfall can be expected. Rainfall varies from 400 to 700 mm in the north to 250 mm in the south. Today, the northern part of this area is intensively used for grain cultivation, whereas the southern part is more steppe-like. Although very few trees grow in the area, Hillman[3] has shown that the region has the potential to be covered by deciduous oak park woodland under modern climatic conditions in the absence of deforestation, grazing and cultivation. The dominating presence of oak charcoal, the presence of acorns and the charcoal of small oak branches in archaeological deposits of the Early Bronze Age Tell Mozan (fig. 1) samples suggest that the oak *Rosaceae* park woodland had a more southern distribution than at present.[4] At Tell Leilan (fig. 1), besides riverine vegetation probably deriving from the Jarrah terraces, deciduous *Quercus* (oak) is dominant in the Early Bronze Age samples, which supports the results from Mozan. The *Rosaceae* species typically belong to this oak park woodland as well. Preliminary anthracological research at Leilan, however, also indicated that the samples contain more *Pistacia* (oriental terebinth) and *Amygdalus* (almond) than at Mozan. This indicates that the terebinth-

1. KOLARS and MITCHELL, 1991.
2. WILKINSON, 2002a; HOLE, 1997; DECKERS and RIEHL, 2007a and 2007b.
3. MOORE *et al.*, 2000: fig. 3.7.
4. DECKERS, 2005; DECKERS and RIEHL, 2007b.

almond woodland steppe was closer to the site than at Mozan, where few such indicators were found. The local presence of *Pistacia* and *Amygdalus* in the Tell Leilan surroundings is indicated by the presence of small branches in the Early Bronze Age archaeological deposits.[5] From Tell Chagar Bazar a few new botanical samples also indicate the presence of deciduous oak,[6] besides the previously found *Fraxinus* sp. (ash), *Ulmus* sp. (elm), *Platanus* sp. (plane), and *Populus* sp. (poplar).[7]

The present study will focus on settlement and land-use patterns in the 3rd to 2nd millennia BC. Although several studies have been published about the recognition and modeling of Early Bronze Age settlement patterns, land use and communities within the Upper Khabur plain, with each model providing new insights,[8] there are still contradictions between the presently available studies. For example, while Wilkinson[9] came to the conclusion that there was barely enough land to support Early Bronze Age inhabitants on the plain, anthracological research indicates the presence of deciduous oak park woodland in the Tell Mozan, Leilan and Chagar Bazar region.[10] Moreover, settlement hierarchy and land use studies are sometimes undertaken with selective data sets,[11] which represent reasonable models, but could be amended by basic data. In many land use models, subsistence patterns are reduced to the consumption of cereals, with no inclusion of archaeobotanical data. We decided to work with a maximalistic dataset by using archaeobotanical data to model the subsistence pattern in greater detail and by using all visible tell sites and features despite the fact that some of them may not belong to the 3rd millennium BC. Our approach in modeling landscapes around the settlements will be "conventional", *i.e.* based on settlements as a whole with little space for individual decision-making, as opposed to recently published agent-based models[12] that have the potential to provide dynamic outcomes.

DATA SET AND METHODOLOGY

It has been demonstrated in several publications that satellite imagery has great potential for finding archaeological and landscape features in the Near East.[13] Within this study tells, lower towns, river systems and "hollow ways" have also been mapped with satellite imagery in a GIS system.

CORONA images with a resolution of *ca* 2 to 3 m were used in this study. The negatives from the following missions were processed and printed on photo paper: D025-055 1105-1FWD, 5 Nov 68, D025-011, 1102-1 AFT, 11 Dec 67 and D025-003, 1108-1AFT, 6 Dec 69. They were subsequently scanned and geo-referenced. Landsat-7ETM$^+$ three band maps at 14.3 m resolution were used for this. Moreover, the area is also covered by six SRTM (*Shuttle Radar Topography Mission*) one-degree-tiles (36-38 N, 38-41E) of three arc-second resolution (*i.e.* 90 m horizontal resolution, ± 6 m 90% relative vertical accuracy) which were imported into the GIS system.

Although the automated technique for finding tells on SRTM data developed by Menze *et al.*[14] is a rapid method to gain an overview of tell sites, the visual assessment of CORONA images still provides more complete results. Therefore, within this study, tells, as well as wadis, were manually digitized from georeferenced CORONA images. As a control, the tells and wadis identified were superimposed on SRTM images. Although a lot of archaeological survey work has been undertaken in the Upper Khabur area over the last 60 years,[15] not all tells have been systematically surveyed. Tells in the area are classic multi-period occupations which formed at least in the 6th to 4th millennia BC, attained their maximum settled area in the 3rd millennium BC, and showed relatively little occupation on the main mounds after the late 3rd millennium BC.[16] In the Tell Beydar and western Upper Khabur region for example, survey showed that nearly all 3rd millennium occupations occurred on tells. In the 3rd millennium BC, very few small, low sites were occupied, unlike in other periods.[17] Overall, 908 tells were identified within this study and we will assume that their maximum extent dates to the 3rd millennium BC. It should be noted however, that survey in the eastern part of the Khabur basin indicated that the greatest extent of settlement took place in the early 2nd millennium BC (Early Middle Bronze Age).[18] A comparison of the

5. Unpublished data Deckers.
6. *Ibid.*
7. MALLOWAN, 1947: 15.
8. *e.g.*, GOLDHAUSEN and RICCI, 2005; WILKINSON, 1997; WILKINSON *et al.*, 2007.
9. WILKINSON, 2003: 122, fig. 6:18.
10. DECKERS and RIEHL, 2007b and unpublished data Deckers.
11. *e.g.*, GOLDHAUSEN and RICCI, 2005; WILKINSON, 2003.
12. *e.g.*, WILKINSON *et al.*, 2007.
13. *e.g.*, VERHOEVEN and DALES, 1994; KENNEDY, 1998; WILKINSON, 2003; KOUCHOUKOS, 2001; DONOGHUE *et al.*, 2002; PHILIP *et al.*, 2002; UR, 2003 and 2005; HRITZ, 2004; POURNELLE, 2003; HRITZ and WILKINSON, 2006.
14. MENZE *et al.*, 2005.
15. See *e.g.*, VAN LIERE et LAUFFRAY, 1954; MEIJER, 1986; LYONNET, 1996, 1997 and 1998; LYONNET (éd.), 2000; WILKINSON, 2000b and 2002b; WILKINSON and BARBANES, 2000; RISVET, 2005.
16. WILKINSON and TUCKER, 1995: 51; WILKINSON, 2002b.
17. WILKINSON, 2002b.
18. RISVET, 2005.

archaeological survey result maps by Risvet[19] with our own remote sensing data indicates that we have an overestimation of 3rd millennium BC settlements in this area, especially towards the south. This, however, will not have great impact on our general conclusion regarding 3rd millennium BC land use, since those sites were mostly relatively small (see below).

A number of analyses have been undertaken on this database to gain insight into the settlement pattern and land use. First, a location analysis was performed to understand the relation between tells and rivers within this area. Secondly, hollow ways were followed to determine whether they crossed over watersheds or interfluves. This analysis was done on the exaggerated 3D-SRTM data overlain by the CORONA satellite images with the mapped radial lines. And thirdly, sustaining areas were calculated for the mapped settlements. Based on the mapped tell and lower town surfaces, we can calculate Early Bronze Age sustaining areas based on realistic population density values, food consumption patterns and yields within this climatic zone. As mentioned above, we assume that all tells and lower towns had their maximal expansion during the Early Bronze Age, which may result in some overestimation of the area under cultivation. On the other hand, we also realize that some parts of tells may be covered by later alluvium, since typical tell locations within this area are along rivers and, therefore, we also realize that in some cases, we may underestimate the area under cultivation. Furthermore, in our calculations, we suppose that the tell sites supported a significant sedentary population.

Estimates of population densities in Near Eastern sites based on contemporary and historical settlements provide values varying between 100 and 200 persons / ha.[20] Pfälzner,[21] however, argues that, based on ethnographical demography for prevailing house types and sizes, population density values between 300 and 400 persons / ha are more realistic, for example, at the Early Bronze Age archaeological sites Tell Selenkahiye, Tell Halawi A and Tell Chuera.

FAO data from 2005 suggests a considerably high caloric intake of 3057 kcal / capita / day, indicating a strong deviation from the average consumption of 2200 kcal for a normal population with 15% under 12 years old and under the assumption of moderate labor. Since we will use annual yields from the period between 1961 and 1970 for reasons explained below, it is also interesting to look at the 1970 calory intake pattern of *ca* 2342 kcal / capita / day (table 1), which resembles normal population consumption patterns. The diet of the Syrian population in 1970 was, according to FAO, composed of 89.1% plant food, whereas only 10.9 % was derived from animal products, with 3.8% from milk and 3.2% from meat. Of the 89.1% plant products, 54.4% were from cereals, 9.6% were from oil plants and nuts, 4.6% from pulses, 4.5% from fruits and almost 9.2% from sugar cane. If we assume, for further calculations, that the composition of the diet in the Bronze Age was similar to today concerning the percentage proportion of carbohydrates, proteins and fats, the 9.2% of carbohydrates mainly from sugar cane (*Saccharum* sp.) should be replaced by items which were available to the Bronze Age population such as cereals and fruits.

As mentioned above, more than half of the calory input today is derived from cereals, particularly wheat (53.3%), while barley is mostly cultivated for animal feed and seed. We may thus assume that a more or less healthy diet may cover the required calories by even more than 50% of cereals, while slightly more than 18% comes from other cultivated and wild plant food.

The estimation of ancient yields is a difficult task, complicated by numerous methodological problems as for example the interpretation of barley yields from ancient texts. Interpretations of ancient texts present highly different values, and in some cases it is difficult to evaluate whether the documents refer to yields of one year or more.[22] Documented barley returns at Mari for example are mistrusted, because the purpose of the texts is unclear.[23] Assyriologists also suggest that the higher average yields at Mari in comparison to southern Mesopotamian agriculture may be due to a larger size of the Mari *ikû*.[24] Another criterion discussed with yield calculations is their disparity. However, looking at modern yield variation between 1990 and 2004, this fact is not astonishing. Table 2 shows the strong variability in barley yields of the Near East even today. The maximum yields per hectare in Syria, Lebanon and Israel reflect above all differences in water availability during the growing season of the cereal. Looking at the minimum yields as recorded in the Mari tablet minimum yields are indeed comparatively high; however mean and maximum yields, lay well within the modern yield ranges of Lebanon and Israel. Maximum yields at Nuzi that is considered as both irrigation and rain fed land,[25] were similar to those in Mari that is considered as irrigation land, but there was much more

19. RISVET, 2005.
20. ADAMS, 1981; KRAMER, 1980.
21. PFÄLZNER, 2001.
22. POWELL, 1985.
23. VAN KOPPEN, 2001: 482.
24. *Ikû* is a unit of land measurement that equals 0.35 ha.
25. ZACCAGNINI, 1990.

Table 1 – *Food consumption and yields in Syria based on FAO data.*

Food type	Mean yield (1961-1970) in kg/ha	Food consumption (kcal/capita/day) in 1970 (2 342,65 kcal)	Food consumption (1970 in %)	Hypothetical food consumption (kcal/capita/day) in a 2 200 kcal diet	Hypothetical food consumption (kg/capita/year) in a 2 200 kcal diet
Wheat	740,347	1 249,95	53,356	1 173,837	113,169
Barley	792,323	21,91	0,935	20,576	0,000
Millet	798,400	8,26	0,353	7,757	0,018
Rice, paddy	2 215,170	65,68	2,804	61,681	12,115
Broad beans, horse beans, dry	1 086,054	1,51	0,064	1,418	1,169
Chick peas	716,135	0,00	0,000	0,000	1,576
Lentils	720,043	0,00	0,000	0,000	2,629
Peas, dry	950,747	0,72	0,031	0,676	0,247
Bitter vetch	748,450	0,00	0,000	0,000	0,000
Grass pea	748,450	0,00	0,000	0,000	0,000
Other pulses	no data	105,44	4,501	99,019	15,627
Linseed	642,189	0,00	0,000	0,000	0,000
Safflor	827,599	0,00	0,000	0,000	0,000
Other oil plants and nuts	no data	225,40	9,622	211,675	29,288
Figs	2 512,367	0,00	0,000	0,000	1,156
Grapes	3 150,948	42,65	1,821	40,053	11,077
Other fruits	3 007,860	62,73	2,678	58,910	94,852
Vegetables	6 852,940	63,18	2,697	59,333	54,369
Tubers, roots and cane	10 652,590	239,30	10,215	224,728	258,436
Spices and others	700,500	0,94	0,040	0,883	5,590
Mammal meat		69,56	2,969	65,324	10,557
Poultry		6,48	0,277	6,085	5,480
Fish and sea food		2,11	0,090	1,982	1,587
Milk, whole, fresh		89,46	3,819	84,013	70,318
Bird eggs (incl. hen eggs)		11,11	0,474	10,433	5,934
Other animal fats		76,26	3,255	71,616	1,224

data on low yields, resulting in particularly low minimum and mean yields. There may be a bias in the low yield data for the Nuzi area because barley proportions are considered to have constituted up about 80% of the cereal crops[26] and should have been a dominant part of subsistence and also surplus production.

It was concluded that supplementary irrigation of at least wheat was used at the fortified agricultural production centre of Tell Sabi Abyad on the Balikh River.[27] Although the relation of seed corn to yield is considered to be amongst the best during the period, the barley yield reconstructed for the site is comparatively low. It is plausible to assume that in fact not barley, but other crops were irrigated.

A source of more or less exact yields, compared to those available in ancient texts, is the statistical database of the FAO,[28] which covers a time range from 1961-2004. It has to be taken into account that since 1990 an agricultural revolution has taken place in Syria with the introduction of high-yielding varieties of crop plants, chemical fertilizers, herbicides and intensified irrigation.[29] Therefore, within this study we used yield values as given by FAO for the period between 1961 and 1970 (table 1). We decided to use this data because it represents the only complete and, for this period, most reliable information source useful in our calculations.

Yields also depend on the amount of seed that has been sown. Wilkinson and Tucker[30] mentioned that today *ca* 4% of seed is necessary to secure the next crops. This is in accordance with 3rd and 2nd millennia BC texts. For example yield ratios for Mari were 1:29,[31] which would have required *ca* 3.4% of seeds. Exceptionally low seed-to-yield ratios (1:1.5 and 1:8) were mentioned in texts from Nuzi which means that yields

26. *Ibid.*
27. WIGGERMANN, 2000.
28. http://faostat.fao.org/site/395/default.aspx
29. MAZID *et al.*, 2003.
30. WILKINSON and TUCKER, 1995: 84.
31. VAN KOPPEN, 2001: 482-485.

Table 2 – *Barley yields for different periods and regions in the Near East;* ***1****, FAO Statistics Division 2006;* ***2****, 1* ikû *= 0,35 ha, which would assume that the Mari* ikû *equals the Babylonian* ikû*, 1* gur *= 120 litres at Mari, in contrast to the more common 300 litres in Akkadian systems: values according to van Koppen*[32] *and Lafont;*[33] ***3****, volume weight according to Lewy;*[34] ***4****, according to Zaccagnini,*[35] imēru *surface (1* imēru*: 1* ikû *= 1: 5; 1* imēru *= 18 ha),* imēru *volume (1* imēru *= 67 litres);* ***5****, according to Wiggermann.*[36]

Region of cultivated barley and origin of source	Min. yield	Max. yield	Mean yield
Modern barley yield (t/ha) Syria 1990-2004 **(1)**	0,2	1,5	0,7
Modern barley yield (kg/ha) Syria 1990-2004 **(1)**	160,0	1 500,0	654,0
Modern barley yield (t/ha) Lebanon 1990-2004 **(1)**	1,1	3	1,7
Modern barley yield (kg/ha) Lebanon 1990-2004 **(1)**	1 080,0	2 950,0	1 734,0
Modern barley yield (t/ha) Israel 1990-2004 **(1)**	0,2	2,1	0,9
Modern barley yield (kg/ha) Israel 1990-2004 **(1)**	180,0	2 140,0	855,0
ARMT 23 426 Mari district (water damaged) (*gur/ikû*) **(2)**	0,7	1,7	1,2
ARMT 23 426 Mari district (water damaged) (kg/ha); H. vulgare **(3)**	144,0	349,7	246,9
ARMT 23 426 Mari district (water damaged) (kg/ha); H. distichion **(3)**	168,0	408,0	288,0
ARMT 23 591 Naser district (*gur/ikû*) **(2)**	4,9	7,5	5,7
ARMT 23 591 Naser district (kg/ha); H. vulgare	1 008,0	1 542,9	1 172,6
ARMT 23 591 Naser district (kg/ha); H. distichion	1 176,0	1 800,0	1 368,0
ARMT 24 3 Zaqum, Bab-nahlim, Der (*gur/ ikû*) **(2)**	6	8	7
ARMT 24 3 Zaqum, Bab-nahlim, Der (kg/ha); H. vulgare	1 234,3	1 645,7	1 440,0
ARMT 24 3 Zaqum, Bab-nahlim, Der (kg/ha); H. distichion	1 440,0	1 920,0	1 680,0
ZIEGLER, 1999:49 note 300 fields of princesses (*gur/ikû*) **(2)**	7,5	9,3	8,4
ZIEGLER, 1999:49 note 300 fields of princesses (kg/ha); H. vulgare	1 542,90	1 913,10	1 728,00
ZIEGLER, 1999:49 note 300 fields of princesses (kg/ha); H. distichion	1 800,0	2 232,0	2 016,0
Nuzi (*imēru/imēru*) **(4)**	2	705	108,5
Nuzi (kg/ha); H. vulgare	4,5	1 574,5	242,3
Nuzi (kg/ha); H. distichion	5,2	1 836,9	282,7
Tell Sabi Abyad (kg/ha); with supplementary irrigation **(5)**			421,0

must have been very low, or that exceptionally high numbers of seeds were sown.[37] Within our calculations of the necessary cultivated land, we added an extra 6% for the necessary seed.

Even nowadays, with sophisticated agricultural technology, crop yields are very variable from one year to the next, and can only partially be explained as a result of economic preferences.[38] Even under the application of irrigation, there is a strong correlation between yield and soil moisture, which is apparent by comparing regional averages of annual precipitation anomalies[39] with the yields of barley and wheat provided by FAO (see fig. 2). In consideration of the distinctive Holocene climate history in the Near East, we may certainly expect lower yields during times of increased aridity even when irrigation was practiced. The comparison of the precipitation anomalies with mainly non-irrigated barley yields between 1961 and 2003 shows a strong correlation between peaks in precipitation and yield.

Comparing archaeobotanical data[40] with modern consumption data includes various methodological problems, mainly due to the fact that the representativity of the assemblage is unknown concerning the amount of taxa used for the diet, as well as their proportions. If we assume that most of the yield was consumed, and that the archaeobotanical dataset of crop remains is a product of garbage accumulation, storage and

32. VAN KOPPEN, 2001.
33. LAFONT, 2000.
34. LEWY, 1944.
35. ZACCAGNINI, 1990.
36. WIGGERMANN, 2000.
37. ZACCAGNINI, 1975 and 1990.
38. See also effects of moist and dry years on the Syrian economy as outlined by WIRTH, 1971.
39. ZHANG *et al.*, 2005.
40. RIEHL and KÜMMEL, 2005.

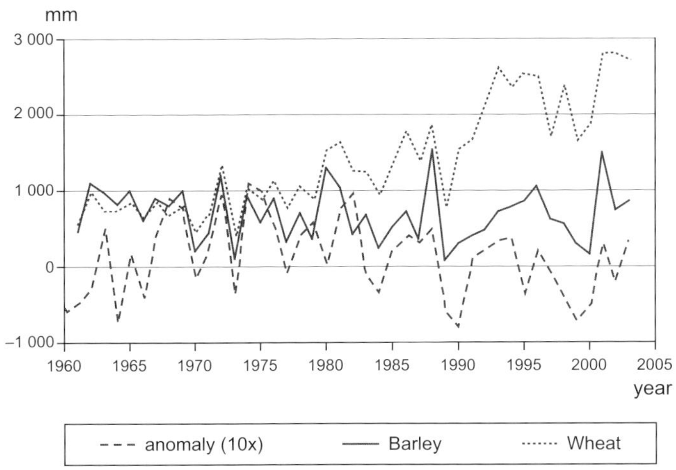

Fig. 2 – *Comparison of precipitation anomalies with barley and wheat yields between 1961 and 2003.*

accidental burning of seeds, we have a dataset that is not 100% equivalent to the proportions of ancient crop yields. This is a general problem in archaeobotanical research and has to be tolerated when working with archaeobotanical data. Additionally, the general difference of preservation in different contexts results in either over- or under-representation of some species, as is seen in some of the oil plants. If they are preserved, vegetative remains from fruits, tubers and roots are usually not identified, and thus are generally under-represented in the archaeobotanical record.

As already mentioned many of the modern crops are new additions and were not available to Bronze Age people. In our database, crops today used for sugar production and consumption that account for a large part of the calorific input (10.2%) need to be substituted when considering the ancient assemblage. These may well be represented by cereals, which occur in much higher proportions in the archaeobotanical assemblage than in the modern diet (see table 3).

For the calculation of the cultivated surface according to the archaeobotanical data, we used the following formula, according to several assumptions detailed below.

$$L_{(A,B,Z)} = [P_{(S \bullet D)}] \bullet \left[\frac{Z \bullet C_{(A,B)} \bullet 365}{Y_{(A)} \bullet K_{(A)} \bullet 1000} \right]$$

The variables are L (land needed for cultivation in ha / year), A (crop species), B (archaeological site), Z (calorific need in kcal / capita / day), P (expected population in capita) in S (settled area as calculated in ha) under D (assumed population in capita / ha), Y (modern mean yield in kg / ha), C (crop percentage), $K_{(A)}$ (kcal / 100 g of crop A)

RESULTS AND DISCUSSION

TELLS AND THEIR HYDROLOGICAL CONTEXT

For this study, tells were considered to be located along a stream if they were located less than 250 m from a current or relict stream bed. Of the 908 tells identified within the research area, 71.6% were positively identified to be located along an identifiable (relict) stream, while only 28.4% were not (fig. 1). It should be mentioned that there may well be some (relict) streams that were not mapped, and therefore, that even more tells may have been located along (relict) streams, as is also suggested by unmapped depressions on the SRTM images. Our calculations are, however, based on interpretations of the CORONA images with clearly visible streams, which therefore represent a minimum. Moreover, the Wadi-Radd area is generally swampy but mostly lacking clear channels that were mapped, so that the moist context of some tells may be somewhat underrepresented on figure 1. It however needs to be investigated whether waterlogged conditions also occurred there during the Early Bronze Age.

Therefore, watercourses constituted a major axis along which human settlement in the Upper Khabur has been founded (fig. 3). This strong alignment of tells along wadis has been observed by Stein and Wattenmaker[41] in the Tell Leilan region. Although not all these wadis and relict wadis had perennial flow, water could have been obtained by digging into the wadi bed or by the construction of a dam.

The inhabitant's relationship with the rivers may have been ambiguous. In fact, the watercourses must have constituted a latent threat to the very existence of settlements and harvests on the alluvial plain. Although rainfall mainly occurs between October and April within this area, maximal discharge of the Jaghjagh in the sixties took place between October and May.[42] During the spring especially, floods could have been very damaging. For example, at the village Mozan no water flows within the barely visible wadi, but about 25 years ago it contained water; to such an extent that the village was often flooded, houses were damaged and fields just north of the

41. STEIN and WATTENMAKER, 1990.
42. KERBE, 1987: 718.

Table 3 – *Calculations of cultivation surfaces according to ancient crop proportions as archaeobotanically identified at Tell Mozan and Tell Brak for the different periods.*

Food type	Crop proportions at EBA Tell Mozan	Crop proportions at EBA Tell Brak	Crop proportions at MBA Tell Mozan	Crop proportions at MBA Tell Brak	Kcal/100g of the crop	Hypothetical amounts of crops in kg/year/capita at EBA Tell Mozan based on a 2200 kcal diet	Hypothetical amounts of crops in kg/year/capita at EBA Tell Brak based on a 2200 kcal diet	Hypothetical amounts of crops in kg/year/capita at MBA Tell Mozan based on a 2200 kcal diet	Hypothetical amounts of crops in kg/year/capita at MBA Tell Brak based on a 2200 kcal diet	Added extra 6% seed for EBA Tell Mozan	Added extra 6% seed for EBA Tell Brak	Added extra 6% seed for MBA Tell Mozan	Added extra 6% seed for MBA Tell Brak	Mean yield (1961-1970) in kg/ha	Cultivation surface (ha) needed per capita/year at EBA Tell Mozan based on a 2200 kcal diet	Cultivation surface (ha) needed per capita/year at EBA Tell Brak based on a 2200 kcal diet	Cultivation surface (ha) needed per capita/year at MBA Tell Mozan based on a 2200 kcal diet	Cultivation surface (ha) needed per capita/year at MBA Tell Brak based on a 2200 kcal diet
Wheat	43,02	5,58	26,36	4,27	313	97,32	12,63	60,1	9,62	103,16	13,39	63,71	10,19	740,3472	0,139333	0,018083	0,086051	0,013768
Barley	54,97	84,05	66,72	94,76	320	124,35	190,2	152,13	213,33	131,81	201,61	161,25	226,13	792,3227	0,166354	0,254454	0,20352	0,285402
Broad beans, horse beans, dry	0	0	0,17	0	326	0	0	0,39	0	0	0	0,42	0	1 086,054	0	0	0,000385	0
Chick peas	0	0	0,09	0	325	0	0	0,2	0	0	0	0,21	0	716,1349	0	0	0,000292	0
Lentils	1,18	0,02	0,69	0	319	2,66	0,05	1,58	0	2,82	0,07	1,67	0	720,0426	0,003914	0,0001	0,002321	0
Peas, dry	0,12	10,16	0	0	287	0,28	22,99	0	0	0,29	24,36	0	0	950,7468	0,000309	0,025627	0	0
Bitter vetch	0,12	0,04	3,54	0,73	314	0,28	0,09	8,08	0,07	0,29	0,1	8,56	0,08	748,45	0,000392	0,000128	0,011442	0,000104
Grass pea	0,37	0,09	0,52	0,07	314	0,83	0,21	1,18	0,15	0,88	0,22	1,25	0,16	748,45	0,001177	0,000299	0,001674	0,000211
Linseed	0	0,04	0,09	0	372	0	0,09	0,2	0	0	0,1	0,21	0	642,1893	0	0,000149	0,000325	0
Safflor	0	0	0,09	0,16	580	0	0	0,2	0,37	0	0	0,21	0,39	827,599	0	0	0,000252	0,000472
Figs	0,07	0	0,17	0	63	0,17	0	0,39	0	0,18	0	0,42	0	2 512,3668	0,00007	0	0,000166	0
Grapes	0,15	0,02	1,56	0	71	0,33	0,05	3,55	0,01	0,35	0,06	3,76	0,01	3 150,9484	0,000112	0,000018	0,001193	0,000003
															0,311661	**0,298859**	**0,307622**	**0,299961**

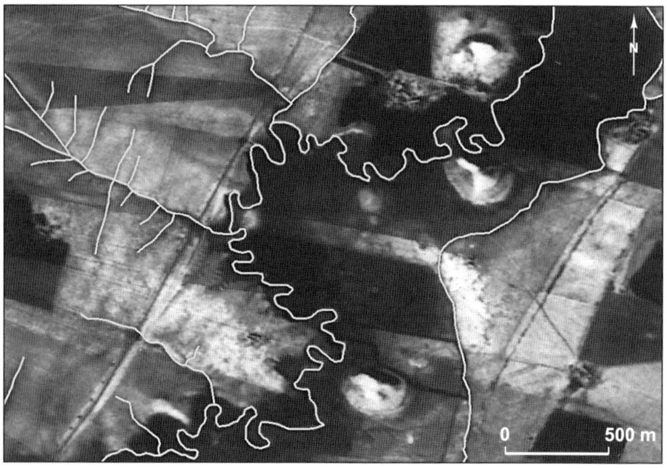

Fig. 3 – *Typical location of tells along wadis in the Upper Khabur Basin.*

village could not be cultivated.[43] The upward construction of settlements, more precisely the development of tells, höyüks and magoulas,[44] may have been to protect against flooding. Table 4 demonstrates this for a few Near Eastern tell sites, where flooding problems occurred and subsequent measures were taken that made the settlement higher and drier. Many tell sites exhibit flooding evidence between anthropogenic layers, such as Hayez Höyük,[45] Değirmendere, Imikusşaği Höyük,[46] Köşkerbaba Höyük, Tell al-Muqayyer[47] and Tell Fara[48] emphasizing the typical fluviatile setting of numerous tell sites. The high situation

43. Elderly men in village, personal communication.
44. Respectively Arabic, Turkish and Greek words for settlement mound.
45. YAKAR, 1985: 321.
46. SEVIN and KÖROĞLU, 1986; SEVIN and DERIN, 1987; KÖROĞLU, 2001.
47. WOOLLEY, 1955.
48. MARTIN and ERNST, 1988.

Table 4 – *List of archaeological sites with flooding evidence and the following countermeasures.*

Site	Location	Problem	Measures	Reference
Yassihöyük (Turkey)	Sakarya river plain	river destroyed part of settlement	site was raised with fill	MARSH, 1999
Platia Magoula Zarkou (Greece)	floodplain	aggradation and flooding	human accumulation rate of occupation deposits overtook natural sedimentation	VAN ANDEL *et al.*, 1995: 138
Aşıklı Höyük (Turkey)	Melendiz River	flooding	artificial fill	ESIN, 1999: 19
Tell Jerablus (Syria)	Euphrates floodplain	flooding	wall and fill	PELTENBURG, 1999: 98-103
Tell Bederi (Syria)	Khabur	founded only 1.5 m above Khabur	needed artificial raising	ERGENZINGER, 1991: 43
Tell-ed-Deir (Iraq)	former Euphrates channel	flooding	heaped up mounds with sand from the free parts of the site	PAEPE and BAETEMAN, 1978
Babylon (Iraq)	Euphrates	high water weakened its mud-brick foundations	whole center of city was raised by up to five meters	COLE, 1994: 94

of the settlement probably provided protection against normal flooding but could not, however, protect against extreme floods. Further research is necessary to investigate the 28.4% of tells that may not have been located along a stream in the Upper Khabur. As mentioned above, some (palaeo) channels may have been overlooked from the CORONA images. However, some tells were probably not located along a stream, suggesting that also other factors played a role in the formation of tells.

The modeled May discharge of the Euphrates by Bryson for the last 10,000 years[49] is particularly interesting for the observed settlement pattern in the Upper Khabur: from 2000 BC onwards, modeled May discharge increased considerably, causing more extreme floods. This coincides with the time that a reduced occupation on tell sites took place.[50] Perhaps the typical riverine locations became too dangerous from then on.

RADIAL LINES

While Wilkinson,[51] Wilkinson and Tucker[52] and Ur[53] interpret the radial lines as "hollow ways" that became depressed by prolonged periods of use by men and animals, McClellan[54] suggested that they served as channels for collecting, storing and distributing water. The radial lines around the Bronze Age centers imply that these features were formed at this time and were used as long as the tell was in use.[55]

Within this study, all 1823 hollow ways identified were analyzed to determine whether they were leading to a topographic depression or whether they disregarded topography in ways that it would be impossible for water to be transported over longer distances. In 44.1% of the cases, watersheds were crossed, indicating that these radial lines were unlikely to be related to water harvesting (see example for the Brak region fig. 4). Similarly, Wilkinson and Tucker[56] have observed that hollow ways cross watersheds. Therefore, if they were canals, complex engineering works would be required. In 30.82% of the cases, however, the radial lines lead to a topographic depression, while in 25.1% of the cases, it was impossible to decide this with the use of 3D-data. It is well-known that some hollow ways are occupied by wadis or gullies for long distances, whereas others accommodate wadis for a short interval and cause kinks in the channel pattern. They do not, however, appear as natural wadis but have become adopted by wadis at low points.[57] Therefore, the hydraulic aspect of some hollow ways is more likely a secondary effect.

While McClellan highlighted the problem of interpreting the fade-out points of some of the radial features as hollow ways, Wilkinson[58] and Ur[59] argue convincingly that their discontinuation indicates the limits of cultivation for a site and therefore provides the interface between the arable and pasture zone. More precisely, animals and people could not walk freely through fields that were under cultivation and therefore they always used the same tracks at the boundaries of fields. Where the fields discontinued, the flocks could disperse. Based on these fade-out points of the radial features, the area under

49. RIEHL *et al.*, 2008, fig. 1.
50. WILKINSON, 2003.
51. WILKINSON, 1990 and 1993.
52. WILKINSON and TUCKER, 1995.
53. UR, 2004.
54. MCCLELLAN, 2000.
55. WILKINSON, 1990 and 1993; WILKINSON and TUCKER, 1995; UR, 2003 and 2004.
56. WILKINSON and TUCKER, 1995.
57. *Ibid.*: 25
58. WILKINSON, 1993 and 1994: 492-493.
59. UR, 2004.

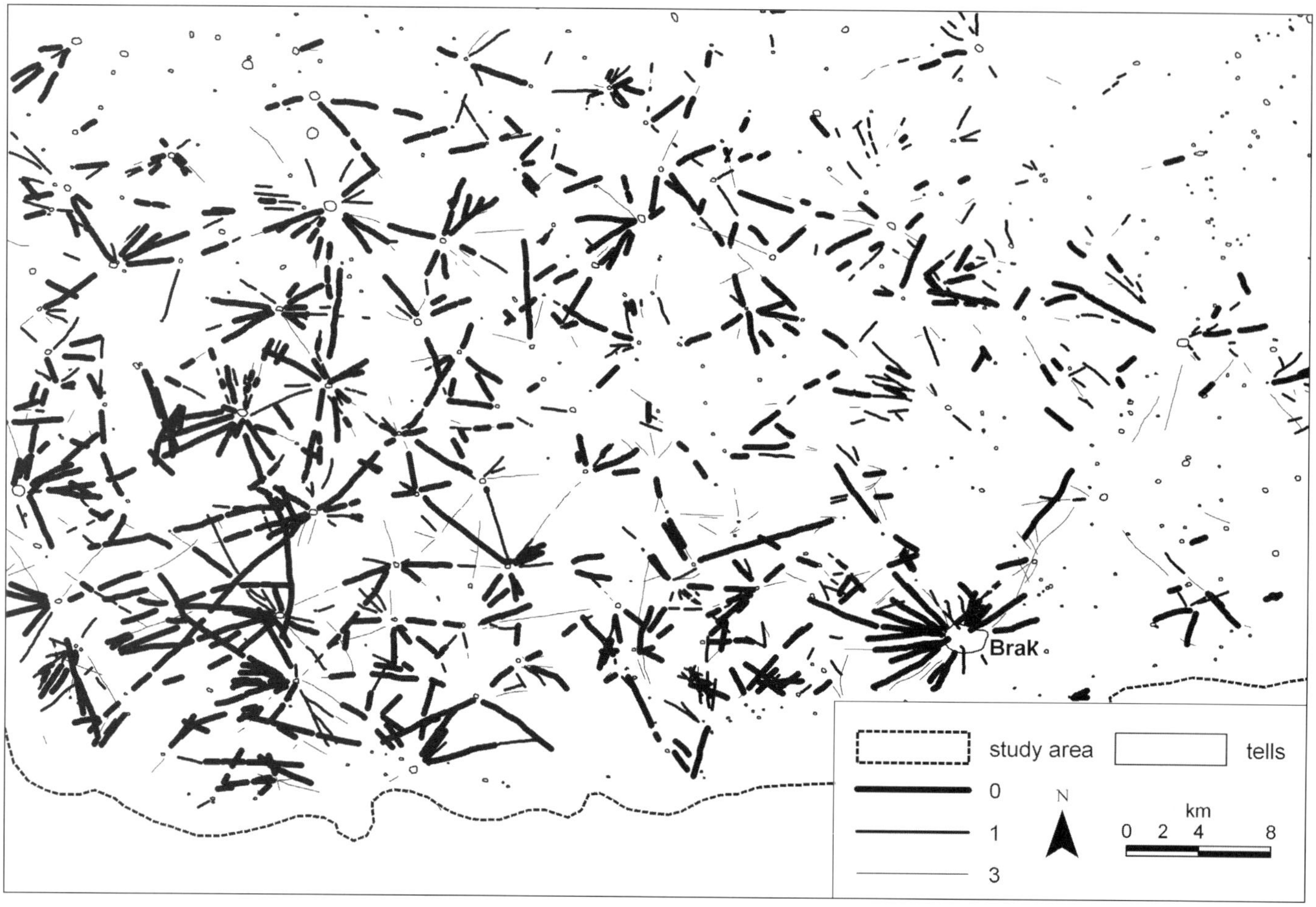

Fig. 4 – *Radial lines and their crossing of topographic boundaries for the Western Upper Khabur Basin.* ***0****, crossing of topographic boundaries;* ***1****, leading to a depression;* ***3****, not identifiable.*

cultivation could be mapped for some sites.[60] This was, however, not possible for all sites. In order to check this hypothesis, another method was applied to gain an insight into the areas under cultivation, which will be detailed below.

AN ESTIMATION OF EARLY BRONZE AGE SUSTAINING AREAS

From the hypothetical amounts of crops in kg / year / capita for the different sites (Mozan and Brak) and periods (Early and Middle Bronze Age) in our study area (table 3), the amount of land necessary to support the inhabitants was calculated from the site size (*e.g.*, fig. 5 and table 5). The GIS-based calculation of the settled area is 2347.4 ha, while our study area covers 663109.8 ha, making 660762.4 ha of cultivable land. (see *e.g.*, for EBA Mozan fig. 6a-d).

The results indicate that with normal population densities of between 100 and 400 persons / ha and continuous cultivation, more than half of the study area was free of cultivation and there would have been room for pasture and oak park woodland vegetation (table 5). This contradicts what has been thus far suggested: that the Upper Khabur basin was so densely inhabited during the Early Bronze Age that it could have hardly provided enough food for the inhabitants, especially towards the south, implying that there was not much space for woodland and firewood.[61] These interpretations are based on an incomplete tell database and a non-dynamic catchment radius of 3 km around

60. WILKINSON, 1994.

61. WILKINSON 2000a and 2003.

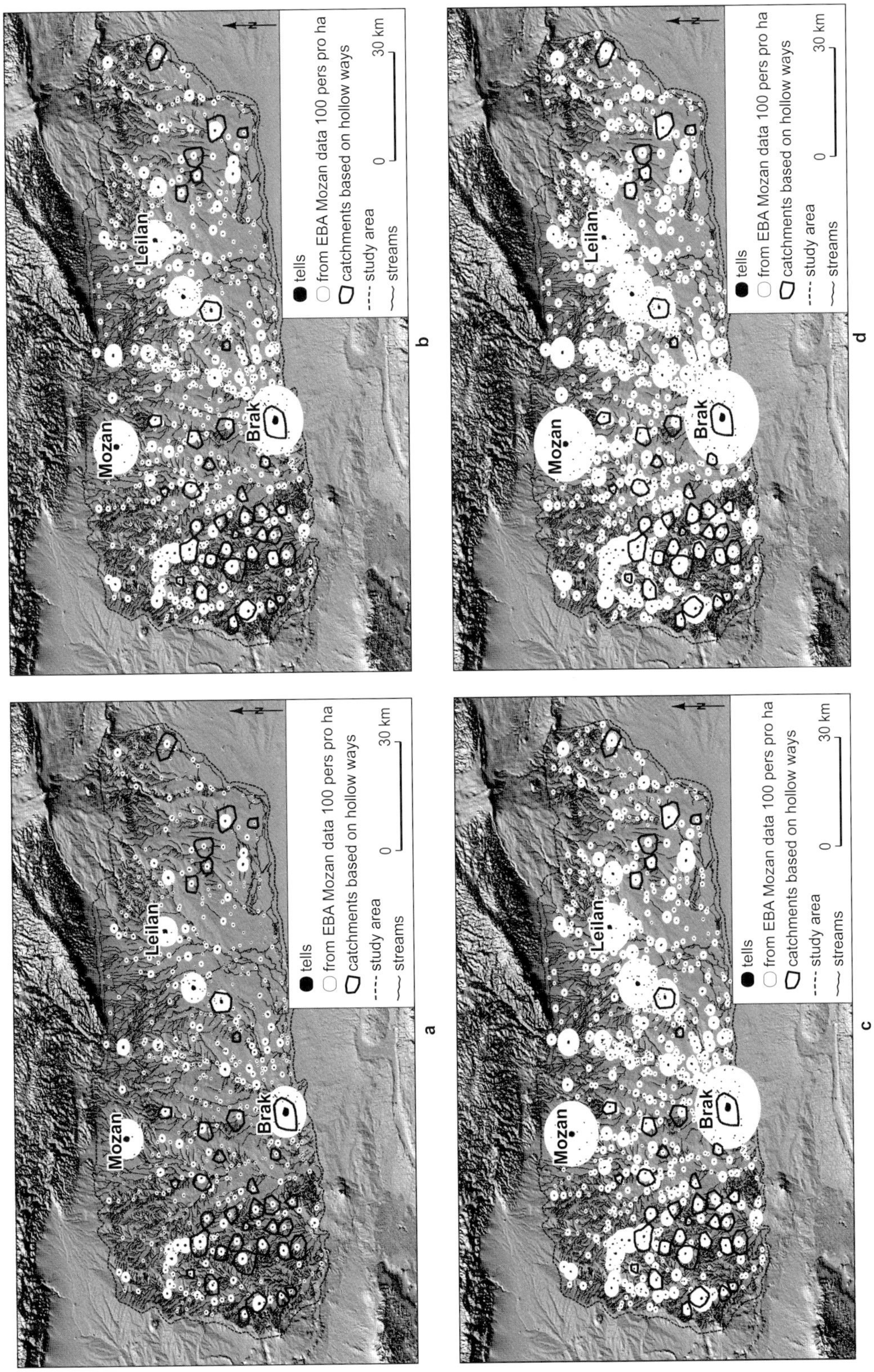

Fig. 5 – *Estimation of land necessary to support inhabitants with crops according to proportions archaeobotanically established at EBA Tell Mozan, based on the assumption of site densities of* ***a****, 100 pers. / ha;* ***b****, 200 pers. / ha;* ***c****, 300 pers. / ha;* ***d****, 400 pers. / ha. An estimation of catchment zones according to fade-out points of hollow ways is also indicated.*

Table 5 – *Cultivation surface (ha) calculated for the different sites and periods under consideration of different diets and population densities.*

Assumed population density (capita/ha)	Total site surface area in ha	Total amount of cultivable land in ha (study area minus site surfaces)	Land cultivated in ha after data from EBA Mozan	% of land under cultivation after data from EBA Mozan	Land cultivated in ha after data from EBA Brak	% of land under cultivation after data from EBA Brak	Land cultivated in ha after data from MBA Mozan	% of land under cultivation after data from MBA Mozan	Land cultivated in ha after data from MBA Brak	% of land under cultivation after data from MBA Brak
100	2 347,4	66 0762,4	73 159,3	**11,1**	70 154,2	**10,6**	72 211,3	**10,9**	70 412,8	**10,7**
200	2 347,4	66 0762,4	146 318,6	**22,1**	140 308,4	**21,2**	144 422,6	**21,9**	140 825,6	**21,3**
300	2 347,4	66 0762,4	219 477,9	**33,2**	210 462,6	**31,9**	216 633,9	**32,8**	211 238,3	**32,0**
400	2 347,4	66 0762,4	292 637,2	**44,3**	280 616,8	**42,5**	288 845,2	**43,7**	281 651,1	**42,6**
600	2 347,4	66 0762,4	438 955,9	**66,4**	420 925,2	**63,7**	433 267,8	**65,6**	422 476,7	**63,9**

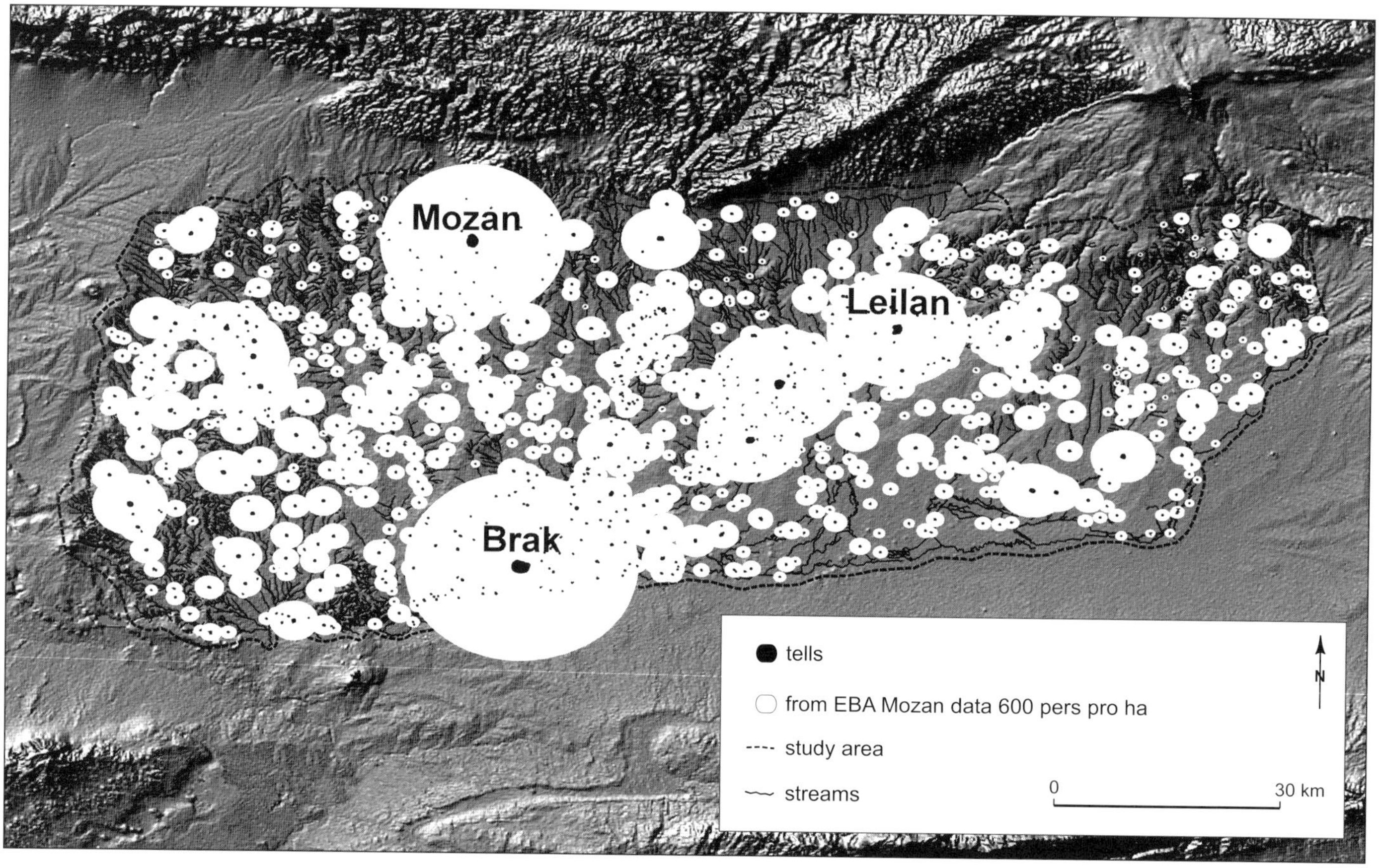

Fig. 6 – *Estimation of land necessary to support inhabitants with crops according to proportions established from archaeobotanical data at EBA Tell Mozan, based on the assumption that the people of each ha of settlement space needed to provide food for 600 persons.*

each mapped site. If however fallowing was in use during the 3rd millennium BC, for which no evidence exists,[62] the amount of agricultural land needs to be doubled and respectively less for park woodland would have remained.

62. Earliest textual evidence is dated to the early 1st millennium BC, see WILKINSON, 1997.

As mentioned above, fade-out points of hollow ways could be another approach in order to gain an insight into the area under cultivation, although this was not possible for every site. If we compare the results based on calculations for continuous cultivation with those of fade-out points of hollow ways (fig. 5), we observe similarities in many cases, especially with the 300 capita / ha values. Differences between the calculated cultivation zones and those delineated by hollow way fade-out points may also be due to the possibility that subordinate sites delivered surplus production for larger centres. Ur and Wilkinson[63] found indications of this through a detailed study of the Tell Beydar region. As for Tell Brak, however, the difference in results by the two methods may be caused by the fact that it was difficult to decide the exact limits of the site based on the CORONA images.[64]

SETTLEMENT DISRUPTIONS OF THE LATE 3RD MILLENNIUM BC

It is well-known that significant social changes took place at the end of the 3rd millennium BC in Northern Syria. During this period, some settlements reduced in size, while others contain evidence of destructions and were deserted.[65] Although the 3rd millennium BC chronology is still in the process of refinement,[66] current evidence from several North Mesopotamian archaeological sites that were excavated indicates that the settlement disruptions did not all take place at the same time. Therefore, we need to look for multiple cause and effect explanations for the observed pattern.[67] A factor that may have played a role in the observed changes at the end of the 3rd millennium BC, is that *ca* 2300 BC a Southern Mesopotamian, Akkadian intervention took place in Syria. Within the Akkadian empire, all Southern Mesopotamian city-states were united and a new imperial administration came into rule. The best evidence for Akkadian control over North-eastern Syria was found at Tell Brak where an Akkadian administrative building was found, possibly related to the collection of tribute from the region and provision of the troops. While some consider Tell Mozan to have been a powerful Akkadian client- or ally-state,[68] others interpret it as a Hurrian city, geographically separated from Tell Brak that was located in the Akkadian realm.[69] The Akkadian kings probably ceded control of the Upper Khabur area during the reign of Sar-kali-sarri (2175-2150 BC), and social turmoil in the north both preceded and followed their departure.[70] Soon afterwards, the kings of Urkiš (Tell Mozan) seized power in Northern Mesopotamia and formed the "kingdom of Urkiš and Nawar." [71]

Although the archaeological evidence of Greater Mesopotamia suggests a diachronic settlement pattern disruption, a survey in the Tell Leilan region within our study area demonstrates a rather synchronous sharp reduction in the number of settlements for the period between 2200 and 1900 BC.[72] This desertion of sites has often been correlated with droughts. The $\delta^{18}O$ isotopic record at Soreq Cave in Israel[73] shows the occurrence of a dry peak between 2150 and 2050 BC after a moist phase between 2550 and 2150 BC. A less obvious, dry phase between 2300 and 2100 BC is also visible within the Lake Van isotopic record from Eastern Turkey,[74] thus suggesting a number of considerable climate fluctuations in the wider area of investigation. Stable carbon isotope evidence from cereal finds in different Bronze Age sites reveals a generally increased water-stress on crop plants during the Middle Bronze Age.[75] Wilkinson[76] argues that the large-scale Early Bronze Age societies in Northern Syria were particularly prone to longer periods of drought and that the late 3rd millennium urban collapse may have resulted from environmental deterioration caused by urban societies. At Tell Mozan where settlement continued throughout the period between 2200 and 1900 BC, in relation to the newly established kingdom of Urkiš and Nawar, archaeobotanical evidence for climatic aridification is only weak. With continued high deciduous oak proportions and ubiquities in the charcoal assemblages, there are hardly any changes visible in the strata between 2250 and 2100 BC. The seed remains from strata dated to 2100-2000 BC indicate increased soil moisture, probably related to a short moist peak at about this time, but with no clear evidence of irrigation. Only from the Middle Bronze Age (2000-1800 BC strata) onwards, a reduction in the ubiquity and fragment percentages of deciduous oak and a shift in the crop assemblage from mainly free-threshing wheat to barley can be observed at Mozan. The reduction in oak percentages may be a result of both anthropogenic impact on the vegetation and climatic drying.[77] Recent surveys have shown that the Tell Leilan region was again densely resettled for 200 years, from 1900 BC onwards.[78] During that period the Upper Khabur region was considered to be of strategic

63. UR and WILKINSON, 2008.
64. Hamoukar is located just outside the study area.
65. AKKERMANS and SCHWARTZ, 2003.
66. See also the ARCANE project.
67. See KUZUCUOĞLU et MARRO, 2007.
68. AKKERMANS and SCHWARTZ, 2003.
69. BUCCELLATI, 1999.
70. FRAYNE, 1995.
71. FRAYNE, 1997: 457-464.
72. STAUBWASSER and WEISS, 2006; RISVET, 2005.
73. BAR-MATTHEWS and KAUFMAN, 1998.
74. WICK *et al.*, 2003.
75. RIEHL *et al.*, 2008.
76. WILKINSON, 1997.
77. DECKERS and RIEHL, 2007b.
78. STAUBWASSER and WEISS, 2006; RISVET, 2005.

importance for the kings of Mari, Ešnunna, Ekallātum, Susa, Babylon and Aleppo.[79] Eighteenth century BC texts from Leilan suggest that the countryside was a patchwork of land used for agriculture and pasture.[80]

If we want to investigate the impact of tribute on the environment, we could increase the population that needs to be fed. Figure 6 is based on the assumption that the people had to produce about double their own needs to meet the requirements of taxation, *i.e.* calculations are based on the assumption of 600 persons / ha. If this was the case, our calculations indicate that the resource limits must have been reached in many areas, leaving only up to less than 40% of the area uncultivated (table 3, fig. 6). Thus, overexploitation related to Akkadian control's demand of tribute (*ca* 2300 BC) may have played a role in the observed collapse of the Upper Khabur. The Mozan anthracological evidence however,[81] with continued high percentages of deciduous oak during this period, suggests that it is rather unlikely that such a tribute situation with intensified agriculture could have been reached there under the Akkadians. As mentioned above, Mozan may have been more the exception than the rule at that time due to the powerful position of Urkiš, which is even more visible shortly afterwards when it seized power over Northern Mesopotamia.

CONCLUSION

Our satellite GIS-based settlement study indicates that the settlement pattern of the Upper Khabur is strongly oriented towards wadis. The typical site form of tells in this area for the 3rd millennium BC may be related to its location along streams "living higher means drier." There may be a causal link between the reduced use of tell settlements after 2000 BC, and the modeled, coinciding increased flooding.

The radial lines are unlikely to be irrigation features, though a lot of them could be hollow ways that indicate the interface between the arable and pasture zone. This has been supported through a comparison of site exploitation catchments based on site size, calorific needs and archaeobotanical evidence assuming normal population densities, with site catchments based on hollow ways.

Resource exploitation calculations indicate, contrary to previous studies, that the Upper Khabur probably was not overexploited during the mid 3rd millennium BC. This also has been suggested by recent botanical studies, which indicate the presence of open oak park woodland. Land overexploitation related to a large tribute may have played a role in the observed settlement discontinuation *ca* 2200 BC.

ACKNOWLEDGEMENTS

Many thanks are due to the "Landesstiftung Baden-Württemberg", the Belgian Vocational Society (Belgische Stichting Roeping) and the Deutsche Forschungsgemeinschaft for supporting this study. G. Wang spent a lot of work digitizing satellite images, for which we are sincerely grateful. Many thanks are due to the reviewers for helping to improve the quality of the manuscript.

***Katleen* DECKERS**
University of Tübingen
Zentrum für Naturwissenschaftliche Archäologie
Rümelinstraße 23, D-72070 Tübingen
GERMANY
katleen.deckers@uni-tuebingen.de

***Simone* RIEHL**
University of Tübingen
Zentrum für Naturwissenschaftliche Archäologie
Rümelinstraße 23, D-72070 Tübingen
GERMANY
simone.riehl@uni-tuebingen.de

79. RISVET, 2005: 116.
80. *Ibid.*: 132.
81. DECKERS and RIEHL, 2007b.

BIBLIOGRAPHY

ADAMS R. MCC.

1981 *Heartland of Cities: Surveys of Ancient Settlement and Land Use on the Central Floodplain of the Euphrates*. Chicago: Chicago University Press.

AKKERMANS P.M.M.G. and SCHWARTZ G.

2003 *The Archaeology of Syria. From complex hunter-gatherers to early urban societies (c. 16000-300 BC)*. Cambridge: Cambridge University Press.

BAR-MATTHEWS M. and KAUFMAN A.

1998 Middle to Late Holocene (6,500 yr. Period) palaeoclimate in the Eastern Mediterranean region from stable isotopic composition of speleothems from Soreq Cave, Israel. *In*: ISSAR A.S. and BROWN N. (eds.), *Water, environment and society in times of climatic change*: 203-214. Dordrecht: Kluwer Academic Publishers.

BUCCELLATI G.

1999 Urkesh and the question of Early Hurrian Urbanism. *In*: HUDSON M. and BARUCH A. (eds.), *Urbanization and land ownership in the ancient Near East*. Cambridge: Harvard University (*Peabody Museum Bulletin* 7).

COLE S.W.

1994 Marsh formation in the Borsippa region and the course of the Lower Euphrates. *Journal of Near Eastern Studies* 53,2: 81-109.

DECKERS K.

2005 Anthracological research at the archaeological site of Emar on the Middle Euphrates, Syria. *Paléorient* 31,2: 153-168.

DECKERS K. and RIEHL S.

2007a Fluvial environmental contexts for archaeological sites in the Upper Khabur Basin (Northeastern Syria). *Quaternary Research* 67,3: 337-348.

2007b An evaluation of botanical assemblages from the 3rd to 2nd Millennium BC in Northern Syria. *In*: KUZUCUOĞLU C. et MARRO C. (éd.), *Sociétés humaines et changement climatique à la fin du troisième millénaire: une crise a-t-elle eu lieu en Haute Mésopotamie ? Actes du colloque de Lyon, 5-8 décembre 2005*: 481-502. Istanbul-Paris: Institut Français d'Études Anatoliennes-Georges Dumezil-De Boccard (*Varia Anatolica* 19).

DONOGHUE D.N.M., PHILIP G., GALIATSATOS N. and BECK A.

2002 Satellite imagery for archaeological applications: A case study from the Orontes Valley, Syria. *In*: BEWLEY R.H. and RACZOWSKI W. (eds.), *Aerial archaeology: Developing future practice*: 211-223. Amsterdam: IOS.

ERGENZINGER P.

1991 Geomorphologische Untersuchungen im Unterlauf des Khabur. *In*: KÜHNE H. (Hrsg.), *Die rezente Umwelt von Tall Seh Hamad und Daten zur Umweltrekonstruktion der Assyrischen Stadt Dur-Katlimu*: 25-50. Berlin: Reimer.

ESIN U.

1999 Archaeological evidence from the Aşıklı excavations for climatic fluctuations in Central Anatolia during the Early Holocene 10/9. Mill. B.P. *Anadolu Araştirmadiri* 15: 1-26.

FRAYNE D.R.

1995 *Sargonic and Gutian periods (2234-2113 BC)*. Toronto: University of Toronto Press (*Royal inscriptions of Mesopotamia* 2).

1997 *Ur III Period (2112-2004 BC)*. Toronto: University of Toronto Press (*Royal inscriptions of Mesopotamia* 3/2).

GOLDHAUSEN M. and RICCI A.

2005 Political centralisation in the Syrian Jezira during the 3rd millennium: A case study in settlement hierarchy. *Altorientalische Forschungen* 32,1: 132-157.

HOLE F.

1997 Evidence for Mid-Holocene environmental change in the Western Khabur drainage, Northeastern Syria. *In*: DALFES H.N., KUKLA G. and WEISS H. (eds.), *Third millennium BC climate change and Old World collapse*: 39-66. Heidelberg: Springer (*NATO ASI Series I, Global Environmental Change* 49).

HRITZ C.

2004 The hidden landscape of Southern Mesopotamia. *Akkadica* 125,1: 93-106.

HRITZ C. and WILKINSON T.J.

2006 Using Shuttle Radar Topography to map ancient water channels. *Antiquity* 80: 415-424.

KENNEDY D.

1998 Declassified satellite photographs and archaeology in the Midle East: Case studies from Turkey. *Antiquity* 72: 553-561.

KERBE J.

1987 *Climat, hydrologie et aménagements hydro-agricoles de Syrie*. Bordeaux: Université de Bordeaux I. Thèse non publiée.

KOLARS J.F. and MITCHELL W.A.

1991 *The Euphrates River and the Southeast Anatolia Development Project*. Illinois: Southern Illinois University Press.

KOUCHOUKOS N.

2001 Satellite images and Near Eastern landscapes. *Near Eastern Archaeology* 64: 1-2, 80-91.

KRAMER C.

1980 Estimating prehistoric populations: an ethnoarchaeological approach. *In*: BARRELET M.-T. (éd.), *L'archéologie de l'Iraq: Perspectives et limites de l'interprétation anthropologique des documents*: 315-334. Paris: Éditions du CNRS.

KUZUCUOĞLU C. et MARRO C. (éd.)

2007 *Sociétés humaines et changement climatique à la fin du troisième millénaire: une crise a-t-elle eu lieu en Haute Mésopotamie ? Actes du colloque de Lyon, 5-8 décembre 2005*. Istanbul-Paris: Institut Français d'Études Anatoliennes-Georges Dumezil-De Boccard (*Varia Anatolica* 19).

LAFONT B.

2000 Irrigation agriculture at Mari. *In*: JAS R.M. (ed.), *Rainfall and agriculture in northern Mesopotamia*: 129-146. Leiden: Nederlands Instituut voor het Nabije Oosten (*MOS Studies*).

LEWY B.

1944 Assyro-Babylonian and Israelite measures of capacity and rates of seeding. *Journal of the American Oriental Society* 64: 65-73.

LYONNET B.

1996 Settlement pattern in the Upper Khabur (NE Syria) from the Achaemenids to the Abbasid Period: Methods and preliminary results from a survey. *In*: BARTL K. and HAUSER S.R. (eds.), *Continuity and change in Northern Mesopotamia from the Hellenistic to the Early Islamic Period*: 349-354. Berlin: Dietrich Reimer Verlag (*Berliner Beiträge zum Vorderen Orient* 17).

1997 Questions sur l´origine des porteurs de pots en Haute-Mésopotamie, du VIe au milieu du IIe millénaire. *In*: BARRELET M.T. et CHARPIN D. (éd.), *Recueil d'études à la mémoire de Marie-Thérèse Barrelet*: 133-144. Paris: SEPOA (*Florilegium marianum* 3).

1998 La prospection archéologique de la partie occidentale du Haut-Khabur (Syrie du Nord-Est): méthodes, résultats et questions autour de l´occupation aux IIIe et IIe millénaires av. n.e. *In*: DURAND J.M. (éd.), *Mari et les hourites*: 363-376. Paris: ERC (*Amurru* 1).

LYONNET B. (éd.)

2000 *Prospection archéologique du Haut-Khabur Occidental (Syrie du N.E.) Vol. 1*. Beyrouth: Institut Français d'Archéologie du Proche-Orient.

MALLOWAN E.L.

1947 Excavations at Brak and Chagar Bazar. *Iraq* 9: 1-266.

MARSH B.

1999 Alluvial burial of Gordion, an Iron-Age city in Anatolia. *Journal of Field Archaeology* 26,2: 163-175.

MARTIN H. P. and ERNST H.

1988 *A reconstruction of the ancient Mesopotamian city of Shuruppak*. Birmingham: Chris Martin.

MAZID A., TUTWILER R.N., AL-AHMED H. and MAYA F.

2003 *Impact of modern agricultural technologies on Durum wheat production in Syria*. Aleppo: ICARDA (*Integrated natural resource management research report series* 3).

MCCLELLAN T.L.

2000 Bronze age water harvesting in North Syria. *In*: ROUAULT O. et WÄFLER M. (éd.), *La Djéziré et l´Euphrate syriens de la protohistoire à la fin du IIe millénaire av. J.-C. Tendances dans l´interprétation historique des données nouvelles*: 137-155. Turnhout: Brepols (*Subartu* 7).

MEIJER D.J.W.

1986 *A survey in Northeastern Syria*. Istanbul: Nederlands Historisch-Archeologisch Instituut.

MENZE B.H., UR J.A. and SHERRATT A.G.

2005 Tell spotting – surveying Near Eastern settlement mounds from space. *In*: S. DEQUAL (ed.), *Proceedings of the 20th CIPA International Symposium (Paris, 26 September – 01 October 2005)*: 458-462. Torino (http://cipa.icomos.org/fileadmin/papers/Torino2005/458.pdf).

MOORE A.M.T., HILLMAN G.C. and LEGGE A.J.

2000 *Village on the Euphrates. From foraging to farming at Abu Hureyra*. Oxford: Oxford University Press.

PAEPE R. and BAETEMAN C.

1978 Fluvial system between Tell ed-Der and Tell Abu Habbah. *In*: DE MEYER L. (ed.), *Tell ed-Der II*: 37-56. Leuwen: Peeters.

PELTENBURG E.

1999 Tell Jerablus Tahtani 1992-1996: A summary. *In*: DEL OLMO LETE M. and. MONTERO FENOLLOS J.-L (eds.), *Archaeology of the Upper Syrian Euphrates. The Tishreen Dam Area. Proceedings of the International Symposium held at Barcelona, Jan. 28th-30th 1998*: 97-105. Barcelona: Ausa (*Aula Orientalis Supplementa* 15).

PFÄLZNER P.

2001 *Haus und Haushalt. Wohnformen des dritten Jahrtausends vor Christus in Nordmesopotamien*. Mainz am Rhein: Verlag Philipp von Zabern (*Damaszener Forschungen* 9).

PHILIP G., DONOGHUE D., BECK A. and GALIATSATOS N.

2002 CORONA satellite photography: an archaeological application for the Middle East. *Antiquity* 76,291: 109-118.

POURNELLE J.R.

2003 *Marshland of cities: Deltaic landscapes and the evolution of early Mesopotamian civilization*. San Diego: University of California. Unpublished PhD.

POWELL M.

1985 Salt, seed and yields in Sumerian agriculture. A critique of the theory of progressive salinisation. *Zeitschrift für Assyriologie und Vorderasiatische Archäologie* 75,1: 7-38.

RIEHL S. and KÜMMEL C.

2005 *Archaeobotanical database of Eastern Mediterranean and Near Eastern Sites* 2005. (http://www.cuminum.de/archaeobotany/).

RIEHL S., BRYSON R. A. and PUSTOVOYTOV K.

2008 Changing growing conditions for crops during the Near Eastern Bronze Age (3000-1200 BC): The stable carbon isotope evidence. *Journal of Archaeological Science* 35,4: 1011-1022.

RISVET L.

2005 *Settlement, economy, and society in the Tell Leilan region, Syria, 3000-1000 BC*. Cambridge: King's College. Unpublished PhD.

SEVIN V. and DERIN Z.

1987 Imikuşaği Kazıları, 1985, VIII. *Kazı Sonuçları Toplantısı* 1: 181-204.

SEVIN V. and KÖROĞLU K.

1986 Imikuşaği Kazıları, 1984, VII. *Kazı Sonuçları Toplantısı* 5: 163-179.

STAUBWASSER M. and WEISS H.

2006 Introduction: Holocene climate and cultural evolution in Late Prehistoric-Early Historic West Asia. *Quaternary Research* 66: 372-387.

STEIN G. and WATTENMAKER P.

1990 The 1987 Tell Leilan Regional Survey: Preliminary report. *In*: MILLER N. (ed.), *Economy and Settlement in the Near East. Analyses of Ancient Sites and Materials*: 5-18. Philadelphia: University of Pennsylvania, University Museum (*MASCA Research Papers in Science and Archaeology Supplement* 7).

UR J.

2003 CORONA satellite photography and ancient road networks: A northern Mesopotamian case study. *Antiquity* 77: 102-115.

2004 *Urbanism and Society in the Third Millennium Upper Khabur Basin*. Chicago: University of Chicago. Unpublished PhD.

2005 Sennacherib's Northern Assyrian canals: New insights from satellite imagery and aerial photography. *Iraq* 67: 317-245.

UR J. and WILKINSON T.J.

2008 Settlement and economic landscapes of Tell Beydar and its hinterland. *Subartu* 21: 305-327.

VAN ANDEL T.H., GALLIS K. and TOUFEXIS G.

1995 Early Neolithic farming in a Thessalian river landscape, Greece. *In:* LEWIN J., MACKLIN M.G. and WOODWARD J. (eds.), *Mediterranean Quaternary river environments*: 131-143. Rotterdam: A.A. Balkema.

VAN KOPPEN F.

2001 The organisation of institutional agriculture in Mari. *Journal of the Economic and Social History of the Orient* 44,4: 451-504.

VAN LIERE W.J. et LAUFFRAY J.

1954 Nouvelle prospection archéologique dans la Haute Jezireh syrienne. *Les Annales Archéologiques de Syrie* 4-5 : 129-148.

VERHOEVEN K. and DALES L.

1994 Remote sensing and geographical information systems (GIS) for archaeological research applied in Mesopotamia. *In*: GASCHE H., TANRET M., JANSSEN C. et DEGRAEVE A. (éd.), *Cinquante-deux réflexions sur le Proche-Orient ancien*: 519-539. Ghent: Peeters.

WICK L., LEMCKE G. and STURM M.

2003 Evidence of Late glacial and Holocene climatic change and human impact in Eastern Anatolia: High-resolution pollen, charcoal, isotopic and geochemical records from the laminated sediments of Lake Van, Turkey. *The Holocene* 13,5: 665-675.

WIGGERMANN F.A.M.

2000 Agriculture in the northern Balikh valley: the case of Middle Assyrian Tell Sabi Abyad. *In*: JAS R.M. (ed.), *Rainfall and agriculture in northern Mesopotamia*: 171-232. Leiden-Istanbul: Nederlands Instituut voor het Nabije Oosten (*MOS Studies* 3).

WILKINSON T.J.

1990 The development of settlement in the North Jazira. *Iraq* 52: 49-62.

1993 Linear hollows in the Jazira, Upper Mesopotamia. *Antiquity* 67: 548-562.

1994 The structure and dynamics of dry-farming states in Upper Mesopotamia. *Current Anthropology* 35,5: 483-505.

1997 Environmental fluctuations, agricultural production and collapse: A view from Bronze Age Mesopotamia. *In*: DALFES H.N., KUKLA G. and WEISS H. (eds.), *Third Millennium BC climate change and Old World collapse*: 67-106. Berlin / Heidelberg: Springer (*NATO ASI Series* 149).

2000a Settlement and land use in the zone of uncertainty in Upper Mesopotamia. *In*: JAS R.M. (ed.), *Rainfall and Agriculture in Northern Mesopotamia. Proceedings of the third Mos Symposium*: 3-35. Leiden-Istanbul: Nederlands Instituut voor het Nabije Oosten (*MOS Studies* 3).

2000b Archaeological survey of the Tell Beydar region, Syria, 1997: A Preliminary Report. *In*: VAN LERBERGHE K. and VOET G. (eds.), *Tell Beydar: Environmental and Technical Studies*: 1-37. Turnhout: Brepols (*Subartu* 6).

2002a Physical and cultural landscapes of the Hamoukar area. *Akkadica* 123: 89-103.

2002b The settlement transition of the second millennium BC in the Western Khabur. *In*: WERR L.A.-G., CURTIS J., MARTIN H., MCMAHON A., OATES J. and READE J. (eds.), *Of Pots and Plans. Paper on the archaeology of Mesopotamia and Syria presented to David Oates in Honour of his 75th Birthday*: 261-372. London: Nabu Publications.

2003 *Archaeological landscapes of the Near East*. Tucson: University of Arizona Press.

WILKINSON T.J. and BARBANES E.

2000 Settlement patterns in the Syrian Jazira during the Iron Age. *In*: BUNNENS G. (ed.), *Essays on Syria in the Iron Age*: 397-422. Leuwen: Peeters (*Ancient Near Eastern Studies Supplement* 7).

WILKINSON TJ. and TUCKER D.J.

1995 *Settlement development in the North Jazira, Iraq. A study of the archaeological landscape*. Warminster: Aris and Philips.

WILKINSON T.J, CHRISTIANSEN J.H., UR J., WIDELL M. and ALTAWEEL M.

2007 Urbanization within a dynamic environment: Modeling Bronze Age communities in Upper Mesopotamia. *American Anthropologist* 109,1: 52-68.

WIRTH E.

1971 *Syrien: eine geographische Landeskunde*. Darmstadt: Wissenschaftliche Buchgesellschaft (*Wissenschaftliche Landeskunden* 4/5).

WOOLLEY C.L.

1955 *The early periods: A report on the sites and objects prior in date to the 3. Dynasty of Ur discovered in the course of the excavations*. Philadelphia: British Museum London and Museum of the University of the University of Pennsylvania.

YAKAR J.

1985 *The later prehistory of Anatolia: The Chalcolithic and Early Bronze* Age. Oxford (*BAR Int. Ser.* 268).

ZACCAGNINI C.

1975 The yields of the fields at Nuzi. *Oriens Antiquus* 14: 181-225.

1990 Again on the yields of the fields at Nuzi. *Bulletin of Sumerian Agriculture* 5: 201-217.

ZHANG X, AGUILAR E., SENSOY S., MELKONYAN H., TAGIYEVA U., AHMED N., KUTALADZE N., RAHIMZADEH F., TAGHIPOUR A., HANTOSH T.H., ALBERT P., SEMAW.I M., ALI M.K., AL-SHABIBI M.S., AL-OULAN Z., ZATARI T., KHELET I.A.D., HAMOUD S., SAGIR R., DEMIRCAN M., EKEN M., ADIGUZEL M., ALEXANDER L., PETERSON T.C., and WALLIS T.

2005 Trends in Middle East climate extreme indices from 1950 to 2003. *Journal of Geophysical Research* 110,22: D22104.1-D22104.12.

RECENSIONS

CHAZAN M. and HORWITZ L.K. (eds). 2007. Holon. A Lower Paleolithic Site in Israel. *Cambridge (Mass.): Harvard University, Peabody Museum of Archaeology and Ethnology* **(American School of Prehistoric Research Bulletin** *50). Par Christophe GRIGGO.*[1]

Cette monographie sur le site paléolithique inférieur d'Holon (Israël) est dédiée à la mémoire de Tamar Yizraeli Noy, conservatrice au département de Préhistoire du Musée d'Israël.

Le site d'Holon, datant du Paléolithique inférieur et découvert en 1951 par Y. Kaplan, archéologue au Département des Antiquités d'Israël, est un site de plein air, situé dans le sud de la zone urbaine de Tel Aviv à proximité de la rivière Ayalon. La construction d'une usine métallurgique et son agrandissement sont à l'origine de fouilles de sauvetage, d'abord en 1963 et en 1964 puis en 1970, l'ensemble étant mené sous la direction de T. Noy. Ces fouilles, réalisées sur une surface de plus de 270 m^2, ont permis de recueillir une industrie riche en bifaces, outils sur éclats et galets taillés ainsi que de nombreux vestiges osseux de proboscidiens, de bovidés, de cervidés et de carnivores. Cependant, l'ensemble de ces artéfacts n'avaient fait l'objet que d'études préliminaires, publiées seulement dans des rapports de fouilles.

Ce n'est qu'en 1994 que M. Chazan et L.K. Horwitz demandèrent à T. Noy l'autorisation de reprendre l'étude exhaustive du site d'Holon. Pour mener à bien un tel projet, plus de 30 ans après les fouilles, ils rassemblèrent une équipe pluridisciplinaire. Le résultat final correspond à la publication de cette monographie qui présente des analyses détaillées des vestiges lithiques et osseux mais aussi du contexte géoarchéologique, afin de reconstituer les activités des Hommes fossiles et les divers processus mis en place lors de la formation du site d'Holon.

Le chapitre 1 est consacré à la présentation de l'historique des fouilles d'Holon. On prend ainsi conscience du rôle de T. Noy. Sans son intervention rapide, ce site du Paléolithique inférieur aurait pu être détruit en quelques heures. Elle a su mener à bien cette opération de sauvetage, constituant ainsi une première pour l'archéologie préhistorique d'Israël. De nombreuses photographies d'ambiances de fouilles et surtout de coupes stratigraphiques illustrent ce chapitre, permettant ainsi de se faire une bonne idée du contexte. Cependant, la qualité médiocre de reproduction de la plupart d'entre elles les rend peu lisibles.

Dans le chapitre 2, M. Netser présente la séquence quaternaire de la plaine côtière du sud de l'Israël dans laquelle se trouve le site d'Holon. Les cartes et schémas, cette fois-ci de bonne qualité, permettent de bien comprendre la succession des différents événements, surtout d'origine fluviatile, qui ont abouti à la formation du site.

Les résultats des datations par TL et ERS sont présentés et commentés par N. Porat dans le chapitre 3. Il propose un âge de 205 ka, ce qui place le site d'Holon dans le stade isotopique 7 vers la fin du Paléolithique inférieur. Par conséquent, il est contemporain de Tabun E et de Yabroud 1 – couche 18/19.

L'industrie lithique est présentée par M. Chazan dans les chapitres 4 et 5, qui correspondent respectivement aux études typologique et technologique. Au total 1468 artéfacts lithiques sont analysés. Ils comprennent 100 bifaces, 39 choppers, 1179 éclats (387 sont retouchés et 147 correspondent à des Nahr Ibrahim) et 150 nucléus. Pour obtenir ces différents décomptes, il m'a fallu compiler des données réparties dans plusieurs tableaux du chapitre 5. Je regrette donc que ces décomptes ne soient pas clairement présentés dès le début du chapitre 4 afin d'avoir rapidement une idée de la composition lithique. Nous regrettons également la pauvreté des dessins représentant cette industrie, aussi bien en terme de qualité que de quantité. Les tracés sont épais et les ombrages peu précis. Seuls 7 bifaces, 4 choppers, 10 éclats retouchés et 15 Nahr Ibrahim sont représentés ce qui est bien trop peu. En outre, l'analyse des bifaces est essentiellement morphométrique et aucune observation n'est faite en ce qui concerne une analyse techno-morpho-fonctionnelle. Cette dernière approche aurait permis à M. Chazan d'aller bien plus loin dans la caractérisation

1. Christophe GRIGGO – Université Joseph Fourier Grenoble I, UMR 6636 LAMPEA, Institut Dolomieu, 15 rue Maurice Gignoux, 38031 Grenoble cedex – FRANCE – Christophe.Griggo@ujf-grenoble.fr

chrono-culturelle de l'industrie d'Holon. En effet, parvenir à définir les modes d'utilisation ou les types de gestes associés à des bifaces apportent bien plus d'informations sur les capacités techniques et cognitives des hommes qui les ont fabriqués que lorsque l'on se limite à donner leurs formes, leurs dimensions et leurs poids moyens.

Quoi qu'il en soit, M. Chazan attribue l'industrie d'Holon, riche en grattoirs et en bifaces surtout cordiformes, aux industries du Paléolithique inférieur récent du Levant se rapprochant de l'Acheuléo-Yabroudien.

Les analyses fauniques d'abord paléontologiques puis taphonomiques ont, quant à elles, été effectuées par plusieurs spécialistes. Elles se caractérisent toutes par une présentation un peu « scolaire » mais qui a l'avantage d'être complète et rigoureuse. En effet, pour chacun des taxons ou facteurs taphonomiques, les auteurs font un bilan des connaissances pour la région du Levant, puis ils décrivent le matériel avant de proposer leurs interprétations.

Dans le chapitre 6, H. Monchot et L.K. Horwitz donnent une présentation générale de l'assemblage faunique du site. Parmi les 1569 restes recueillis, 573 sont déterminés d'un point de vue taxonomique. Le daim de Mésopotamie est l'espèce qui a livré le plus grand nombre de vestiges. Viennent ensuite l'aurochs, l'éléphant antique et l'hippopotame. Seuls quelques restes indiquent la présence de la gazelle, du cerf, du sanglier ainsi que d'une tortue d'eau douce. Pour chaque taxon, les classes d'âge et le sexe ratio sont analysés afin de donner un nombre minimum d'individus (NMI) précis. Cependant au cours de cet exercice, les auteurs ont commis une erreur : contrairement à ce qu'ils semblent penser, pour le daim, il est tout à fait possible de préciser le sexe des animaux à partir des bois, car seuls les mâles en portent : « because these (the antlers) were fragments, it was impossible to determine the sex of the animals » (p. 87).

Les chapitres suivants concernent les études paléontologiques de chacun des groupes zoologiques :

- la tortue d'eau douce (*Mauremys caspica*) dans le chapitre 7 par G. Hartman et L.K. Horwitz,
- les artiodactyles, autres que les cervidés, dans le chapitre 8 par L.K. Horwitz et H. Monchot,
- les cervidés dans le chapitre 9 par A.M. Lister,
- les proboscidiens dans le chapitre 10 par P. David et A.M. Lister.

Pour chacune de ces études, les textes sont complétés par un grand nombre de tableaux et de figures dont de nombreuses photos. Après avoir argumenté les déterminations paléontologiques, les auteurs s'intéressent à la position biostratigraphique d'Holon et à son paléoenvironnement. Ils confirment ainsi l'attribution au stade isotopique 7, alors que le climat était chaud et sec et que l'environnement était constitué d'une mosaïque de zones herbeuses et de zones buissonnantes avec un point d'eau à proximité.

Dans le chapitre 11, H. Monchot et L.K. Horwitz réalisent une analyse taphonomique exhaustive afin de préciser le ou les agent(s) responsable(s) de l'accumulation osseuse et de reconstituer les différents événements susceptibles de l'avoir modifiée. Les étudiants en archéozoologie trouveront là une synthèse relativement complète des principaux outils développés dans cette discipline et pourront s'inspirer de la rigueur avec laquelle ils sont utilisés sur le site d'Holon. Les auteurs s'intéressent à l'homogénéité de l'assemblage, aux états de conservations et à la fragmentation des os, aux représentations squelettiques en fonction de la localisation anatomique ou de leur potentiel nutritif, à la densité des os en relation avec le taux de survie ou avec un transport fluviatile et aux modifications osseuses liées aux activités des carnivores ou des Hommes. En ce qui concerne ce dernier point, ils réfutent l'existence de trois outils sur os qui avaient été retenus, dans un premier temps par T. Noy. H. Monchot et L.K. Horwitz concluent que les Hommes et plus généralement les carnivores, par leurs activités de prédation et/ou de charognage, ont joué un rôle important dans la constitution de l'accumulation osseuse dans le site d'Holon.

Dans le chapitre 12, M. Chazan, H. Monchot et L.K. Horwitz se sont associés pour faire une étude de la répartition spatiale des artéfacts lithiques et des restes fauniques afin de déterminer le mode d'occupation à Holon. Il s'agit là d'une étude purement statistique. Il faudra donc se contenter des deux plans généraux présentés dans le chapitre 1 pour d'autres informations. Il en découle que le site, qui ne semble constitué que d'un seul niveau archéologique, correspond à un palimpseste qui s'est formé sur les marges de la rivière Ayalon. Holon ne constituait pas un camp de base mais offrait, pour les Hommes du Paléolithique inférieur, un environnement idéal pour la chasse ou la consommation de charogne.

En guise de conclusion dans le chapitre 13, L.K. Horwitz et M. Chazan proposent une synthèse des principaux résultats obtenus et élargissent la discussion en les comparant aux autres sites levantins du Paléolithique inférieur récent.

Cette monographie va contribuer très rapidement à placer le site d'Holon parmi les sites de référence pour le Paléolithique inférieur récent du Levant et cela pour plusieurs raisons. Tout d'abord, Holon est un site de plein air et il vient ainsi compléter nos connaissances le plus souvent obtenues dans des sites en grotte ou sous abri ; ensuite, les analyses sont exhaustives, claires et bien argumentées. Nous pouvons regretter que l'analyse du matériel lithique se soit surtout focalisée sur des approches typologiques et morphométriques. En revanche, l'étude de la faune est un modèle du genre.

Le fait le plus important qui est mis indirectement en avant dans cette monographie est la contribution des fouilles de sauvetage dans la compréhension des modes de vie et des organisations sociales des Hommes de la préhistoire. En effet, lorsque celles-ci sont menées avec la rigueur de T. Noy, qui a pris soin d'enregistrer le maximum d'informations, il est possible, même après quelques dizaines d'années, d'obtenir des résultats quantitatifs et qualitatifs qui peuvent rivaliser avec ceux provenant de fouilles programmées.

***BARKER G., GILBERTSON D. and MATTINGLY D. (eds.). 2007.* Archaeology and Desertification. The Wadi Faynan Landscape Survey, Southern Jordan. *Oxford: Oxbow Books (*Wadi Faynan Series Volume *2*, Levant Supplementary Series *6) (*Supplementary CD*). By Alison V.G. BETTS.*[1]**

This is a massive volume of very timely published research based on fieldwork undertaken in the late 1990s. The volume represents an extremely well structured research report that brings together archaeologists, environmental scientists and historians in what can only be described as a landmark study. Unfortunately, given the resources and personnel required for such an undertaking, it is unlikely that many other teams will be able to repeat the exercise elsewhere in the Near East, but it clearly demonstrates the value of such an approach, particularly as a broad background to more detailed archaeological work.

The stated aim of the study was to contribute to present day desertification debates by providing a long-term perspective on the relationship between environmental change and human history. The area selected for study, the arid lands of southern Jordan, was known to be a place where the developmental history had been characterized by patterns of changing land use and exploitation in a marginal environment. A study with such a strong emphasis on environmental science in relation to patterns of land use and settlement is fairly unusual in the Middle East. Perhaps the best parallel is the seminal study of the Negev by Evenari *et al.*[2], which has provided important reference data for many archaeologists trying to understand and interpret ancient subsistence strategies in the southern Levant. The study area of Wadi Faynan is unusual in that the presence of copper ores was a particular magnet for human activity despite the increasingly difficult nature of the environment. This was in part due to climatic deterioration, but was also the result of toxic pollutants created through the mining process itself. As it happens, in the end the research was only able to offer a broad brush contribution to desertification debates, but it has provided a very significant long-term study of economic and social changes set against natural and man-made environmental deterioration that is of much value to scholars seeking to interpret the developmental history of the southern Levant.

The form that the volume takes is a substantial methodological introduction followed by a chronological discussion of the changing patterns of human occupation in the wadi set against environmental data. The book comes with appendices in a supplementary CD containing GIS files, a gazetteer of sites illustrated in colour and individual catalogues of lithics, ceramics, bones and other finds. The studies presented in the volume are detailed and, importantly, reflective. There is discussion of the development of methodologies as the team faced practical constraints, and as one group of specialists identified questions that required further study by another. For the settlement history of the wadi, the authors were also able to draw on more detailed studies of others from excavated sites for most key periods from the Early Neolithic to the Roman / Byzantine era. While the survey extensively recorded the distribution of sites throughout all periods, this volume should not be read as an archaeological report as this was never the intention of the authors. Data collection was inclusive but not exhaustive. Nevertheless, the resulting record is in itself a valuable tool for archaeologists in that it provides detail on almost the full range of sites of any given period. This includes the smaller hamlets, activity sites, fields, canals and sherd scatters that are often neglected in site based studies. The scope of the volume means that the archaeological evidence is sometimes presented in relatively general terms but each chronological chapter contains a well researched and useful summary of the broader Near Eastern historical background. These background

1. Alison V.G. BETTS – Archaeology A14, University of Sydney, Dept. of Archaeology, NSW 2006, Sydney – AUSTRALIA – alison.betts@usyd.edu.au

2. EVENARI *et al.*, 1971.

introductions provide a big picture account that does not and, given the scope of the work, cannot address the nuances of debate in each period. This is understandable in terms of broad historical outline but where environmental variability has been suggested as an instrument of change in land use patterns, more detailed discussion might be useful. An example of this is the summary treatment of hypotheses concerning the role of the environment in the PPNB / PPNC transition. There has been much discussion regarding this in published literature and conflicting environmental and archaeological evidence has been presented. In a volume that specifically addresses climatic variability and its impact on human land use, a more detailed treatment of the evidence might have been appropriate. The authors note that small-scale changes in volume or seasonality of precipitation would not register in the coarse-grained level of data available to them, but given the significant presence of Neolithic sites in the Wadi Faynan, the topic should merit some debate. Similarly, in discussing the abandonment of walled sites by EBIV, Kenyon's long dismissed Amorite hypothesis is mentioned, yet evidence for a "major drought event" is referred to but not further incorporated in analysis of the evidence from Wadi Faynan. For scholars trying better to understand the complex and fragmentary environmental evidence for the ancient Levant, this may have been helpful.

The chronological range of the work is from the Paleolithic to the Ottoman period and includes a section on ethnoarchaeology. This is not only valuable in itself but also helps to inform interpretation of archaeological data from earlier periods. A particularly important aspect of the work overall is its focus on the environmental impact of mining. The geology of the area is complex, with a great variety of different rock outcroppings, all heavily displaced by the faulting and uplifting associated with the formation of the Rift Valley. Copper ores occur in quantities suitable for exploitation, but a variety of other minerals are also present including arsenic, lead and an array of "toxic, noxious and poisonous elements". The diseases associated with mining copper and lead, and ancillary exposure to the other toxic minerals include a wide array of debilitating and potentially lethal complaints. Wadi Faynan was a copper mining area from as early as the Chalcolithic but the pollution from mineral extraction activities reached its peak in the Roman / Byzantine era. Soils were poisoned with high levels of copper, lead and other rarer metals, and these minerals would have been entering the food chain. Mortuary evidence from Roman / Byzantine times indicates high mortality levels, while the potential ill-effects of long term environmental contamination are still a serious issue to day. Mining activity also had other damaging environmental impacts. Pollen analysis shows that although the area was still well wooded in the Iron Age, by Roman times wood fuel for the furnaces was being brought in from the plateau at the top of the Rift valley, presumably because the local supply was largely exhausted. The removal of tree cover left the landscape at significantly greater risk of erosion and soil depletion. Mining came to an end at some time in the later Byzantine period although it is not known whether this was due to economic considerations or possibly as a result of the Arab conquest.

An interesting aspect of the work, and one that would be relevant to others aiming to undertake similar approaches is how far the research program was, in the end, able to address the stated questions of desertification, environmental change and human history. The work was able to show variations in patterns of land use and settlement through time, and provide a local model for environmental change. It was perhaps less successful in making clear and solid links between the one and the other. In the prehistoric periods environmental conditions may have been a major factor in the nature of human settlement, but economic and political considerations became increasingly significant in later times. In addition, the presence of minerals in commercial quantities in the Wadi Faynan masked the realities of the marginal subsistence potential of the region. In summing up their evidence, and following a useful discussion of the theoretical background, the authors admit that the questions they address are too complex to be solved within the framework of the fieldwork they were able to undertake. They admit that "many of the interpretations... must remain qualitative initial estimates". They also state that:

> "the evidence is... not capable of sustaining detailed discussion about the significance of past climates, and climate change, for the human populations..., nor about the importance of rapid habitat change for human settlement, nor about the magnitudes of impacts of human activities upon ecosystems."

What then has this book achieved? The value of this important work lies in its scope and structure. The authors' own assessment of their work is that the project established a localized environmental stratigraphy, the start of an OSL based chronostratigraphy and demonstrated conclusively that the study area experienced profound climatic, geomorphic, and ecological changes that brought about dramatic landscape changes. Apart from the data on the toxic effects of mining activity, the overall sequence of environmental change and related patterns of settlement in Wadi Faynan broadly matches that of the surrounding wider region. The data presented in this volume can act as a valuable model for interpretation of other sub-regions of the southern Levant.

In short the volume is an extremely well organized and clearly presented multi-disciplinary regional study that is of value for the individual findings of the various scientific teams and collectively as a model for long-term change and development in the region. It is a fine example of a well planned and integrated research program.

BIBLIOGRAPHY

EVENARI M., SHANA L. and TADMOR N.
1971 *The Negev. The Challenge of a Desert.* Cambridge (Mass.): Harvard University Press.

SCHYLE D. 2007. **Ramat Tamar and Metzad Mazal. The Early Neolithic Economy of Flint Mining and Production of Bifacials Southwest of the Dead Sea.** *Berlin : ex oriente. By Leslie A. QUINTERO.*[1]

Archaeological studies today usually consider resource acquisition and tool-production trajectories as important aspects of lithic analyses. Conceptually, we owe much to the joint impetus of processualism and the *chaîne opératoire*, of course, but we owe equally to the early initiatives of field archaeologists and geologists, regardless of their (or our own) continental orientations. The monumental works of Lech and Schild, and the many others in the Eastern European confederacy of mine and quarry researchers, were instrumental in helping to frame modern approaches to lithic industrial studies. *Unconventional Archaeology*[2] and *5000 Jahre Feuersteinbergbau*[3] continue to inspire even those of us working in other parts of the world. My particular inspirational moment was 16 years ago in Berlin when the late Wolfgang Taute of Cologne University presented his insightful assessments of Ramat Tamar and Metzad Mazal, the Neolithic flint mine and accompanying tool-production sites located southwest of the Dead Sea, to a group of international scholars. We were assembled for the Proceedings of the First Workshop on PPN Chipped Lithic Industries, lithic archaeologists working on the stone-tool technologies of the Neolithic of the Levant. This meeting marked a shift in our thinking, a fragile consensus on the need for new approaches to behavioral studies of ancient tool technologies. The framing of chronologies and culture histories, which had for so long occupied the field, was slowing shifting from center stage. From this encounter our excitement grew with expanding views of the potential of stone-tool research, and the insights into economic organizations and ancient behaviors that resource acquisition and tool-production studies in particular provide. At that moment, Taute, perhaps more than any other scholar, in my view, exemplified a better way to understand the past.

Taute's Berlin presentation was envisioned as preliminary to his final assessment of his data. Sadly, he died before he could complete his study. The volume discussed here, which is dedicated to Taute, is the result of Schyle's extraordinary task to complete Taute's analyses. The volume presents the lithic assemblages from the two sites that were explored and reported preliminarily by Taute,[4] and is augmented with Schyle's own perspectives. The site of Ramat Tamar is notable as the first Neolithic flint mine to be discovered and studied in the Near East (outside of Egypt). It also functioned as a tool-blank production area. Metzad Mazal is a nearby living area and flint-knapping workshop where flint blanks from Ramat Tamar were further processed and configured, primarily into axe / adze preforms and finished tools.

In many ways, this volume illustrates the continuing growth of our ideas (and perhaps some of the growing pains) since that first meeting in Berlin, ideas that at their best are changing lithic research in the Levant across the spectrum of prehistory. Here, Schyle's goal was twofold. First he proposed to finish

1. Leslie QUINTERO – Lithic Technology Laboratory, Department of Anthropology, University of California, Riverside CA 92521-0418 – USA – quintero@ucr.edu

2. SCHILD, 1980.

3. WEISGERBER *et al.*, 1980.

4. TAUTE, 1981 and 1994.

Taute's resource analysis of the mine site, to complete Taute's technological study of the axe / adze production strategies at the two sites, and to determine the production organization underlying the industries at the two sites (and associated find spots in the vicinity). Next, he devised a strategy for a more comprehensive study of the economic organization of the combined enterprise.

Commendably, the first set of goals was configured and ultimately achieved within the maturing framework of lithic analysis that emerged from the Berlin workshop and subsequent meetings. That is, processual / *chaîne opératoire* methods were used to assess the industrial activities that occurred at the mine, the nature of the assemblages that were created at the two sites, and how these sites likely were related to one another. Of particular importance was an understanding of the production and use-life stages of the bifacial tools at the sites. The work was greatly facilitated by refitting studies of bifacial tools and cores (initiated by Taute and completed by Schyle) and by a replicative mining project conducted at Ramat Tamar undertaken by colleagues.[5] In the case of the final economic study, mathematical extrapolations were made from the data bases in order to project possible underlying economic strategies and industrial decisions of the Neolithic miners and knappers. The first set of goals was initiated and partially completed by Taute, the second goal was Schyle's alone.

It should be noted that the investigation of the sites and extraction of data were quite varied. The open-cast mine at Ramat Tamar was extensively excavated by Taute and his team and an impressive assemblage was recovered, but only a small portion of it was transported to Cologne by Taute for his study. The axe / adze production site of Metzad Mazal received only initial exploration and analysis by Taute, although he intended to return for further investigation. While Schyle made use of Taute's notes and analyses, his own data assessment was conducted on a random sample of that portion of Taute's original sample that was stored in Cologne. This unfortunate sampling constraint handicapped some of the ensuing economic extrapolations. However, he also drew on the mining replication conducted at Ramat Tamar which added much necessary pragmatic information and insight into the energetics of open-cast mining.

Fundamentally, our understanding of stone industries requires not only recovery of tools, but of their production detritus as well, in order that the organization and processes of production, and the intent of the craftspeople can be understood. The key to understand these clues, of necessity, is analogy with experiments in tool production. Even refitting studies ultimately rely on an understanding of process provided by doing the task itself. In this way, technologically meaningful attributes become apparent and can be separated from the nearly endless, meaningless information. As is the case in most studies, including the one discussed here, researchers rely on the expanding accumulation of technological data from these baseline replicative efforts to understand their own assemblages. In this volume, stages and modes of tool production are revealed largely from extensive technological attribute studies. Hence the careful reader will appreciate the detail supplied here (including 51 tables, 94 plates, 8 figures), although some might wonder at the shortage of standard lithic terminology, as commonly used internationally.[6]

Even under the best of circumstances, economic analyses that focus on product output and scale of work are difficult. Here the study is severely handicapped by the uncertainties inherent in the minimal exposure of Metzad Mazal and the previously mentioned sampling strictures, and by working from Taute's records, which were at times incomplete. Also, there are no radiocarbon dates to bracket the time depth of the flint mining and production at Ramat Tamar. Consequently, the economic interpretations that are offered regarding the scale and output of work, and how these relate to other Neolithic sites and lithic industries, are not well grounded and extend far afield. Comparisons with estimates of Neolithic wood-cutting efforts at the Neolithic townsite of 'Ain Ghazal, without an understanding of its axe / adze assemblage, and without experimental information on the use-lives of axes and adzes, and their functions, to arrive at a scale of axe / adze production at the workshops at Ramat Tamar and Metzad Mazal are not verifiable. The even broader interpretations of resource procurement and exchange within Neolithic society of the southern Levant, as offered in the volume, are simply conjectural. More useful for understanding the industrial efforts and organizational dynamics of the Levantine sites would be comparisons to the many similar flint mines and axe / adze-production sites so well documented for the Neolithic of Europe, such as those discussed in *Flint Mining*.[7]

Nonetheless, Ramat Tamar remains a fabulous and unique find. As a result of several studies conducted there to understand the exposure of the flint bed and its means of access, the mechanisms of flint extraction there seem well understood and clearly described. The huge effort of extracting flint without metal tools to remove the massive hard-rock overburden is

5. See BARKAI *et al.*, 2007.

6. *e.g.*, TIXIER *et al.*, 1980; INIZAN *et al.*, 1992; YERKES and KARDULIAS, 1993.

7. LECH, 1995.

certain to impress. The mining enterprise is an accomplishment that has to inspire Neolithic scholars today. Together with the information from the axe / adze production workshops at Ramat Tamar and Metzad Mazal, the grand endeavor and organization of work described in this volume offers a rare glimpse into what must have been a ubiquitous Neolithic pattern of exploitation of flint and woodland resources, a view that we seldom are privileged to explore.

BIBLIOGRAPHY

BARKAI R., GOPHER A. and WEINER J.

2007 Quarrying Flint at Neolithic Ramat Tamar: An Experiment. *In*: ASTRUC L., BINDER D. et BRIOIS F. (éd.), *Systèmes techniques et communautés du Néolithique précéramique au Proche-Orient. Actes du 5ᵉ colloque international, Fréjus, du 29 février au 5 mars 2004*: 25-32. Antibes: Éditions APDCA.

INIZAN M.-L., ROCHE H. and TIXIER J.

1992 *Technology of Knapped Stone (Préhistoire de la pierre taillée 3)*. Meudon: C.R.E.P.

LECH J. (ed.)

1995 Special theme: Flint Mining: dedicated to the Seventh International Flint Symposium – Poland 1995. *Archaeologia Polona* 33.

SCHILD R. (ed.)

1980 *Unconventional Archaeology: New Approaches and Goals in Polish Archaeology*. Wrocaw: Polska Akademia Nauk, Instytut Historii Kultury Materialnej.

TAUTE W.

1981 Mesad Mazzal, ein Siedlungsplatz des präkeramischen Neolithikums südlich des Toten Meeres (Vorbericht). *In*: FREY W. und UERPMANN H.-P. (Hrsg.), *Beiträge zur Umweltgeschichte des Vorderen Orients*: 236-256. Wiesbaden: Ludwig Reichert (*Beihefte zum Tübinger Atlas des Vorderen Orients, Reihe A, Naturwissenschaften* 8).

1994 Pre-Pottery Neolithic Flint Mining and Flint Workshop Activities Southwest of the Dead Sea, Israel (Ramat Tamar and Mesad Mazzal). *In*: GEBEL H.G. and KOZLOWSKI S.K. (eds.), *Neolithic Chipped Stone Industries of the Fertile Crescent*: 495-509. Berlin: ex oriente (*Studies in Early Near Eastern Production, Subsistence, and Environment* 1).

TIXIER J., INIZAN M.-L. et ROCHE H.

1980 *Préhistoire de la pierre taillée 1 : Terminologie et technologie*. Antibes: C.R.E.P.

WEISGERBER G., SLOTTA R. and WEINER J. (Hrsg.)

1980 *5000 Jahre Feuersteinbergbau: die Suche nach dem Stahl der Steinzeit*. Bochum: Deutsches Bergbau-Museum.

YERKES R.W. and KARDULIAS P.N.

1993 Recent Developments in the Analysis of Lithic Artifacts. *Journal of Archaeological Research* 1: 89-119.

UERPMANN H.-P., UERPMANN M. and ABBOUD JASIM S.A. (eds). 2006. **Funeral Monuments and Human Remains from Jebel al-Buhais.** *Government of Sharjah: Department of Culture and Information* (**The Archaeology of Jebel al-Buhais** *1). Par Anne-marie TILLIER.*[1]

Ce premier volume d'une série annoncée est consacré aux travaux conduits dans l'Émirat de Sharjah au nord de la péninsule d'Oman, sous la responsabilité du Département des Antiquités et dirigés par Sabbah Abboud Jasim. Les fouilles dans le secteur du Jebel al-Buhais ont débuté en 1994 et ont impliqué rapidement la collaboration de Hans-Peter et Margarethe Uerpmann de l'Institut de Pré-Proto-histoire et d'Archéologie médiévale de l'université de Tübingen. Elles ont permis de mettre au jour de nombreuses structures funéraires couvrant plusieurs périodes d'occupation, depuis la fin du Néolithique jusqu'à l'époque hellénistique. L'exploration archéologique du Sud-Est arabique ayant une histoire récente, toute contribution à notre connaissance du peuplement ancien de cette aire géographique est la bienvenue. Cet ouvrage collectif richement illustré, source d'informations nouvelles sur la chronologie archéologique et les pratiques funéraires dans la région, devrait inciter à l'ouverture de quelques pistes de réflexion.

L'ouvrage débute par un bref chapitre consacré par S.A. Jasim à la présentation archéologique des ensembles les plus récents ; ces derniers couvrent une période allant de l'âge du Bronze à celui du Fer et sont identifiés comme plusieurs unités. Cette présentation archéologique est complétée par des

1. Anne-marie TILLIER – CNRS-UMR 5199 PACEA, Université Bordeaux 1, Bâtiment B 8 – 1ᵉʳ étage, Avenue des Facultés, 33405 Talence Cedex – FRANCE – am.tillier@anthropologie.u-bordeaux1.fr

données figurant dans un second chapitre rédigé par l'équipe allemande impliquée dans le projet. On peut regretter ce cloisonnement qui ne facilite pas la lecture.

Pour l'âge du Bronze, il s'agit toujours de constructions circulaires cloisonnées dont la richesse en vestiges archéologiques varie de l'une à l'autre. Quand ils existent, les restes humains sont fragmentaires (à l'exception du cas de la tombe BHS 64). Pour la période du Wadi Suq, calée entre 2000 et 1300 BC (appellation reprise par les auteurs d'après une proposition de S. Cleuziou en 1981), la majorité de la documentation peut être rapportée à la période dite classique, à savoir la plus ancienne : elle concerne des tombes creusées et aménagées dans le sol. L'exemple le plus original est représenté par la structure BHS 66 en forme de trèfle avec quatre chambres qui a été réutilisée secondairement. Riches en offrandes funéraires, ces chambres ont livré une documentation anthropologique variable d'une chambre à l'autre, allant de deux adultes pour la chambre A à douze pour la chambre C.

Enfin à l'âge du Fer, les pratiques funéraires seraient plus diversifiées, impliquant soit la réutilisation de structures anciennes, le creusement de simples fosses ou bien encore l'emploi de cavités rocheuses naturelles. Relevant de cette dernière catégorie, deux structures (BHS 78 et 85) retiennent plus particulièrement l'attention par l'importance des vestiges humains qu'elles contiennent au nombre desquels des squelettes en connexion anatomique, couchés sur le côté gauche ou droit, avec les membres fléchis (voire hyperfléchis pour les inférieurs dans quelques cas).

Le deuxième chapitre comporte également une étude anthropologique succincte ; ceci s'explique en partie par le fait que le plus souvent (BHS 3, 8, 12, 64, 66 et 78) les vestiges humains ont été laissés *in situ* et les informations proviennent donc d'observations sur le terrain. Il en est autrement pour la documentation issue du site BHS 85 pour lequel on dispose d'une étude postfouille. Le lecteur, qui peut volontiers admettre la limitation de l'étude compte tenu des contraintes évoquées ci-dessus, découvre pourtant que la répartition chronologique des vestiges n'est pas aussi claire puisqu'il y a dans certains cas des dépôts intrusifs. Il reste par ailleurs ébaubi par le traitement suivi. En effet, on peut regretter que la méthodologie employée pour conduire l'analyse anthropologique se réfère à des ouvrages anciens et fasse abstraction de toute référence récente. Pour estimer l'âge des adultes, les problèmes inhérents à l'emploi de la synostose des sutures crâniennes (et encore plus quand l'observation se limite à la face exocrânienne) sont connus depuis longtemps[2]. De la même façon, un long cheminement méthodologique[3] a accompagné la recherche de critères fiables pour l'estimation du sexe, depuis le manuel des chercheurs hongrois Acsádi et Nemeskéri datant de 1970 auquel se réfèrent les auteurs. Que dire aussi de la volonté d'estimer la stature des sujets à partir des droites de régression de Pearson publiées en 1899 plutôt qu'une référence au travail de Sjøvold (1990) par exemple. Ce chapitre se termine par une partie consacrée à l'évaluation des caries, hypoplasies de l'émail dentaire, diverses atteintes du parodonte et quelques informations sont rassemblées sur les marqueurs d'activités. Si aucune précision n'est fournie quant aux méthodes de cotation, les auteurs semblent pour cette investigation plus au fait des travaux récents.

Le troisième chapitre de l'ouvrage en constitue la pièce maîtresse et elle est vouée aux tombes néolithiques de l'ensemble BHS 18, situées près du mont Hajar, un ensemble qui représenterait le plus ancien cimetière trouvé au Sud-Est de l'Arabie. Il s'agit en fait du travail de doctorat défendu par H. Kiesewetter à l'université de Tübingen qui porte sur la documentation issue des fouilles conduites entre 1996 et 2000, celle relevant des fouilles plus récentes (2001-2005) devant faire l'objet d'une monographie ultérieure. L'analyse intéresse 365 des 400 squelettes datés du Ve millénaire BC et lui sont adossés un catalogue des tombes ainsi qu'une annexe composée de huit tableaux synthétiques portant aussi bien sur le type de sépulture, le mode de dépôt du corps et la conservation des squelettes que sur des données classiques liées à l'étude anthropologique des squelettes (mesures crâniennes et infra-crâniennes, distribution par âge et par sexe, inventaire des variations anatomiques non métriques ou caractères discrets).

Bien illustrée, la présentation des pratiques funéraires, s'appuyant sur une interprétation ostéo-archéologique de chaque dépôt (qui pourrait être plus poussée), éclaire le lecteur sur la distribution, sur quelques centaines d'années et près de 100 m^2, de sépultures primaires (individuelles ou multiples), de la présence de remaniements liés à l'intrusion de nouveaux défunts (sépultures plurielles) et de sépultures secondaires. Aucune sélection ne se remarque selon l'âge ou le sexe au moment du décès, bien que la partie septentrionale montre une concentration des enfants. Seules les sépultures primaires, toutes orientées vers l'est, sont accompagnées de parures sans sélection apparente liée au sexe (en dehors des perles de cornaline associées aux femmes) ou à l'âge. La sépulture multiple la plus intrigante contient cinq individus reposant tous en position fléchie sur le côté, orientés identiquement, pratiquement dans les bras les uns des autres.

2. *e.g.*, MASSET, 1971 et 1989 ; UBELAKER, 1987 ; SCHMITT 2002.

3. *e.g.*, CUNHA et VAN VARK, 1991 ; BRUZEK, 2002 ; BRUZEK *et al.*, 1994.

La densité des données biologiques mises à la disposition, qu'il s'agisse de paramètres classiques (caractères ostéométriques, variations anatomiques non-métriques, lésions osseuses dégénératives, pathologiques ou traumatiques) ou culturels (trépanations), représente l'une des premières études monographiques consacrées aux populations néolithiques parcourant la péninsule d'Oman. Identifiant deux types de morphologie crânienne, l'auteur suggère l'existence à BHS 18 d'un groupe assez hétérogène et cette partition s'identifie également au niveau des pratiques funéraires, l'un (le type "round-headed") étant majoritairement représenté dans les sépultures secondaires.

Du point de vue démographique, l'auteur constate qu'un tiers des individus est composé de sujets immatures et que pour les adultes, une fréquence de décès plus élevée s'observe pour les jeunes (entre 20 et 34 ans) et les très âgés (au-delà de 50 ans). On peut regretter, lors du traitement des caractères non métriques, que l'auteur privilégie la fréquence générale de ces traits et la question de leur répartition selon le groupe ("long-headed", "round-headed"), au détriment d'une interprétation de leur occurrence dans les sépultures multiples, ce qui serait plus instructif pour la compréhension de ces regroupements.

Une première comparaison faite avec les sites déjà connus aux Émirats arabes unis et en Oman, contemporains ou plus récents, tels Umm al-Quwain ou le cimetière de RH 5, permet de dégager quelques conclusions. Aucune différence apparente ne s'observe entre RH 5 et BHS 18 quant au taux de mortalité. Les groupes représentés au Jebel al-Buhais se distingueraient par une santé buccale relativement bonne, un taux faible d'indicateurs de stress non spécifiques (*cribra orbitalia*, hypoplasies de l'émail dentaire) et de marqueurs de maladies infectieuses, ce qui amène H. Kiesewetter à privilégier un état de santé relativement correct, que l'auteur met en relation avec la nutrition, laissant de côté d'autres facteurs de causalité. Elle souligne l'incidence des trépanations sur trois des sujets et la haute fréquence de lésions traumatiques (14% des sujets de plus de 15 ans et majoritairement des sujets masculins avec fractures du crâne), qu'elle n'hésite pas à associer à des rivalités internes.

De toute évidence on le voit, cette troisième partie consacrée à l'ensemble néolithique, avec une analyse consciencieuse, est la contribution la plus enrichissante de l'ouvrage, même si on peut regretter quelques déductions un peu hâtives dans l'interprétation des données. Elle offre une panoplie de réflexions qui suscite l'intérêt. Il faudra attendre l'étude exhaustive du site pour approfondir l'interprétation de la diversité biologique illustrée et saisir son origine.

BIBLIOGRAPHIE

ACSÁDI G. and NEMESKÉRI J.

1970 *Human life Span and Mortality*. Budapest: Akademia Kiado.

BRUZEK J.

2002 A method for visual determination of sex, using the human hip bone. *American Journal of Physical Anthropology* 117: 157-168.

BRUZEK J., MURAIL P., HOUËT F. and CLEUVENOT E.

1994 Inter- and intra-observer error in pelvic measurements and its implication for the methods of sex determination. *Anthropologie Brno* 32: 215-223.

CUNHA E. and VAN VARK G.N.

1991 The construction of sex discriminant function from a large collection of skulls. *International Journal of Anthropology* 16,1: 53-66.

MASSET C.

1971 Erreurs systématiques dans la détermination de l'âge par les sutures crâniennes. *Bulletins et Mémoires de la Société d'Anthropologie de Paris* 12,7 : 85-105.

1989 Age estimation on the basis of cranial sutures. *In*: ISCAN M.Y. (ed.), *Age markers in the human skeleton*: 71-103. Springfield: Charles C. Thomas.

PEARSON K.

1899 On the reconstruction of the stature of prehistoric races. Mathematic contribution to the theory of evolution. *Philosophical Transactions of the Royal Society Series A* 192: 169-244.

SCHMITT A.

2002 Estimation de l'âge au décès des sujets adultes à partir du squelette. Des raisons d'espérer. *Bulletins et Mémoires de la Société d'Anthropologie de Paris* 12,1-2 : 51-73.

SJØVOLD T.

1990 Estimation of stature from long bones utilizing the line of organic correlation. *Human Evolution* 5,5: 431-447.

UBELAKER D.H.

1987 Estimating Age at death from human skeletons: An overview. *Journal of Forensic Science* 32: 1254-1263.

PELTENBURG E. (dir.). 2006. The Chalcolithic Cemetery of Souskiou-Vathyrkakas, Cyprus. Investigations of Four Missions from 1950 to 1997. Vol. 1. ***Nicosie: Département des Antiquités. Par Sabine FOURRIER[1] et Françoise LE MORT[2].***

ARCHÉOLOGIE (S.F.)

La nécropole de Souskiou-*Vathyrkakas* représente l'un des sites majeurs et, paradoxalement, l'un des moins bien connus du Chalcolithique chypriote. De fait, contrairement à celle d'autres périodes de l'histoire de l'île, l'archéologie du Chalcolithique a essentiellement porté sur des habitats, qui ont livré quelques tombes, sous ou à proximité des maisons. La nécropole de *Vathyrkakas-Cemetery 1*, séparée de l'habitat de *Laona* par un ravin, apparaissait donc tout à fait exceptionnelle. C'était le plus ancien exemple de zone réservée pour les morts, strictement distincte du lieu des vivants, et l'on y pratiquait des inhumations multiples et non plus individuelles. Son pillage intensif, au moins depuis les années 1950, a alimenté les collections privées en objets exceptionnels (vases et figurines, pendentifs en picrolite), pour lesquels on ne dispose que de catalogues[3], sans contexte de découverte. L'intérêt du site, et la menace constante des pillages, au moins depuis les années 1950, ont entraîné plusieurs fouilles, de sauvetage puis programmées, qui sont toutes restées inédites ou seulement connues par des rapports préliminaires. En 1950-1951, l'expédition anglaise, alors active sur le site de l'ancienne Paphos, y a fait quelques sondages sous la direction de J.H. Iliffe et T.B. Mitford. Devant la poursuite des fouilles illégales, le Département des Antiquités a ensuite confié à la nouvelle mission allemande en charge de l'exploration de Kouklia, sous la direction de F.G. Maier, de réaliser une opération de sauvetage[4]. Le site continuant à se dégrader, le Département des Antiquités a organisé une fouille à la fin de la même année, puis en 1991 et de 1994 à 1997, sous la direction de D. Christou[5].

Ce volume de synthèse vient donc combler un vide important. Il témoigne de la réussite de la politique chypriote en faveur de la publication de fouilles restées inédites : l'ouvrage, de belle qualité formelle, est édité dans les collections du Département des Antiquités et sa réalisation a bénéficié d'une aide de la fondation Shelby White-Leon Levy. E. Peltenburg, qui en a coordonné l'étude et la rédaction, était certainement le mieux placé pour mener à bien cette tâche difficile. Dans le cadre du « Lemba Archaeological Project » (LAP) qu'il dirige, il a fouillé plusieurs sites majeurs de l'ouest de l'île (Lemba, Kissonerga[6]) et il est l'un des meilleurs connaisseurs de la préhistoire (du Néolithique à l'Âge du Bronze) dans la région de Paphos. Il fouille, en outre, depuis quelques années, le site de Souskiou-*Laona* (habitat et nécropole), ce qui lui permet d'avoir une connaissance directe du matériel et des structures contemporains de *Vathyrkakas* et, partant, de replacer le *Cemetery 1* dans son environnement immédiat.

E. Peltenburg s'est entouré de différents spécialistes qui suivent les normes du LAP (notamment pour la typologie du mobilier), et il s'est adjoint la collaboration de D. Christou, responsable des fouilles chypriotes, qui signe un chapitre particulier sur ses travaux (chapitre 3). Le volume suit un plan simple : historique des fouilles (chapitre 1), catalogue des tombes (chapitre 2), étude des restes humains (chapitre 4), analyse des différentes catégories de mobilier (chapitres 5-9) et des restes fauniques (chapitres 10-11), synthèse historique (chapitre 12). Des annexes viennent compléter l'ouvrage et permettent une consultation pratique.

Comme les autres publications du LAP, le volume s'inscrit dans les principes de la "New Archaeology", dont il possède les qualités comme les défauts. Ainsi, il fait preuve d'une très grande rigueur scientifique dans la compilation et la présentation des données qui frise parfois la rigidité. Publier les résultats de travaux qui se sont déroulés sur près de 50 ans, dans des conditions de fouille et d'enregistrement très diverses, était une gageure. Les problèmes d'inventaire, exposés p. 9-10, ont été nombreux : certaines tombes ont ainsi été inventoriées plusieurs fois ou fouillées partiellement en plusieurs fois, ce qui crée des doublons. Plutôt que de proposer une numérotation générale harmonisée, l'auteur a choisi de proposer un inventaire le plus large possible, en gardant les numéros supprimés

1. Sabine FOURRIER – UMR 5189 HiSoMA CNRS-Université Lyon 2, Maison de l'Orient et de la Méditerranée Jean Pouilloux, 7 rue Raulin, 69365 Lyon Cedex 07 – FRANCE – sabine.fourrier@mom.fr

2. Françoise LE MORT – UMR 5133 Archéorient CNRS-Université Lyon 2, Maison de l'Orient et de la Méditerranée Jean Pouilloux, 7 rue Raulin, 69365 Lyon Cedex 07 – FRANCE – francoise.le-mort@mom.fr

3. Notamment VAGNETTI, 1980.

4. L'opération a eu lieu en 1972. Deux brefs rapports ont paru : MAIER, 1973 : 193-194 et 1974 : 41-43.

5. De brefs rapports ont paru dans la « Chronique des fouilles et découvertes archéologiques à Chypre », publiée chaque année dans le *BCH* (1973 et 1993). D. Christou a, par ailleurs, présenté un premier bilan des travaux réalisés lors du colloque d'Edimbourg (CHRISTOU, 1989).

6. Les résultats sont publiés dans la collection *Studies in Mediterranean Archaeology* 1985 (vol. 70,1), 1988 (vol. 70,2), 1991 (vol. 70,3), 2003 (vol. 70,4).

(par exemple, p. 17 : 34-36), ou en indiquant les recoupements possibles (p. 11 : 7 et p. 18 : 39). Bien qu'imparfait, ce système a du moins le mérite de permettre les renvois aux publications précédentes et à d'éventuelles corrections. Mais quel besoin d'informer le lecteur de la découverte de paquets de cigarettes (par exemple p. 18 : 39 ou p. 24 : 71), données brutes de fouille, quand la mention de « tombe pillée » ou « stratigraphie bouleversée » suffit amplement ? De même, le choix d'exclure de la publication les objets découverts illégalement est discutable[7] : ils sont d'ailleurs mentionnés dans le chapitre de D. Christou (p. 40-42). Malgré le caractère disparate et souvent partiel des informations disponibles, la publication replace le matériel, les tombes et l'ensemble de la nécropole dans son contexte géographique et chronologique. Cette méthode donne d'évidents résultats et bouleverse certaines interprétations traditionnelles. Ainsi, l'étude du micro-outillage lithique montre qu'il a été déposé intentionnellement dans les tombes : seuls les objets finis sont attestés sur le site, lequel est toujours resté inhabité (p. 133-139). De même, la répartition des *dentalia* par tombe et l'absence de trace d'usure invitent à nuancer leur interprétation traditionnelle comme éléments de colliers : peut-être n'avaient-ils pas tous cette fonction et peut-être les colliers étaient-ils uniquement élaborés à des fins funéraires (p. 92-93 et p. 143). La mise en relation des tombes et de l'habitat suscite des hypothèses intéressantes : les inhumations multiples dans les tombes apparaissent au moment où l'habitat se stabilise et où les maisons sont utilisées sur plusieurs générations. Peut-être la fermeture d'une tombe correspond-elle à l'abandon d'une maison (p. 178). Enfin, replacée dans une évolution diachronique, la nécropole de *Vathyrkakas* apparaît moins isolée (p. 180-181).

La terminologie employée est de temps en temps compliquée à l'envi. Par exemple, le vase SVP 86/20 est assurément exceptionnel et n'a évidemment pas qu'une fonction pratique (p. 112)[8]. Les interprétations vont parfois bien au-delà de ce que les données disponibles permettent. L'usage de quantifications est ainsi discutable pour un lot réduit de céramiques (p. 116-117) : la découverte de deux vases dans une tombe qui n'a livré aucun autre matériel n'autorise pas à conclure que la céramique « représente en l'occurrence une part significative du dépôt funéraire », à moins qu'il ne s'agisse d'une lapalissade. La reconstitution des pratiques funéraires (p. 170-174) est souvent audacieuse. En l'absence de textes ou d'images, il est difficile de parler d'« ancêtres » ou d'« officiants ». L'hypothèse que l'exceptionnelle tombe 73 soit celle d'« officiants », médiateurs entre les « ancêtres » et les vivants (p. 176-177), chargés de faire la place dans les tombes en rangeant soigneusement les ossements (et au besoin en découpant les liaisons anatomiques encore en place) parce qu'on y a découvert des lames de *chert* et des aiguilles en os (pour coudre les éventuels sacs d'os !), suscite les plus grandes réserves.

La qualité et la rareté du matériel publié (vases anthropomorphes et zoomorphes, figurines et pendentifs, perles de faïence – dont l'origine est encore incertaine, spirale en cuivre), la variété et les caractéristiques singulières de certaines tombes (dont la tombe 73, sans doute couverte d'un tumulus, p. 127-128), font de Souskiou-*Vathyrkakas* une nécropole importante, mais non plus exceptionnelle. Cette publication, qui étudie les objets en contexte, permet d'en apprécier l'originalité, en même temps qu'elle insère le site dans une évolution historique où il trouve toute sa place. Désormais sûrement datée du Chalcolithique Moyen (*ca* 3200-2900), la nécropole est fréquentée et abandonnée au même moment que l'habitat de Souskiou-*Laona*. On attend donc avec impatience les résultats des fouilles du LAP sur ce dernier site.

ANTHROPOLOGIE BIOLOGIQUE (F.L.M.)

Malgré les difficultés inhérentes à l'étude des restes humains du cimetière chalcolithique de Souskiou-*Vathyrkakas*, fouillé en plusieurs étapes sur une longue période, avec des méthodes très diverses et ayant, en outre, subi de nombreux pillages, ceux-ci figurent en bonne place dans la publication coordonnée par E. Peltenburg.

Le second des douze chapitres de ce volume consiste en un catalogue qui résume l'ensemble des informations disponibles pour chacune des tombes inventoriées. Parmi ces informations, on trouve pour certaines tombes des données relatives aux restes humains : état de conservation, disposition, nombre d'individus. Il s'agit de données succinctes, notamment en ce qui concerne la position des os mais, compte tenu des pillages et des perturbations subies par les tombes, on ne pouvait s'attendre à mieux. À la lecture de ce catalogue, il apparaît que les tombes peuvent soit contenir un seul sujet, soit, plus fréquemment, regrouper plusieurs individus. Dans ce dernier cas, il aurait été théoriquement intéressant de connaître en détail la

7. Les raisons invoquées dans l'introduction (p. 1-2) sont d'ordre divers : malgré l'avis d'autres spécialistes, la provenance de ces objets n'est pas sûre ; ils n'ont pas leur place dans une publication qui vise à la mise en contexte du matériel ; cela encouragerait le commerce illicite.

8. Ce qui donne en langage conceptuel : "Pottery vessels, in addition to serving a wide range of utilitarian functions, played important roles in constructing and negotiating identities of individuals during multiple stages of the life course."

position relative des différents corps afin de pouvoir discuter du caractère simultané ou successif des dépôts. Malheureusement, les éléments nécessaires à ce type de discussion ont vraisemblablement été en grande partie détruits par les pillages. La présentation des données relatives au nombre de sujets inhumés et à leur identité biologique (âge, sexe), sous la forme de deux rubriques distinctes (résultats issus de l'analyse des dents et de celle des ossements), est conforme à la structure du chapitre sur les restes humains (chapitre 4) mais n'en demeure pas moins curieuse et propre à entraîner des répétitions. Il aurait, me semble-t-il, été plus judicieux de fournir dans ce catalogue des données biologiques plus synthétiques.

Le chapitre 4, consacré à la population inhumée, est divisé en deux parties: la première (par D.A. Lunt et M.E. Watt) concerne les dents, la seconde (par Z. Parras) porte sur les restes osseux.

Le sous-chapitre "The human dentitions" comprend une présentation de l'état de conservation du matériel et des méthodes utilisées pour le dénombrement et l'estimation de l'âge au décès, un exposé des résultats ainsi qu'une analyse des pathologies dentaires et de certaines variations morphologiques (incisives en pelle et taurodontisme). Les auteurs précisent que le matériel est érodé et très fragmenté et que l'usure des dents est importante, ce qui a restreint les possibilités d'analyse. Ils font preuve de prudence dans l'énoncé des résultats en raison des conditions de fouille et des biais que celles-ci ont pu introduire, notamment en ce qui concerne le nombre minimum d'individus qui est estimé à 187. Les lésions carieuses ont pu être recherchées sur 1351 dents permanentes, le taux de caries étant de 2,4%. Les auteurs soulignent le fait qu'en raison de l'état des dents et notamment de l'érosion *post mortem* qu'elles ont subie, ce taux est vraisemblablement inférieur au taux réel de caries dans la population. Comparé aux données disponibles pour le Chalcolithique chypriote, il apparaît proche de celui observé pour la période 4 de Kissonerga-*Mosphilia* (2,3%) mais inférieur à celui observé à Lemba-*Lakkous* (5,3%).

Le sous-chapitre "Osteological evidence" comprend également l'état de conservation des restes osseux et les méthodes d'étude utilisées. Les résultats concernent l'estimation du nombre minimum d'individus et de l'âge au décès, la détermination du sexe, la démographie de la population inhumée et la stature. Les ossements sont, comme les restes dentaires, incomplets et très fragmentés et leur étude s'en est trouvée restreinte. À partir de ces vestiges osseux, le nombre minimum d'individus est estimé à 170. La répartition de ces individus dans les différentes classes d'âge traduit une sous-représentation des enfants de moins de 7 ans. Le profil de mortalité de la population est comparé avec ceux des populations inhumées dans les deux établissements chalcolithiques chypriotes de Kissonerga-*Mosphilia* et Lemba-*Lakous*, et dans un cimetière du Bronze ancien de Turquie, Karataş-Semayük. Ces comparaisons font apparaître des différences importantes entre Souskiou-*Vathyrkakas*, Kissonerga-*Mosphilia* et Lemba-*Lakous,* où les enfants âgés de moins de 7 ans sont proportionnellement beaucoup plus nombreux et où on note une plus grande proximité avec le profil de mortalité du cimetière de Karataş-Semayük. Les phénomènes de conservation différentielle, les perturbations subies par les tombes et un éventuel prélèvement incomplet des ossements lors des fouilles ne semblent pas suffisants pour expliquer cette sous-représentation des enfants de moins de 7 ans à Souskiou-*Vathyrkakas*. L'hypothèse selon laquelle les enfants « manquants » auraient été enterrés ailleurs (dans les espaces domestiques ?) est soulevée.

Le sous-chapitre traitant des restes osseux comprend également une petite partie relative au mode de dépôt des corps. Compte tenu des perturbations subies par une grande partie des tombes, de l'absence de photographies des ossements *in situ* et de l'absence fréquente de relevés, seules des hypothèses peuvent être émises. Sur la base des catégories d'ossements présentes dans les tombes « intactes » (7) et « partiellement pillées » (13), il semblerait que la majorité d'entre elles aient contenu des sépultures primaires.

Malgré les explications fournies (p. 53: note 1), on peut regretter cette subdivision du chapitre sur les restes humains en deux parties totalement distinctes. Une présentation commune des résultats obtenus à l'issue des deux études séparées qui ont été conduites aurait, me semble-t-il, été préférable. Quoi qu'il en soit, ce volume constitue une contribution fondamentale à la connaissance des pratiques funéraires et de la biologie des populations chalcolithiques chypriotes et, plus largement, des populations préhistoriques de l'île. La série de Souskiou-*Vathyrkakas* est, en effet, celle qui comprend le plus grand nombre d'individus, après celle de Khirokitia (VIIe-VIe millénaire av. J.-C.). L'on ne peut que remercier D.A. Lunt, Z. Parras et M.E. Watt, ainsi que l'ensemble des auteurs de cet ouvrage, de s'être attelés à l'étude complexe d'un matériel ingrat et d'avoir ainsi fourni à la communauté scientifique quantité de données nouvelles sur les populations préhistoriques chypriotes.

BIBLIOGRAPHIE

CHRISTOU D.
1989 The Chalcolithic Cemetery 1 at Souskiou-*Vathyrkakas*. *In*: PELTENBURG E. (ed.), *Early Society in Cyprus*: 82-94. Edinburgh: Edinburgh University Press.

MAIER F.G.
1973 Excavations at Kouklia (Palaepaphos). Sixth Preliminary Report: Seasons 1971 and 1972. *Report of the Department of Antiquities of Cyprus*: 186-201.
1974 Ausgrabungen in Alt-Paphos. Sechster vorläufiger Bericht: Grabungskampagne 1971 und 1972. *Archäologischer Anzeiger*: 28-51.

VAGNETTI L.
1980 Figurines and Minor Objects from a Chalcolithic Cemetery at Souskiou-*Vathyrkakas*. *Studi Micenei ed Egeo-Anatolici* 21: 17-72.

ASTRUC L., BINDER D. et BRIOIS F (éd.). 2007. **Systèmes techniques et communautés du Néolithique précéramique au Proche-Orient.** *Antibes : Éditions APDCA. Par Jean VAQUER.*[1]

Cet ouvrage constitue le recueil des actes du cinquième colloque international de l'organisation *PPN Chipped Stone Industries in the Near-East* qui s'est tenu à Fréjus en 2004. Il édite une trentaine de contributions sous forme d'articles courts et bien illustrés.

Dans la première section se trouve un article qui propose les résultats d'une révision des collections des stratigraphies en grottes qui ont permis la construction du cadre chronologique et culturel de l'Épipaléolithique du Levant sud. Des éléments du début du Néolithique figurent dans la plupart de ces collections témoignant de fréquentations qui d'après A. Belfer-Cohen et N. Goring-Morris seraient liées à des activités symboliques sur les sites perçus comme immémoriaux. Deux approches expérimentales sont consacrées respectivement à l'extraction du silex à Ramat Tamar (R. Barkai *et al.*) ou à la fabrication et à la fonction des herminettes taillées de type Mureybet (J.-A. Sanchez Priego et M. Brenet). La première qui cerne bien l'importance de la texture de la roche encaissante dans la conduite de l'exploitation en carrière achoppe sur les témoins concrets d'outillages conservés qui, hormis une masse en basalte, auraient été en bois végétal, voire en roche locale comme le suggère P. Wilke le modérateur de cette session. La seconde qui intègre beaucoup mieux la documentation archéologique réelle et les incidences des duplications débouche sur des propositions convaincantes au sujet de l'utilisation de ces herminettes dans le domaine de l'architecture en pierre et en bois.

La seconde section regroupe cinq contributions concernant la typologie des industries taillées du PPNA du Levant sud. Il faut retenir en premier lieu une révision critique de la subdivision chronologique Khiamien-Sultanien présentée par G. Sayej qui paraît salutaire au regard de la diversité des assemblages de cette région et des multiples biais qui peuvent affecter les séries dans les sites à occupations de longue durée ou de grande extension comme ceux présentés : Iraq-ed Dubb (I. Kuijt et N.B. Goodale) et Dhra' (N.B. Goodale *et al.*). Les voies proposées pour affiner la perception qu'il s'agisse de l'analyse fonctionnelle de certains outils très typés mais à vocation multiple comme les pointes d'El Khiam (S. Smith) ou des analyses à haute résolution de la composition des outillages aux échelles intra-site ou inter-sites en prenant en compte des séries aux contextes mieux définis paraissent pragmatiques car elles débouchent sur des interprétations privilégiant les aspects paléo-économiques par rapports aux aspects culturels (D. Kaufman et D. Nadel).

Les aspects technologiques sont au cœur de la réflexion sur la période de transition entre le PPNA et le PPNB. Les contributeurs semblent d'accord pour considérer le Levant nord comme la zone la plus précoce pour les changements qui caractérisent la transition PPNA-PPNB. Pour F. Abbès, c'est dès le Mureybétien que les populations du Moyen Euphrate ont joué un rôle essentiel dans l'invention et la systématisation des débitages bipolaires sur des nucléus naviformes étroits pour la production de lames à forme pointue prédéterminée qui étaient transformées en armatures à Mureybet et en couteaux à Jerf el Ahmar. Le PPNB ancien bien attesté dans cette région s'est développé aussi dans le bassin de l'Oronte vers 8400 av. J.-C. d'après les données de Tell Ain el-Kerkh présentées par M. Arimura qui souligne quelques caractéristiques de cette période. Les réflexions présentées par une équipe de tracéologues (J.-J. Ibáñez *et al.*) sur les multiples évolutions techniques

1. Jean VAQUER – CNRS, UMR 5608 TRACES, 36 allée Jules Guesdes, 31000 Toulouse – FRANCE – vaquerjean@aol.com

survenues au cours du PPN proposent des corrélations avec d'autres sous-systèmes techniques liés aux activités de subsistance ou de fabrication. C'est ainsi que de nombreuses innovations apparues précocement n'ont pris leur essor que lors du démarrage réel de l'économie de production dans un cadre social qui mettait en avant la programmation, la prévision et les échanges.

La section consacrée aux industries du PPNB moyen et récent apporte des données nouvelles sur des problématiques particulières. La présence de microlithes géométriques larges, probables éléments de faucilles, dans des séries de la fin du PPNB comme celle de Tor at-Tareeq en Jordanie est envisagée par D. Olzewski comme un apport du Nord Levant. L'étude de P. Wilke *et al.* des débitages bipolaires sur nucléus naviformes en orthoquartzite à Ain Jamman et dans d'autres sites du sud de la Jordanie est présentée comme un exemple d'adaptation réussie. La documentation sur cette méthode de taille est d'ailleurs considérablement enrichie grâce aux énormes séries lithiques du site d'Yiftahel présentées par Y. Garfinkel. Des précisions sur l'évolution typologique des industries lithiques du PPNB sont apportées par la présentation des séries de la base de la stratigraphie du site de Beidha qui sont comparées à celles du site de Shkarat Msaied et replacés dans un contexte d'interactions plus large par C. Hoffmann Jensen. Pour Chypre c'est un tableau évolutif sur toute la durée du Néolithique acéramique qui est dressé par C. McCartney. Elle propose notamment de reconnaître à partir de séries de surface une étape intermédiaire entre celle d'Akrotiri et celle corrélée au PPNB ancien du continent qui rendrait compte du syncrétisme entre les aspects locaux et les apports exogènes tout au long du PPN chypriote.

Les problématiques liées à l'exploitation et à la diffusion de l'obsidienne sont développées à partir de la reconnaissance des variétés exploitées et de leur origine par rapport aux points de découvertes. Ce type d'approche fondé sur le classement par couleur et texture et validé par des analyses physico-chimiques aboutit à des résultats révélant des sélections différentes selon les périodes, voire les objectifs aussi bien dans des sites proches des sources comme Musular (présenté par N. Kayacan et M. Özbasaran) que dans des sites récepteurs situés en première couronne des réseaux de diffusion comme Akarçay Tepe (présenté par O. Maeda), voire un site placé en limite des 300 km considéré par C. Renfrew comme le seuil théorique d'approvisionnement régulier (cas du site de Magzaliyah présenté par E. Healey). Les corrélations entre les centres d'exploitation identifiés et les options techniques développées sont au cœur des avancées de la recherche, comme le démontre D. Binder dans une magistrale synthèse sur les débitages par pression de la zone anatolienne portant aussi bien sur l'obsidienne que sur les silex de haute qualité. Il existe des dynamiques très différentes entre l'Anatolie orientale où les débitages par pression évoluent très rapidement entre 8500 et 7500 av J.-C. et l'Anatolie centrale moins tournée vers le renouveau et les échanges à grande distance à la même époque.

Le dernier chapitre thématique est consacré aux réserves et caches de supports lithiques au PPN et même au EPN de Çatalhöyük. Qu'il s'agisse de groupes de produits bruts sélectionnés ou lots mixtes comportant aussi des outils façonnés, ces concentrations forment des ensembles qui suscitent de multiples problèmes d'interprétation afférant non seulement au domaine techno-économique mais aussi aux aspects sociaux et symboliques. Un tour d'horizon général des cas de figure reprenant et discutant les propositions de l'article de L. Astruc *et al.*[2] est dressé par O. Barsilai et A.N. Goring-Morris à partir du commentaire des trouvailles des sites PPNB de Nahal Issaron et de Kfar Hahoresh. Il aboutit à des propositions cohérentes, fondées à la fois sur la nature des assemblages et leurs contextes de découverte pour distinguer des stocks ou des réserves prévues initialement pour une utilisation différée et des caches ou des dépôts sacrifiés à des fins symboliques ou propitiatoires. Ces deux grandes catégories se déclinent en de multiples variantes dans les exemples présentés. Sur le site mortuaire et cultuel de Kfar Hahoresh, une concentration de produits a permis le remontage d'un débitage bipolaire qui permet de restituer les actions accomplies par un tailleur peu expérimenté dont toute la production a été laissée sur place. Certains lots de supports de très haute qualité technique comme le groupe de lames bipolaires du site de Motza présenté par H. Khalaily *et al.* ou celui de Ain Ghazal présenté par J.-D. Karnes et L. Quintero correspondent à des productions spécialisées et sélectionnées en vue d'échanges. Ce sont plutôt des interprétations d'ordre symbolique qui sont mises en avant par F. Briois pour les petits dépôts de lames de Shillourokambos plus ou moins liés à l'environnement de sépultures. D'autres groupes de pièces en obsidienne intimement liés aux habitations de Çatalhöyük sont interprétées par T. Carter comme relevant d'un ensemble des pratiques rituelles « domestiques ». Ce sont des propositions de modélisation sur le fonctionnement socio-économique de l'exploitation et de la diffusion des pièces en obsidienne d'Anatolie orientale vers le Nord Levant qui sont proposées par L. Astruc *et al.* de façon stimulante à partir de la trouvaille d'un lot de lamelles importées retrouvé à Sabi Abyad II et des données techno-fonctionnelles générales sur le rôle de l'obsidienne dans la région du haut bassin de l'Euphrate.

Cet ouvrage, qui expose des approches technologiques de la taille de la pierre très variées dans leurs objectifs et leur mode

2. ASTRUC *et al.*, 2003.

de connaissance, a le grand mérite de dépasser les aspects méthodologiques ou strictement fonctionnels en établissant des corrélations avec les autres catégories de faits non seulement chronologiques et culturels mais aussi économiques, sociaux voire idéologiques sur l'ensemble de la zone d'émergence du Néolithique du Proche Orient. Il s'adresse de ce fait à un lectorat qui dépasse largement le cercle des lithiciens du domaine considéré.

BIBLIOGRAPHIE

ASTRUC L., ABBÈS F., IBÁÑEZ ESTÉVEZ J.J. et GONZÁLE URQUIJO J.
2003 « Dépôts », « réserves » et « caches » de matériel lithique taillé au Néolithique précéramique au Proche-Orient : quelle gestion de l'outillage ? *Paléorient* 29,1 : 59-77.

HANSEN S. 2007. **Bilder vom Menschen der Steinzeit. Untersuchungen zur anthropomorphen Plastik der Jungsteinzeit und Kupferzeit in Südosteuropa.** *Mainz: Deutsches Archäologisches Institut, Eurasien Abteilung. Verlag Philipp von Zabern (***Archäologie in Eurasien, Band** *20). Par Alain LE BRUN*[1].

S. Hansen aborde l'étude des figurines anthropomorphes du Néolithique et du Chalcolithique de l'Europe du Sud-Est de quatre points de vue différents. D'un point de vue stylistique, il rend compte des proportions du corps, du rendu plastique de ses diverses parties, des décors… D'un point de vue typologique, le personnage est représenté debout, les bras étendus à l'horizontale ou bien assis, les bras repliés sur la poitrine… Sont également pris en compte le contexte dans lequel ces objets ont été découverts ainsi que leur distribution spatiale et chronologique. Il en conclut que cette plastique est un phénomène chronologiquement et spatialement défini, lié à un certain ordre économique et social. Quant aux caractéristiques formelles, elles lui apparaissent comme « eine Verbindung von Allgemeinem und Besonderem », comme une conjonction du général et du particulier. Il est sans aucun doute superflu de relever la banalité du propos. Il l'est beaucoup moins de s'arrêter sur cette formulation, plus précisément sur l'emploi du terme « particulier ». Ce n'est pas là engager une querelle sémantique. Il en va, en fait, de bien autre chose car ce « particulier » ne renvoie pas à l'individu. Au contraire, il lui fait écran si bien que l'être humain pris dans son individualité se trouve de la sorte escamoté comme une colombe sous le coup de la baguette magique d'un prestidigitateur. On aboutit ainsi au paradoxe de ce livre, à savoir que dans un texte traitant de la représentation que l'être humain se fait de lui-même, question donc qui le touche au plus près, ce dernier est désespérément absent. Par conséquent, les développements peuvent succéder aux développements, la machine tourne à vide.

Reste la base de travail que constitue le catalogue[2] présenté région géographique par région géographique et qui occupe la majeure partie du premier volume de ce gros ouvrage, près de 310 de ses 540 pages complétées par 202 illustrations et 9 tableaux et enrichies par les 532 planches de dessins ou de photographies réunies dans un second volume.

BIBLIOGRAPHIE

RONEN A.
1995 Core, Periphery and Ideology in Aceramic Cyprus. *Quartär* 45-46: 177-206.

1. Alain LE BRUN – CNRS UMR 7041 ArScAn, Maison de l'Archéologie et de l'Ethnologie René-Ginouvès, 21 allée de l'Université – 92023 Nanterre Cedex – FRANCE – alr.lebrun@wanadoo.fr

2. La note 358 p. 82 appelle une clarification : « Asprots » est un terme créé par A. Ronen (1995) pour désigner les colons néolithiques de l'île de Chypre dans lesquels il propose de voir les membres d'une secte qui auraient recherché l'isolement de l'île pour mieux vivre leurs croyances. Ce terme n'a donc rien à voir avec l'Atlas des Sites du Proche-Orient contrairement à ce qui est suggéré.

OUVRAGES REÇUS

ALIZADEH A. 2008. *The Development of a Prehistoric Regional Center in Lowland Susiana, Southwestern Iran. Final Report in the Last Six Seasons of Excavations, 1972-1978.* Chicago : University of Chicago (*Oriental Institute Publications* 130). 395 p.

ARIMATSU Y., MIKUNI H., OGAWA Y. and NISHIAKI Y. 2009. *Catalogue of Archaeological Materials in the Department of Archaeology of Western Asia, Part 9. West Asian pottery collected by the Tokyo University Iraq-Iran Archaeological Expedition.* Tokyo: The University Museum, The University of Tokyo (Material Reports 76). 207 p.

BEAUNE S. A. DE. 2008. *L'homme et l'outil. L'invention technique durant la préhistoire.* Paris : CNRS Éditions (*Collection Le passé recomposé*). 166 p.

CREVECŒUR I. 2008. *Étude anthropologique du squelette du Paléolithique supérieur de Nazlet Khater 2 (Égypte). Apport à la comprehension de la variabilité passée des hommes modernes.* Leuven : Leuven University Press (*Egyptian Prehistory Monographs* 8). 319 p.

HAYDEN B. 2008. *L'homme et l'inégalité. L'invention de la hiérarchie durant la préhistoire.* Paris : CNRS Éditions (*Collection Le passé recomposé*). 161 p.

IBÁÑEZ J.J. (ed.) 2008. *Le site néolithique de Tell Mureybet (Syrie du Nord). En hommage à Jacques Cauvin. Vol. I et II.* Oxford : Archaeopress-Maison de l'Orient et de la Méditerranée Jean Pouilloux (*BAR Int. Ser.* 1843). 731 p.

LAUFFRAY J. 2008. *Fouilles de Byblos Tome VI. L'urbanisme et l'architecture. Vol. 1. De l'époque proto-urbaine à l'occupation amorite (de l'Énéolithique à l'âge du Bronze II). Vol. 2. Atlas de plans et coupes.* Beyrouth : Presses de l'IFPO (*BAH* 182). Vol. 1: 518 p. Vol. 2 (en coffret): 19 plans.

SAGONA A. and ZIMANSKY P. 2009. *Ancient Turkey.* London-New York: Routledge. 420 p.

SCHELTEMA G. 2008. *Megalithic Jordan. An introduction and field guide.* Amman : ACOR (*American Center of Oriental Research Occasional Publication* 6). 144 p.

VALLA F. 2008. *L'homme et l'habitat. L'invention de la maison durant la préhistoire.* Paris : CNRS Éditions (*Collection Le passé recomposé*). 141 p.

VILA E., GOURICHON L., CHOYKE A.M. and BUITENHUIS H. 2008. *Archaeozoology of the Near East VIII. Actes des huitièmes Rencontres internationales d'Archéozoologie de l'Asie du Sud-Ouest et des régions adjacentes/Proceedings of the eighth international Symposium on the Archaeozoology of southwestern Asia and adjacent areas. Tome I-II.* ASWA VIII – Lyon, 28 juin-1er juillet 2006. Lyon : MOM (*Travaux de la Maison de l'Orient et de la Méditerranée* 49). 646 p.

BASE DE DONNÉES BIBLIOGRAPHIQUES

Bibliographical References database

EN LIGNE/ON LINE

Les références bibliographiques annuelles de ***Paléorient*** – dès à présent plus de 18 000 références interrogeables – sont accessibles sur le web. Cette base de données est actualisée chaque année par G. DOLLFUS.

At the following address you may consult the annual bibliographical references lists published by **Paléorient**: *18,000 references. This database is updated by G. DOLLFUS once a year.*

http://web.mae.u-paris10.fr/paleo_index.htm

Depuis juin 2005 (volume 30,2), la base de données bibliographiques n'est plus publiée que par voie électronique.

From June 2005 (volume 30,2), the annual bibliographical references list is consultable only on the website.

Paléorient est désormais consultable du volume 1 au volume 30 (2004), avec une barrière mobile de quatre ans, sur le portail de diffusion public et gratuit PERSÉE.

The contents of **Paléorient** *is now available on line with a moving wall of four years, presently from Volume 1 (1973) to Volume 30 (2004). This resource can be found on the free public website* PERSÉE.

www.persee.fr

Achevé d'imprimer en juin 2009 par EMD S.A.S. – 53110 Lassay-les-Châteaux
N° d'impression : 21577 – Dépôt légal : juin 2009
Imprimé en France